FUNDAMENTALS OF ECOLOGY

S.K. Agarwal
Head, P.G. Department of Botany,
Government Autonomous College,
Kota

A P H PUBLISHING CORPORATION
4435-36/7, ANSARI ROAD, DARYA GANJ
NEW DELHI-110 002

Published by
S.B. Nangia
A P H Publishing Corporation
4435-36/7, Ansari Road, Darya Ganj
New Delhi-110 002
Ph. : 23274050
E-mail: aphbooks@gmail.com

2026

Typesetting at
Paragon Computers
New Delhi-110 059

Printed at
Balaji Offset
Navin Shahdara, Delhi-32
Ph.: 22324437

Preface

The science of ecology has attained a significant position because of its interdisciplinary nature and utility for human welfare. The problem of scientific management of natural resources which are under great stress from the ever-increasing population and over-exploitation of resources and industrialization, demands a deep insight into the 'ecological order' of nature. People have already begun to look at the subject enthusiastically. If their enthusiasm cuts across the specialized scientists barriers and at the same time provides support to the scientists from the social perspectives, then ecology may move from laboratory to community and *vice-versa*.

The Earth and its environment is the true laboratory of nature. Living organisms in soil, air and water owe their continued survival to the balance of the physical and chemical conditions prevailing on earth and its environment. In view of the ecological factors, mankind has to see that no such circumstances prevail which can force an abnormal and harmful situation resulting in ecological collapse.

In writing this fetschrift on the "Fundamentals of ecology", it was aimed at providing advanced students with a comprehensive survey of the Indian literature because ecological literature is dispersed in a great variety of journals, commonly published in Europe or America. Indian workers often do not have ready access to them. The absence of such a work has, inconvenienced both students and the teachers alike. It is hoped that this attempt will shed some light on the expanding horizons, serious controversy on major concepts by opposing schools of thought, and stimulate others to clarify the subject further.

The large volume of literature on ecology has led to extensive bibliography, which however, covers only a small fraction of the work done in our country in this field. Wherever possible review papers or ones which give further guidance to research priorities, have been cited.

The author extends special gratitude to Prof Frank B. Golley (Athens, USA), Prof Ralph D. Amen (North Carolina), Prof R.H. Whittaker (New York), Dr. Y. Tadaki (Tokyo), Prof R. Misra (Varanasi), Prof L.N. Vyas (Udaipur), Prof P.S. Dubey (Ujjain), Prof R.R. Das (Gwalior), Prof R.S. Ambasht (Varanasi) and numerous other ecologists for generously supplying the reprints of their research papers, for their suggestions for improvements and clarifications. Credits for illustrations have been scrupulously assigned in the accompanying legends, although the author has not in every case been able to contact their authors for permission, to them my thanks are also due.

The author is confident that the students and the teachers will find this book full of information and a handy guide for further detailed study.

S.K. AGARWAL

Contents

		Page
	Preface	(*iii*)
	Illustrations	(*vii*)
	Tables	(*xiii*)
1.	Introduction	1
2.	Principles of ecology	21
3.	Radiation	39
4.	Temperature	55
5.	Atmosphere - gases and wind	63
6.	Water	75
7.	Soil	87
8.	Topographic factors	113
9.	Biotic factor	117
10.	Pyric factor	133
11.	Autecology	137
12.	Genecology	187
13.	Organisation of communities	197
14.	Community dynamics	233
15.	Ecosystem	253
16.	Production ecology	283
17.	Biogeochemical cycles	309
18.	Systems ecology and ecosystem modelling	321
19.	Biogeography	355
20.	Famous Indian Ecologists	389
	Bibliography	393
	Subject Index	439

Illustrations

		Page
1.	Biology cake layer	4
2.	Spectrum of biological organisation and the scope of ecology	4
3.	Interaction of various components of an ecosystem	22
4.	The inter-relations between various components of the environment and organisms	25
5.	Generalised diagram of the operation of feedback mechanisms in natural ecosystems	26
6.	The relationship of photosynthetic carbon assimilation to ambient carbon dioxide concentration	29
7.	Graphic illustration of the law of tolerance	32
8.	Biomagnification of pesticides in the food chain	35
9.	The general relations among net productivity, entropy and species diversity in ecosystem development	38
10.	Types of radient energy on the earth	40
11.	The electromagnetic spectrum of radiant energy	41
12.	Greenhouse effect on Earth	43
13.	Energy exchange at about noon on a sunny day	44
14.	Energy exchange at night	45

15.	The various streams of energy between a plant and its environment	47
16.	Day and night time surface heat exchange	57
17.	The temperature inversion conditions within the atmosphere	59
18.	Effect of temperature on physiological activity of organisms	60
19.	Effect of temperature on biomass accumulation and decomposition	62
20.	Diagrammatic sketch showing the principle zones of atmosphere along with variations in temperature	64
21.	Effect of abrasive ice particles and strong winds on the branch growth	72
22.	Hydrologic cycle	76
23.	Hydrologic cycle	77
24.	Ecosystem water balance	78
25.	A generalised profile of soil	94
26.	Classification of soil texture	97
27.	Generalised diagram showing the different forms of soil, water and their possible relationship with soil and plant water status	102
28.	Outline map of India showing soil types	111
29.	Altitudinal zonation and distribution of various species of plants on the western and eastern Himalayas	114
30.	Obligatory mutualism in lichen between algae and fungus	121
31.	(a) Leguminous root with bacterial nodules	122
	(b) T.S. leguminous root passing through bacterial nodules	

32. Mutualisic association involving algae : 123

(A) Blue-green algae and Cycas root

(B) Algae (Zoochlorella) and Paramecium

33. Diagramttic representation of vascular Arbuscular Mycorrhiza showing vesicles and arbuscules 124

34. Gastropod shell occupied by a hermit crab with attached hydroids 125

35. Epiphyte Vanda on a tree branch 127

36. Commensalism for food - Termite and Staphylinid beetle 128

37. Commensalism for transport - Phoresis (Shark) and Remora (Sucker fish) 128

38. Total stem parasite (a) Cuscuta on host plant, (b) section through host and parasite 130

39. Balanophora on host roots 131

40. Viscum album on host branch 131

41. Ignition point by lightening in relation to topography 134

42. Biological clock diagrammatic 185

43. Stratification in forest communities, (a) Coniferous forest, and (b) Broad leaved forest 202

44. Life forms in plants 209

45. Three possible ways of dispersion in a community: 1. regular, 2. random, 3. clumped 215

46. Normal frequency diagram 217

47. Relation between canopy cover and basal cover 220

48. Ecotone between two communities 227

49.	Diagrammatic representation of successions	236
50.	A generalised field succession process	238
51.	Diagram showing ecological succession from a pond to a climax forest	243
52.	Succession on rock in Indian arid zone	247
53.	Succession on sany loam in Indian desert	248
54.	Different components of eco-systems	254
54-A	Relative primary production values in percentage for different types of ecosystems	259
55.	A generalised model of an ecosystem to show its structure and function	260
56.	Major linkages between the components of a simple terrestrial ecosystem	261
57.	Food web in a grassland ecosystem	262
58.	Ecological pyramids	264
59.	Fate of solar energy in herbaceous community at Udaipur	269
60.	Energy content (K cal/g dry weight) of *Crotalaria medicaginea* during different phases of growth in root, stem, leaf and seeds at Udaipur	271
61.	Diagrammatic representation of flow of energy	278
62.	Y-shaped energy flow model in an ecosystem	279
63.	Pattens schematic model of energy flow based on the relationship HFTS	280
64.	Primary production model	284
65.(a,b,c)	Climate and grassland production at Udaipur	288

66. Monthly fluctuations in net above ground and below ground plant biomass in the herbaceous community at different physiographic units 290

67. The average whorl heights, branch length and fresh weights of branches and leaves for trees of different girth classes 297

68. A model of the transport of a chemical element within the biosphere 311

69. The global cycle of carbon 313

70. The nitrogen cycle 315

71. The sulphur cycle 318

72. The local cycle of phosphorus 319

73. Element of systems ecology 323

74. Initial input-output model of an ecosystem 328

75. A compartmental conceptual model of a forest ecosystem 329

76. A single compartment of ecosystem model 330

77. A detailed working model for nitrogen cycling in a teak forest 331

78. Symbols used to show various processes in ecosystem models 332

79. Monthly fluctuation of live green biomass 334

80. Dry matter relationship 336

81. Dry matter transfer 337

82. Flow of biomass 338

83. Flow of biomass 338

84. A holon (H) with input (Z) and output (Y) 339

85. Linear system 340

86. Non-linear system 341

87. A simple river basin with three sources of pollution 347

88. Simple air pollution problem with three sources emitting SO_2 351

89. Polar projection map of the world showing major biogeographical regions 356

90. Cosmopolitan distribution of family Papilionaceae 358

91. Discontinuous distribution of family Lecythaceae 359

92. Forest types of India 370

93. Floristic regions of India. 377

Tables

		Page
1.	Percentage absorption of solar radiation by some common surfaces	46
2.	Albedo values of some common surface	46
3.	Morphological and physiological characteristics of heliphytes and sciophytes	50
4.	List of light sensitive seeds	51
5.	Classification of plants according to their photoperiodic requirements	52
6.	Composition of dry air by volume	66
7.	Summary of cloud types	82
8.	Soil skeleton	95
9.	Textural classes of soils	96
10.	Classification of soils	98
11.	USDA revised system of classification of soils	98
12.	Various orders of soils in the USDA 7th approximation	99
13.	Comparative water holding capacities of soils	101
14.	Chemical composition of humus	104
15.	The pH value and its soil reaction	107
16.	Possible biological interactions among populations	118
17.	Biological interactions	119
18.	Some algal - animal symbiotic association	124
19.	Distribution of genus *Stipa*	145

20. Distribution of the genus *Onionis* 145
21. Dicotyledonous flora of India 147
22. Coppicing capacity of some plants 149
23. Average number of seeds per plant 151
24. The number of seeds/fruit in different plant habits growing in the same habitat of *Trianthema portulacastrum* 152
25. Average seed output of *Gisekia pharaceoides* growing in and around Pilani 152
26. Effect of intensity of grazing and seed output 153
27. Viability of certain seeds 155
28. Imbibition by seeds at 28°C 158
29. Percent imbibition after different dry storage period 159
30. Effect of temperature on percent imbibition 160
31. Effect of photoperiod on imbibition percentage in *Indigofera astragalina* seeds 160
32. Effect of sulphuric acid pretreatment on *Idigofera tinctoria* seeds 161
33. Effect of varying soil moisture on germination of different crops 163
34. Effect of soil types on seed germination of *Verbena bipinnatifida* 164
35. Effect of soil types on seed germination of *Indigofera cordifolia* 164
36. Seed germination (%) in 29 days of *Limnanthes* spp at various temperatures 166
37. Germination (%) of *Primula* seeds at different storage period, constant and durnal alternating temperatures 167

38. High temperature thermoresponse of germinating *Portulaca oleracea* at 50°C 168

39. Photoperiodic response of germinating *Nerium oleander* seeds 169

40. Effect of duration of light on seed germination 169

41. Spectral sensitivity of germinating seeds 170

42. Spectral responses of germinating seeds of Asclepiadaceae 170

43. Spectral sensitivity of germinating seeds of *Merramia* spp 171

44. Repeated reversal of germination (%) response by alternating red (600 mu) and far-red (730 mu) radiations 175

45. Showing various types of seed dormancy and most suitable treatment for their removal in some of the investigated Indian plants 178

46. Reproductive capacity of some Indian plants 182

47. Ecotype differentiation of some Indian plants 193

48. Distribution of dicots and monocots in the flora of Gogunda and Prasad 200

49. Life-form spectra of some of the investigated localities in India 211

50. Biological spectrum (%) of various altitudinal zones in Yusmarg, Kashmir 212

51. Species density //ha in different stands of Prasad forest range, Udaipur 218

52. Net primary productivity 258

53. Rate of respiration of some aquatic and terrestrial animals at rest 266

54. Net primary production in the grasslands at Varanasi in terms of energy 270

55. Average energy values (K cal /g dry weight) of dominant vegetation of different grasslands 272

56. Efficiency of energy capture in the grassland at Varanasi as percent half total solar incident radiation 273

57. Comparative energy capture efficiencies of herbaceous communities 274

58. System transfer functions of energy dynamics in different grasslands 276

59. Net primary production in arid and semiarid grasslands of Rajasthan 287

60. Net aboveground biomass production ($g/m^2/yr$) on the two stands : stand 1 and stand 2 289

61. Monthly net positive increase (g/m^2) in above ground and below ground biomass for *Apluda* community at different physiographic units of Kalighati forest block, Udaipur 292

62. Net primary above ground biomass production (g/m^2) by grassland at Jodhpur in 1968-69 293

63. Net above ground biomass production by two shrub species at Kota 294

64. Above ground biomass (dry wt kg/ha) of *Calotropis procera* L at Kota 295

65. Growth attributes of *Zizyphus nummularia* in flat alluvial plains of Jodhpur 296

66. Pattern of foliar dry weight production (in % ± SD) in deciduous trees 298

67. Average above ground biomass (kg/tree) production of tree species in the semiarid regions of Rajasthan 299

68. Above ground biomass (mt/ha) production in different forest stands of the semiarid parts of Rajasthan 301

69. Average increment of non-photosynthetic biomass in decidous trees in semiarid parts of Rajasthan 302

70. Contribution of leaf litter by different species in Koriat forest community (expressed in gms/m^2) from October 1971 to March 1972 304

71. Annual leaf litter production 306

72. Canopy characteristics and pigment estimates in dry deciduous trees at Udaipur 307

73. Global distribution of carbon 312

74. Hierarchy in ecological units 323

75. Policy and cost relationship in waste waterlr treatment 350

76. Particulars of sources of sulphur dioxide 351

77. Examples of different types of discontinuous distribution 360

78. Some endemic taxa 362

1. *Introduction*

Man's activities in ancient to very ancient past have shown in undisputed terms, his understanding of environment and its impacts on his life. He learnt to modify it for a better adjustment. The words 'oikos' (house) and 'logos' (study of) were first together by Reiter (1868) as 'oikologie'. However, he did not made any attempt for a precise definition of this discipline. A year later, Haekel (1869) defined oikologie as the science of treating reciprocal relations of organisms and the external world (Koromondy, 1965). The term oikologie was soon substituted by the term 'ecology' which became much common, more popular and firmly entranched in the literature.

Since Haekel, ecology has been defined variously by different ecologists, placing different degree of emphasis on the components of nature and their interactions. To mention some of them are, by Clements (1916) who considered it as the science of community. Charles Elton (1927) defined it as 'the scientific natural history concerned with the sociology and economics of animals'. Taylor (1936) preferred to call it as 'the science of the relations of all organisms to all their environments'. Allee *et al* (1949) considered ecology as 'the science of inter-relation between living organisms and their environment, including both the physical and biotic environments, and emphasizing inter-species as well as intra-species relations'. Clarke (1954) defined ecology as 'the study of inter-relations of plants and animals with their environment which may include the influences of other plants and animals present as well as those of the physical features'. Woodbury (1954) regarded ecology as 'the science which investigates organisms in relation to their environment : a philosophy in which the world of life is interpreted in terms of natural processes'. Macfadyen (1957) defined ecology as 'a science which concerns itself with the inter-relationships of living organisms, plants and animals, and their environments'. Kendeigh (1961, 1974) defined ecology as 'the study of animals and palnts in relation to each other and to their environment'. Andrewartha (1961) defined ecology as 'the scientific study of the distribution and abundance of organisms'. Lewis and Taylor (1967) defined ecology as 'the study of the way in which individual organisms, populations of some species and communities of populations respond to these changes'. Petrides (1968) defined ecology as 'the study of environmental interactions which control the welfare of living things, regulating their distribution, abundance, production and evolution'. E.P. Odum (1971) defined ecology as 'the study of the structure and function of ecosystems or broadly of nature'. Krebs (1972) defined ecology as 'the scientific study of interactions that determine the distribution and abundance of organisms'. Clark (1973) considered ecology as 'the study of ecosystems or the totality of the reciprocal interactions between living organisms and their physical surroundings'. Pinaka (1974) defined ecology as 'the study of relations between organisms and the totality of the biological and physical factors affecting them or influenced by them'. Southwick (1976) defined ecology as 'the scientific study of the relationships of living organisms with each other and with their environments'. Smith (1977) defined ecology as 'a multidisciplinary science which deals with the organisms and its place to live and which focuses on the ecosystem'. Indian ecologist Prof. R. Misra (1967) defined it as 'the study of interactions of forms, functions and factors'.

In all these definitions of ecology one common thing is reflected that is, 'the inter-relationship between the organisms and their environment', although the ways of expression may differ.

What is Ecology

Ecology is a science. It studies the relation between the organisms and their milieu, as well as the relation between the organisms. The organisms within a habitat can be categorised under the following three classes :

1. Primary producers - plants containing chlorophyll.

2. Consumers - herbivorous and carnivorous animals and parasites.

3. Decomposers - fungi and bacteria.

The milieu or the environment consists of abiotic components which exists under three physical states (a) gases, (b) liquids, and (c) solids.

Misra (1974) has interpreted the abiotic components of the environment as 'stage materials' for the drama of life on our planet, and the organisms as the 'actors', who perform their part on the stage during their life-time. In this drame of life the primary producers carry out photosynthesis in the presence of carbon dioxide, water, chlorophyll and convert radient energy into chemical energy, that binds the molecules of various elements into complex compounds - the source of energy for all the vital processes. Primary consumers eat away plants, while the subsequent consumers may be both herbivorous and carnivorous. The consumers get their energy requirements by oxidative breakdown of the complex organic molecules into simpler forms. When net primary producers and consumers die, the decomposers bring about decomposition of the dead bodies and return the abiotic elements into the environment, which could be utilized again by subsequent generations. This phenomenon is also called as 'biogeochemical cycling' of the abiotic elements of the environment, and is a constant feature of an ecological system.

The very simple relations described above clearly show that both the organisms and the environnment are closely interrelated. The actual inter-relationship between them are far more complex than we can imagine, and ecology is the science dealing with such complex relationships between organisms and their environment.

Status of Ecology

Ecology is one of the several basic divisions of biology that is concerned with principles, that the fundamentals common to all life. Physiology, genetics, embryology and evolution are examples of other basic divisions. We traditionally cut the biology 'layer cake into small pieces in two distinct ways as shown in figure 1.

We may cut it horizontally into what are usually called 'basic divisions' because they are concerned with the fundamentals common to all life. We may also cut the cake vertically into what may be called 'taxonomic divisions', which deal with the study of specific kinds of organisms. Thus ecology is a basic division of biology.

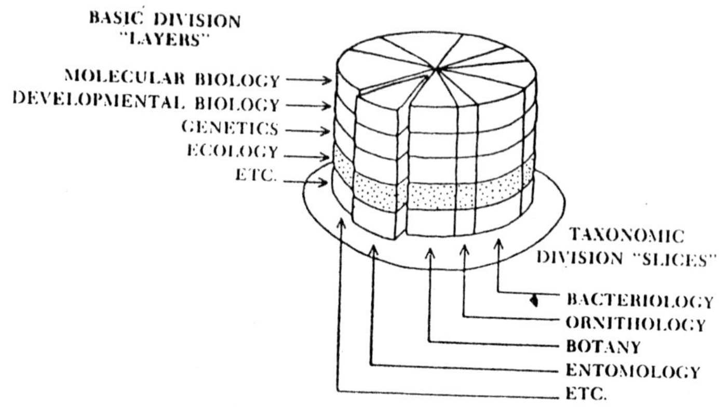

Fig. 1 : The biology "layer cake," illustrating "basic" (horizontal) and "taxonomic" (vertical) divisions. (after Odum, 1971)

Perhaps the best way to delimit modern ecology is to consider it in terms of the concept of levels of organisation (Figure 2).

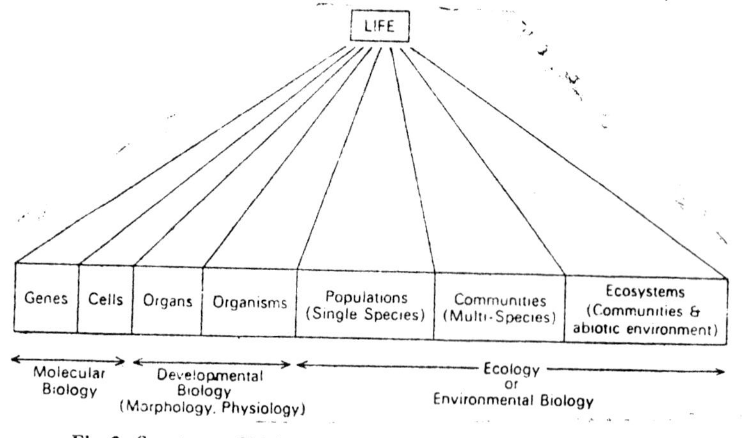

Fig. 2 : Spectrum of biological, organisation and the scope of ecology.

Community, population, organism, organ, cell and gene are widely used terms for several major biotic levels. Interaction with the physical environment (energy and matter) at each level produces characteristic fundamental systems. By a fundamental system we mean 'regulatory interacting and interdependent components forming a unified whole'. Ecology is concerned largely with the system levels beyond that of the organism. In ecology, the term population denotes a group of organisms of any one kind. Likewise, community includes all the populations occupying a given area. The community and the non-living environment function together as an ecological system or ecosystem. Ecosystem is essentially a somewhat more technical term for 'nature'. Finally, the portion of the earth in which ecosystems can operate - that is, the biologically inhabited soil, air and water - is conventially designated as the biosphere.

Some attributes, obviously, become more complex and variable as one proceeds from cells to ecosystems; however, it is an often over looked fact that other attributes become less complex and less variable as we go from the small to the large unit. The reason for this is that a certain amount of integration occurs as smaller units function within larger units. For example, the rate of photosynthesis of whole forest or a whole corn field may be less variable than that of the individual trees or corn plants within the communities, because when one individual or species slows down, another may speed up in a compensatory manner. More specifically we can say that 'homeostatic mechanisms' that dampen oscillations operate all along the line. We are more or less familiar with homeostasis in the individual, as for example, the regulatory mechanisms that keep body temperature in man fairly constant despite fluctuations in the environment. Regulatory mechanisms also operate at the population, community and ecosystem level. For example, we take for granted that the carbon content of the air remains constant, without realizing, perhaps, that it is the integration of organisms and enivironment that maintains the steady conditions despite the large volumes of gases that continually enter and leave the air.

Today, biologists seem to be grouped into two camps. In one are molecular biologists who see all of biology as centered in the areas of chemistry and physics. In the other camp are environmental biologists, who see all of biology in the areas of climatology, systematics, evolution and so on. In many ways, this division is unfortunate, for each group can offer much to the other. Many of the discoveries of molecular biology are relatively meaningless unless they are viewed from ecological point of view, and some of the problems of ecology can be answered only with the help of molecular biology. In reality, biology is a gradient from molecular to the cellular to the ecosystem and each segment of the gradient merges into the other.

Sub Divisions

Sub divisions in ecology, as in any other subject, are useful because they facilitate discussion and understanding as well as suggest profitable ways to specialize within the field of study. Ecological studies have been made with particular reference to plants or animals, hence they have been called 'plant ecology' and 'animal ecology'. In nature, however, it is difficult to study plants or animals alone, as they are always very closely associated and inter-related. Hence, the sub divisions plant ecology and animal ecology are considered vague today. The important branches of ecology are as follows :

1. Autecology - It is the study of individual organism or individual species in relation to the environment. Life history and behaviour as a means of adaptation to the environment are usually emphasized.
2. Synecology - It is the study of groups of organisms which are associated together as a unit in relation to the environment.
3. Habitat ecology - It is dependent upon the nature of habitat for example fresh water ecology, marine ecology, forest ecology, grassland ecology, cropland ecology, desert ecology.
4. Taxonomic ecology - It includes study of organisms belonging to different taxonomic categories and thus the study will be named as algal ecology, fungal ecology, bacterial ecology, insect ecology and so on.
5. Paleo ecology - It is concerned with the study of organisms of the past geological environments.
6. Gene ecology - It deals with genetic make-up of species in relation to their environment. It affords a valuable and reliable data on the origin and inheritance of adaptations in organisms.
7. Space ecology - It is concerned with the development of partially or completely regenerating ecosystems for supporting life during long space flights.

8. Ecosystem ecology - It is the study of community and surrounding non living environment, from structural and functional point of view. The structure is related to species diversity. The more complex the structure, the greater is the species diversity. The function of the ecosystem is related to the flow of energy and the cycling of materials through the structural members of the ecosystem.

9. Conservation ecology - It is concerned with proper management of natural resources such as land, water, air, forest, minerals etc., which are at our disposal today so that these resources are not unnecessarily depleted or destroyed, thus precluding future generations from their opportunity to use and enjoy them.

10. Production ecology - This is concerned with the rate of increase in organic weight in relation to space and time. It is measured in gross and net values. The total organic production is called gross production, while the actual gain in production after deducting the loss due to respiration is called net production.

11. Ecological energetics - It is a recent branch which is attracting the greatest attention of modern ecologists. It deals with the mechanisms and quantity of energy conversions and its flow through organisms within the ecosystem.

Scope of Ecology

In the last few years ecology has grown from an obscure (so far as the general public is concerned) subject, the very name of which had to be explained whenever it was used publicly, to the the great new hope of the world. Its popularity is such that ecology is now a household word, glibly used by news paper writers, ordinary citizens and even politicians, who had never heard of it a few years ago.

People have been badly frightened by the gradual disappearance and destruction of natural resources, and consequences of misuse and lack of respect for man's environment. Some perspective and courageous ecologists have been, for decades, trying to warn man, of the consequences of his headless and wasteful attitude towards his environment, now that these consequences are visible, and smellable, even to the man with no background or training, it is logical to turn to the 'prophets of doom' with cries for help.

In a democracy it is not sufficient just to have a few trained persons, who understand what it's all about; there must be an alert citizenry to insist that knowledge, research and action are properly integrated. We can not afford trial and error procedures in many cases, because many alterations of the environment are not immediately reversible should we discover that the alternation was a mistake. The Copper hill basin in east Tannessee (USA) is a good example of a mistake not easily ractified. Many years ago fumes from smelters were allowed to escape and kill all the forests for many miles around. Although, fumes are no longer present, the area remains virtually a complete desert of raw, red gulleys with little plant cover of any kind, even though man has several times attempted to revegetate the area.

Ecology designates the relationship of organisms with their enviornment and leads to the idea of the 'quality of life', which could be equated in social life by the idea of comfort and wellbeing. However, can we speak of wellbeing and comfort when our natural resources are gradually disappearing and being destroyed ? Some scientists, biologists, and agronomists alarmed by industrial development, which is ceaselessly leading to the profound destruction of the natural environment, have been warning about this in the middle of the twentieth century.

For ecologists, the human being is an integral part of nature and the systems fromed throughout the living world. Destruction and depletion of the external environment is not only destruction of nature but threatens the very existence of man. Konrad Lorenz, winner of the Noble Prize for Medicine in 1973, was one of the first to raise the cry of ecological alarm and to observe the effect of destruction of the environment on plants, animals, man himself and biogeochemical cycles. In his words, 'No one has the right to overturn an ecological balance which belongs to every one', and he added, 'the men of civilization who devastate living nature with blind vandalism threaten their own children with ecological collapse'. Man has the ability of modify the environments, but he is no exception to the rules by which environment control life (Petrides, 1956).

Prof. R. Misra (1976) in his address on ecology and development at the 'All India Symposium on Advancement of Ecology at Muzaffarnagar' pointed out that potent advance in ecology will accure from the attempts to apply the knowledge to the economic development of India and deployment of ecological concepts in redressing or reversing the progress of degradation of the environment. We are all aware of the current problems of environmental deterioration and pollution which are associated with growing human population, exploitation of natural resources, urbanisation, industrialization, illitracy and poverty. Our planners and decision makers are now already sensitized to the need of maintaining the quality of the environment.

Ecology provides us methodology to tackle many of the problems resulting from over exploitation of the resources which constitute our real environment, and which so far eluded us on account of our sectorial approach. Withdrawal of resources, their processing and conversion into goods for use and the resulting waste, are the components of any developmental process, which ultimately make the environment unfit for maintenance of high quality human life.

Man is the only organism of the biosphere, who plans for his future intelligently. Education is his great asset and a knowledge bassed society should be able to regulate the ecosystems of India by judicious planning with necessary inputs of ecology.

Since the ecosystems have faced barbarous exploitation of resources, it is necessary to develop and manage simultaneously a series of artificial ecosystems in different terrains such as hills and mountains, watersheds, plains and land surfaces. These interacting series of systems of a region have to be viable in respect of productivity (crops), protection of environment (forests), compromise multiple use systems (intermixed systems of grass, crop, forest, reservoir, lake etc.).

What ecological factors are we going to give top priority in the perspective planning of the country ? It is obvious that we have to reduce the wastage of our resources by recycling them just as nature does it. Even renewable resources like air, water, plants and animals become non-renewable upon over exploitation, because the base of cycling becomes attenuated. So we have to draw a planning map for the needs of the increasing number in the context of social justice, and better consumption without waste.

—

Ecology provides principal scientific basis for resource management. Whether we are concerned with the production of grains, livestock, timber, wildlife, flowers, experimental organisms, control of pests, prevention of erosion, disposal of wastes, control of radioactive or chemical fallout, deposition of silt, preservation of nature, proper land and water use etc. Thus ecology is the common denominator for the subject areas of agriculture, range management, animal husbandry, forestry, fishries, wildlife management, national parks, erosion, pollution control, oceonography, land and water use planning. The environment controls all organisms - even the genetically improved strains - and ecology is the study of control mechanisms. It is important in planning for all land uses, even including tourism, and to political scientists, to economists, to government administrators, as well as to the public. The full scope of land use and resource planning can not be realised without a basic knowledge of the principals of ecology.

History of Ecology

Interesting enough, although ecology comes from the same root as our word 'economics', the subject that we now call ecology was not given a name until a century later. Man being egocentric, began this type of study in his immediate surroundings. Not until long afterwards did he realize that the man's economics is but a special case of the broader subject. In the words of Wells, Huxley and Wells (1931) 'ecology is really an extension of economics to the whole world of life'. Economics and society might be thought of as the 'ecology of man' in a broad sence.

Though the word 'ecology' was coined only during the last century, the ecological ideas were deep rooted into human history. Man has been aware of the organism - environment relationships ever since he took keen interest in his surroundings. Scriptures of all old civilizations have reference to ecological concepts in a crude manner. The Greek philosopher Theophrastus (4th century BC) has been considered to be the first ecologist (Ramaley, 1940) for his vivid description of inter-relations among organisms and between organisms and their environment.

Ecology as an organised science dates back its origin at the end of the 19th century, when geographers began probing into the details of merely the distribution of flora and fauna. During the same period the relation between structure, function and distribution were also being studied, and this laid the basis of modern ecology. Haberlandt (1884) published his 'Physiological plant anatomy' and Warming (1896) and Schimper (1898) described the physiological foundation of plant geography.

Simultaneously, soil science grew from the concept of Dokuchayev in the year 1889. Soil was conceived, for the first time, to be independent and unique entity, resulting from the combination of climate, living matter, parent materials, relief and time.

The community concept in ecology was introduced by Le Coq. Sendtner and Kerner for plants. This view was later elaborated by Karl Mobius (1877) for animal communities. Notable contributions to plant and animal communities were made by Forbes (1887), Warming (1909), Cowles (1899), Clements (1916) etc.

After 1916, ecological research greately expanded by the establishment of ecological societies and publication of ecological journals. The European ecologists concerned themselves largely with the static approach of classifying vegetation on floristic basis, while their counterparts in America developed the dynamic system of vegetation analysis which emphasizes temporal changes in the community. In fact, Chements (1916) leader of the dynamic viewpoint went as far ahead to compare a unit of vegetation with the animals to a living organism which he called 'complex organism' or 'biome' in order to express the idea that the unit of vegetation is born, grows to adolescence, matures and finally dies even as a living organism does.

The relationship of climate and vegetation to the soil formation was an essential part of Coffey's (1912) classification of North American soils and by 1932, in Britain, Robinson had written 'Soils, their origin and classification' in which he fully realized that neither climate nor geology alone could be a sufficient basis for a classification of soils or an explanation for their genesis.

During the beginning of the present century the population concept came into being, when statistical studies and sampling techniques were employed for solving the community problems. In 1905 Clements text on 'Research methods in ecology' appeared and helped to establish the tradition of measurement, pioneering the use of quadrat and the adoption of instrumental techniques for defining the habitat. Cowles (1911) in his discussion on vegetation cycles, realized the three fold importance of climate, physiography and biota in ecology. Tansley and Ramkin (1911) for the first time used the successional concept in describing vegetation. By 1916 and 1920 Clements was able to produce two works, 'Plant succession' and 'Plant indicators' which contain enormous information and still stimulate the reader.

By 1930's started the era of ecosystem approach to ecology. Tansley (1935) coined the term 'ecosystem' in order to combine plants, animals and environmental complex. The ecosystem forms the structural and functional unit of the biosphere. Development of a similar concept under the name of 'biogeocoenose' by Sukachev (1944) in Russia and the enunciation of trophic-dynamic aspect of Lindeman (1942) and Odum (1957) laid solid foundations of the 'holocoenotic' approach to the study of nature. Margalef (1968) has drawan attention to the unifying principles in ecology and considers maturity of ecosystems as measured by diversity and in terms of energetics. Attempts to analyse more extensive relationships in ecosystems have awaited easy access to computing facilities and the development of the multivariate procedures necessary to define the interactions of the various parameters. Full ecosystem analysis is still in its infancy and is likely to take place in the forefront of ecological advance during the next few decades. Growing population pressure and the need for decisions concerning the management and conservation of natural resources have been the greatest driving forces in the introduction of systems analysis to ecology, the techniques having proved their value to controlled decision-making in the industrial - economic context.

History of Ecology in India

The history of ecology in India is not very different from that of any other country. In Indian writings of Vedic, Epic, and Pauranic etc., we find many references to ecological thought.

Chakra described the importance of Vayu (gases and air), jala (water), desha (topography) and time in regulation of life. Similarly, the concept of Panchatattva (five elements) - Earth (nutrients), water, fire (energy), sky (space) and air (gases) reflect the idea of circulation of materials. Indians have always regarded and respected plants and animals. In this country cutting of a green tree has been considered a crime and planing of a tree a cherity.

We had always recognised the world as one family, which could be equated to the existance of an ecological balance on this planet, as is clear from our age old concept of :

वसुधैव कुटुम्बकम

(Hitopdesh)

We have our age old unbroken links with nature and with life. We must again learn to invoke the energy of growing things and to recognise, as did our ancients in India centuries ago, that one can take from the earth and the atmosphere only so much as one can put back into them, as is clear from the following hymn from Atharva Veda :

यत ते भूमे विंखनामि क्षिप्र तदपि रोहात्।
मा ते मर्म विमृग्वरि मा ते हृदय मर्पिपम्।।

(Atharva Veda. 12 : 1 : 35)

We have recognised the importance of vegetation which supports diverse forms of life in this world, as expressed in the following sanskrit couplet :

यस्या वृक्षा वानस्पत्या ध्रवास्तिष्ठान्ति विश्वहां।
पृथ्वी विश्वा धायसं घृतामच्छा वंदामसि।।

We have also recognised that one organism is the living of another organism throughout this world, and that they eat each other. The following couplet from Mahabharata mentions the importance of the food chain and interdependence of organisms :

बहुभिः मुर्तैः किंजातैः पुत्रधर्मार्थवर्जितैः।
वरमेकः पथि तरूर्यत विश्रश्रमते जनः।।

In India, the earliest contributions to modern ecology were made by British ecologists to the forests and grasslands. Our early ecological studies have been influenced by European thought mainly due to the fact that most of the workers were either Europeans or were trained in Europe. The Indian sub-continent has been a testing ground for most of the ecological concepts developed in relatively advanced temperate countries. The efforts of Indian ecologists have resulted in defining the differences between temperate and tropical situations.

Descriptive Ecology

The Indian sub-continent has a typically forest climate; but the existence of biotic pressure or edaphic conditions maintains a variety of plant communities diffreing in physiognomy, life-form and structure. As a result we get all stages of temporal and spatial community organisations from almost the aquatic through meadows and savannahs to relict or mature forests. These communities show profound seasonal changes in response to the highly seasonal distribution of rainfall and to some extent to the stature of the dominent species.

Inspired by Schimper's and Warming's works attention was focussed in this country on climatic factors; and the expansion of the meteorological department by about that time provided the much needed data on rainfall and temperature. This stimulated studies on correlation of vegetation with altitude, rainfall, aspect and temperature in different regions of the country. The limit of distribution of tropical, subtropical, temperate and alpine type of vegetation began to be recognised. Champion (1936) proposed the classification of vegetation in his "A preliminary survey of the forest types of India and Burma", while four major vegetational zones correspond to temperature, further sub-divisions are based on rainfall. The relative stability of vegetation types was examined by Champion (1939), who initiated much thinking on casual factors in the development of forest communities, and since then most of the works on forests have put successional trends in the fore front.

An important event in the history of ecological studies in India was the publication of a book entitled "Indian forest ecology" by Prof. G.S. Puri in 1960, which presents a comprihensive survey of the vegetation and its environment in this sub-continent. Puri proposed a modification of the Champion's (1936) classification of Indian vegetation. Some of the tropical vegetation types e.g., grassland, tidal forests etc. considered as climatic by Champion, have been placed under new group called edaphic or biotic types. Other changes deal with the nomenclature of the temperate vegetation. According to Puri the climate of the wet and moist temperate forests (of Champion) is not humid and these may, therefore be properly classed with tropical types. Thus, the dry temperate type may only be truly temperate in character. Champion and Seth (1965) proposed a more detailed classification of Indian vegetation.

Analytical Ecology

Analytical and sunthetical studies of herbaceous and mangrove vegetation around Bombay were initiated by Prof. F.R. Bharucha, who introduced the methodology of Zurich Montpellier school of vegetational analysis in this country. The concept of biological spectrum was utilised to deduce the plant climate of the flora of Madras, Matheran and Mahabaleshwar grassland association (Bharucha and Ferreira, 1941 a, b; Bharucha and Dave, 1944). An attempt was made to distinguish distinct plant associations, characterised by indicator species, which could be correlated with habitat conditions. Bharucha and Dubash (1951) discussed the problem of nitrophily and proposed a formulae for numerical assessment of degree of nitrophily, which takes into account frequency, constancy of nitrates and the average nitrate content. Several grades of nitrophily from tolerance of high nitrate concentration in plant tissues to limited absorption of the salt from the substratum have been described. Bharucha and Satyanarayana (1954) described four purely seasonal calcarious associations generally found on walls around Bombay.

A detailed phytosociological analysis of grassland vegetation of Western Ghats was carried out by Bharucha and Shankarnarayana (1958) using Braun-Blanquet system. The study reaveled six major grassland associations and the behaviour and succession of vegetation in relation to biotic factors. Ranganathan (1958) also believes that the Nilgiri grasslands are climatic climax. Majority of the ecologists, agree that apart from the high Himalayan meadows, there is no climatic climax of a grassland in India. The savannahs and the rolling monsoon grasslands as fillings in the forest areas are the products of fire, grazing and other abiotic activities, which are continuously operating and maintaining the degenerated vegetation (Misra, 1959).

Grasslands of Sagar have been subjected to detailed ecological investigation by Pandeya (1961, 1964, 1967) who delimited eight grassland associations distributed according to topography and anthropogenic factors like grazing and burning. Some new phyto sociological concepts have been evolved by Pandeya (1961). The dominance diagram is an improvement over Raunkiaer's Frequency diagram. It incorporates cover of each species in addition to the number of species belonging to a particular frequency class. Likewise, in calculating the 'community coefficient' by frequency index method, coverage by each species has been included and community coefficient by (FXC) ICC method has been evolved.

Raman (1966) described the organisation of hilly and alluvial grassland in Varanasi. Sant (1964,. 1966) described the effect of grazing and seasonal variation in coverage of important grasses and forbs in alluvial grasslands of Varanasi. Grassland communities of Alwar have been studied by Vyas (1964).

Phytosociological exploration of forests have been extended to remote parts of the country by Indian ecologists. Temperate forests of Himalayan region were extensively studied from the point of view of distribution of conifers in the Kulu Himalayas with special reference to geology (Puri, 1950), succession of forest communities in oak-conifer forests of the Bashahar Himalayas and Punjab and Himachal Pradesh (Mohan and Puri, 1955, 1956), and the ecology of humus in conifer forests of the Kulu Himalayas (1951). Contrary to the openion of the earlier workers, Puri has clearly shown that in Western Himalayas oaks form the climatic climax vegetation; whereever oaks have been destroyed, stable conifer communities have developed, which are maintained due to peculiar geological oil or biotic factors.

Sophisticated sampling techniques have been introduced for studying the structure and composition of forest communities. Pandeya *et al* (1967) used Relative Growth Index which includes the analytic -quantitative parameters of density, basal cover and average height of individual species. Agarwal (1977) used Importance value index which includes relative frequency, relative density and relative dominance in terms of basal area. A stand is named after those characteristic species that have higher grades of RGI or IVI, and similar stands are grouped together to arrive at a community.

Experimental Ecology

With the accumulation of ecological information, largely observational and descriptive, the need for precise determination of behaviour and distribution of plants in relation to various environmental factors became evident. This may be recognised as the infusion of experimental approach in ecological thinking in the country.

Their earliest experimental studies are those of Misra (1944) on *Potomogeton Perfoliatus* and Misra and Siva Rao (1948) on *Lindenbergia Polyantha*. Form variations in *P. perfoliatus* are shown to be preconnditioned by carbohydrate supply to the growing shoots. *L. polyantha* Royle and *L. urtaecifolia* Link and Otto have been shown to be ecotypes dependent upon available calcium in soil. Bakshi (1952), Bakshi and Kapil (1954), Mall (1961) and Pandeya (1953) studied morphology and ecology of several herbaceous species. Such studies have been extended by Ramakrishnan (1960, 1961, a, b 1965, a, b) who has reported ecotypic differentiation in *Euphorbia thymifolia* in response to calcium content of the soil. He first established the presence of the green and red ecotypes in this species, the former an obligate calcifuge and the later a facultative calcicole. Such ecotypes have also been shown by him in *Echinochloa colonum, Euphorbia hirta, Setaria glauca* and *Cynodon dactylon. Euphorbia hirta* have been shown to possess three forms : one upright growing in protected area, the other, a prostrate form which appears again in two forms - the compact form of the footpath and the diffuse form of the grazed lands which are interconvertible in reciprocal transplants. Thus the prostrate ecotypes showed purely phenotypic plasticity in producing the ecads in response to biotic factors of grazing and trampling. The ecotypic micropopulations seems to bridge the gap or the discontinuities in form, by means of ecads in nature.

Autecological studies of a few forest trees such as *Boswellia serrata* (Sharma, 1955) and *Tectona grandis* (Bhatia, 1954) and a large number of herbaceous species like *Euphorbia hirta, E. thymifolia, Setaria glauca* (Ramakrishnan, 1960) *Cyperus rotundus*

(Ambasht, 1964; Tripathi, 1965), *Xanthium strumerium* (Kaul 1959), *Anagallis arvensis* (Pandey, 1968), *Alhagi camelorum* (Ambasht, 1963), *Alysicarpus monilifer* (Maurya and Ambasht, 1973) have been done. Prof. L.N. Vyas and his associates have made observations on the seed dormancy in *Indigofera cordifolia* (Vyas and Agarwal, 1972), *I. astragalina* (Agarwal and Vyas, 1970), *I. linnaei* (Vyas and Agarwal, 1970), and *Abrus precatorius* (Vyas and Agarwal, 1973). They have concluded that in *I. cordifloia* and *I. linnaei* seed coat dormancy could be removed by sulphuric acid pretreatment of 10 minutes. In case of *I. astragalina* sulphuric acid pretreatment of 30 minutes is necessary. The seeds of *A. precatorious* have very thick seed coat, these seeds require sulphuric acid pretreatment of 120 minutes. Vyas and Garg (1970) observed higher germination percentage in *Cleome viscosa* when seeds were irradiated with red light. They have reported physiological immaturity of the embryo which causes dormancy in this species. Gibbrelic acid has been observed by them to be promotary for radical growth under red light. Vyas and Shrimal (1972) studied the effect of thiourea, indol acitic acid and ascorbic acid on the germination and growth of seedlings of *Amaranthus spinosus*. they found that the seeds of this plant were negetively photoblastic. They have also reported that IAA and ascorbic acid although by themselves incapable in inducing germination of *A. spinosus* seeds in light were effective in enhancing the capacity of thiourea for inducing germination.

Vyas and Shrimal (1973) reported that seeds of *Celosia argentea* posses dormancy which could be overcome by GA (100 ppm). They have also reported that nitrogenous substances like ammonium nitrate accelerated the germination promoting activity of gibbrelic acid.

Vyas and Agarwal (1970) while studying the germination behaviour of *Verbena bipinnatifida* have reported that a temperature of 25°C, pH of 6.5 and exposure to red light resulted in maximum germination of seeds of this species. They have reported the presence of dormancy due to inactive enzyme system. Vyas and Garg (1971) have studied in detail the resposes of gibberellin of light requiring seeds of *V. bipinnatifida*. They have reported that GA was not only capable in bypassing the light requirement but also to overcome the temperature blocks of germination.

Vyas and Agarwal (1970) made observations on temperature and spectral sensitivity of germinating *Dalbergia sissoo* seeds. They have reported that freshly collected seeds showed a high temperature dormancy which was gradually overcome with age. The maximum promotive effect of blue light in supra-optimal temperature (35°C) have made them to suggest that blue light acted through a second system as well, independently of its action through phytochrome.

Vyas and Garg (1974) observed that the growth promotion in the hypocotyl by GA and IAA was effective while the site of action of inhibition in radicle remained almost ineffective.

Vyas *et al* (1975) studied the reaction of gibberellin induced dark germination of *Borreria stricta* seeds to temperature and nitrogenous compounds. They concluded that GA (400 ppm) caused these light requiring seeds to germinate in total darkness, preheating of these seeds at 60 °C abolished subsequent GA sensitivity. Vyas *et al* (1975) reported that 2,4 -D delayed the initiation of germination, slowed down the rate, lowered the percent germination and retarded chlorophyll formation in *B. stricta*.

Vyas *et al* (1980) while studying morphaction and GA_3 interaction in *Verbena bipinnatifida* seeds have reported that morphactin inhibited germination and seedling growth of this plant. Radicle of all the morphaction treated seedlings became negetively geotropic. Gibberellic acid failed to modify this response.

Production Ecology

Having acquired familiarity with the ecological behaviour, structure and composition of plant communities studies were taken up on primary productivity in aquatic and terrestrial ecostems.

Agarwal (1971) observed that among *Boswellia serrata*, *Butea monosperma, Lannea coromendelica and tectona grandis*, only *B. monosperma* showed a faster rate of annual accumulation of net above ground biomass. This faster rate have been explained on the basis of leaf area, leaf dry weight as well as life-span of leaves - all of which were observed to be maximum in it. Garg (1973) observed simple linear correlation between number of growth rings and CBH and between CBH and tree height and above ground biomass in Kewra-nal forest. The standing crop biomass of the woody vegetation was reported to be 31.45 mt/ha. Ranawat (1975) observed 27.18 mt/ha standing crop biomass in Koriyat forests.

Vyas *et al* (1971 a, b, c), Vyas *et al* (1972), Garg *et al* (1972), Vyas *et al* (1973, a, b, c), Vyas *et al* (1974), Vyas *et al* (1976a, 1977, 1978 a, b, 1979) have made observations on the plant biomass and net production relations in important tree species in the deciduous forests of southern Rajasthan. They have counted the growth rings of the base of the tree trunks and calculated linear correlations between growth rings and CBH, as well as correlation of CBH with tree height, bole biomass, total above ground biomass and leaf area. In addition to such correlations they have also calculated average increment in non photosynthetic above ground biomass with respect to growth rings.

Ranawat and Vyas (1975), Garg and Vyas (1975) and Vyas *et al* (1976b) reported that the observed amount of litter 4.4 t/ha around Udaipur, 4.8 t/ha around Koriat forest block, and 4.45 t/ha in Bansi Forest block was slightly lesser than the expected values as per the formula by Bray and Gorham (1964). The lower values have been correlated with unfavourable climatic conditions of the semiarid zone. They have also concluded that the tree-density as well as leaf size class of the tree species were mainly responsible for the amount of litter fall.

Misra (1970) has tried to understand the input and output of energy and minerals ina tropical deciduous forest comparing the rates of the flow obtained in the ecosystem with those of temperate forests. He concludes that the rapid rates obtained in the tropical ecosystems makes them highly vulnerable to population growth and technological culture. A reduction in the diversity usually brought about by man, exposes the lanscape to intense bleeding of nutrients so that healing of wounds becomes very difficult if not impossible.

2. *Principles of Ecology*

Organisms are components of the environment. Their growth, behaviour and life-histories are all influenced by the environment in which they live and throuugh their life activities they also bring forth change in the environmental complex. Thus, the organisms and their environment are wedded together and are in a state of constant flux. The relationships between the two are based on certain principles, which may be stated as under :

Principles Of Interdependence And Inter-relatedness

The foremost principle is that of interdependence and inter-relatedness, that is, everything is dependent upon and related to everything else.

The environment is a complex of many factors that interact not only with the organisms but among themselves. The environment of a region exhibit short-term (diurnal) or long-term (seasonal) fluctuations. They may also vary locally, to give local environments, and even do so in extremely restricted areas to give micro-enviuronments. Environment is the most fundamental and far reaching of the natural elements which control life. All organisms in our biosphere are closely dependent on it, and their adaptations to face adverse environmental conditions are both direct and indirect. Organisms with specialised adaptations to face their native environment when moved to the new environmental conditions become manifest (Gopal, 1977).

The multiplicity of factors and their interactions produce a large variety of habitats for plant growth, which is conditioned by temporal variations in their elements during the ontogenetic phases of the plant. The requirements for seed germination, seedling growth, flowering, fruiting etc., are all different, and these are derived from ever changing intensities of the factors. The rate of change, the duration of particular intensities and the maximum and minimum values of the environment upon plant growth and transmitted influences on account of exposure of one plant part upon another make us consider its past and spatial variation respectively (Misra, 1959 a).

Interdependence exists between non-living things e.g., solar radiation vaporises water, heats soil and air, causes the wind movement. Water in all its three forms modifies the light and temperature regimes.

Interdependence among the non-living and living components of environment provides a definite structural and functional stystem which is commonly referred to as an ecosystem. Each ecosystem has the same **basic structure**, composed of producer, consumer and decomposer **organisms**, and the same basic functions namely, the **flow** of **energy** and **cycling** of materials. Such an ecosystem is characterised **by its** capacity for self regulation and self maintenance. The system adjusts itself when affected by outside perturbations, but this can be possible only under certain limits (Figure 3).

The interdependence of organisms is essential for several life activities *e.g.*, pollination, dispersal, predation, symbiotism, parasitism, co-operation, compitition etc. Among the organisms the dependence is primarily for food. Directly or indirectly all organisms including man are dependent for their food upon the

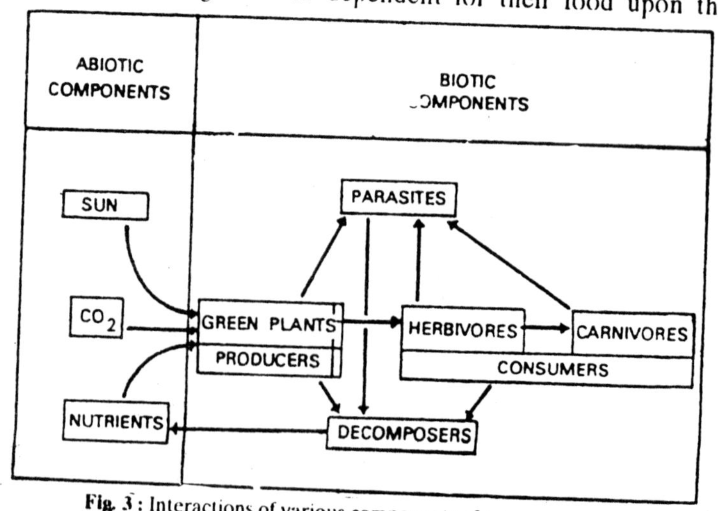

Fig. 3 : Interactions of various components of an ecosystem.

green plants. Plants manufacture their own food material from the very simple molecules using the solar energy. This food passes from one organism to another through the food-chain. Thus, a very large number of organisms become linked with each other forming a network-food-web. In addition to supplying food for animals, plants also provide them with shelter and protection. Consequently, the fauna of a region is dependent upon the vegetation. (Cloudsley-Thompson, 1975). Thus we see that everything is dependent upon and related to everything else.

The interdependence of plants and animals has been discussed by Gundersen and Hastings (1944), who pointed out that plants have become adjusted to the activities of herbivores by various regenerative and protective devices. Because they grow from the base of the leaf, grasses are able to withstand grazing and thus dominate steppe whence other type of plants, which grow at the end of the stem are largely eliminated. Spines in cacti, repellant taste, small or purgative effects protect the desert plants from the herbivores. Animals depend upon plants directly or indirectly for carbohydrates, certain essential aminoacids and vitamins. In their turn, plants are dependent, to a degree, upon the activities of animals through the parts they play in the nitrogen, carbon and phosphorus cycles, through their geological influence in modifying the soil and their role as seed dispersal and pollinating agents. A balanced equilibrium between plants and animals is favourable for both (Cloudsley-Thompson, 1975).

Mutually beneficial interactions have been observed between certain animals and plants of the Indian desert by Sharma (1978). There are many ways in which animals derive benefits from plants, for example, leaves, fruits, seeds, pollen and nectar serve as food for animals and provide them shelter. Animals benefit plants by aiding their pollination, seed dispersal and by proving dung and manure. Burrowing animals enrich the fertility of soil by their droppings and excreta, and the nutrients so released are absorbed by plants roots. Animal burrows promote plant growth by aeration of the otherwise hard soil.

Olson (1972) pointed out that, in its evolution, life has not only responded to, but also has probably caused, major climatic changes. Green plants have polluted the primevel atmosphere with oxygen, thereby making the Earth a suitable environment for habitation by animals.

Principle of Holocoenotic Environment

The term 'environment' means surroundings. All organisms live in some sort of environment. It is a instantaneous menifestation at a given time, of physical, chemical and biological variables, at a given place (Rathore, 1971). The physical environment consists of (*i*) the solid mineral matter on the earth (the lithosphere), (*ii*) the water in the oceans, lakes, rivers and ice-caps (the hydrosphere), (*iii*) the gaseous mixture in the air (the atmosphere), and (*iv*) the radiant solar energy. The position and movement of the earth, and its gravitational force are additional components of the environment. These components result in the variability of magnitude and duration of other environmental factors. The energy interacts with the rocks, water and gaseous components and with the organisms to produce the complex environment with numerous identifiable variables such as heat, light, rain, snow, mist, fog, wind, dust-storm, fire, etc. The environment thus created and maintained by the interactions of these variables functions as a whole unit. Any of these factors cannot be removed or altered without affecting the other factors.

Coper (1926) viewed the vegetation of Earth as a flowing braided stream governed and directed by all environmental factors at all times. Billings (1938) emphasized that successional changes in vegetation cannot be interpreted in terms of one factor, but only by considering the environmental complex as a whole. This principle has been termed that of the 'holocoenotic environment' by Allee and Park (1939), and has been restated and emphasized by Cain (1944). The holocoenotic principle is fundamental to any understanding of environment-organism relationship. Complete explanation of ecological phenomenoa are not possible without it.

The complexity of the relationships between plant and its environment and between the various factors of the environment is almost enough to discourage any attempt at complete analysis and synthesis. In fact, Cain (1944) has stated that such ecological problems not only may be difficult to solve but may really be insolvable in a mathematical sense. However, attempts by ecologists should be and are being made (Lindman, 1942; Major, 1951).

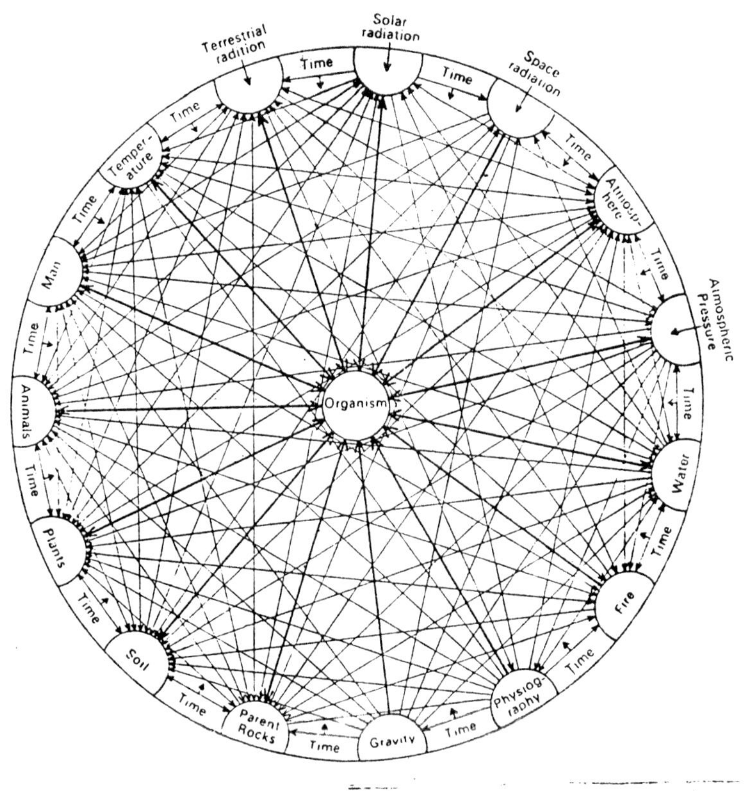

Fig. 4 : The interrelations between various components of the environment and organisms (after Billings, 1938).

It will be of some help in understanding the principle interactions in the environment to understand the holocoenotic concept. Figure 4 is a diagramatic attempt to show such an environment and the interactions between the various factors themselves and between these factors and a plant. The factors in the diagram are large units and there has been some lumping of factors in order to simplify it. Furthermore, the fifteen factors are not of equal weight. Nevertheless, the relations in a complete environment have been shown in the diagram. Some biologists would consider time as an environmental factors, but time light better be considered not as a factor in itself but as a dimension by which all other factors are qualified. Therefore, time has been indicated around the edge of the diagram as affecting all of the reactions within the environment.

Principle of Homoeostasis

Homoeostasis is preservation of an even level. Through species diversity and successional changes the mature systems reach a stage of stability.

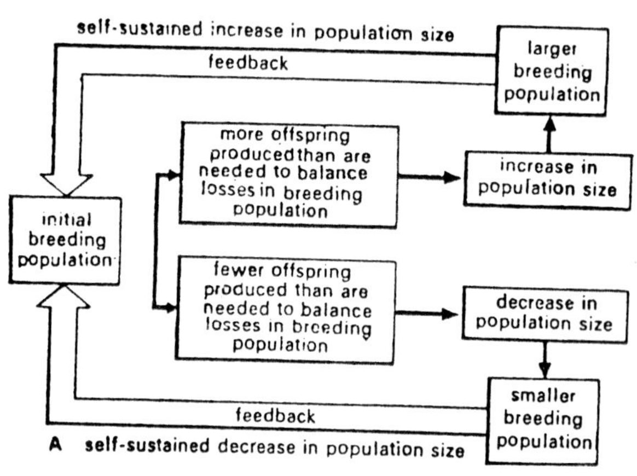

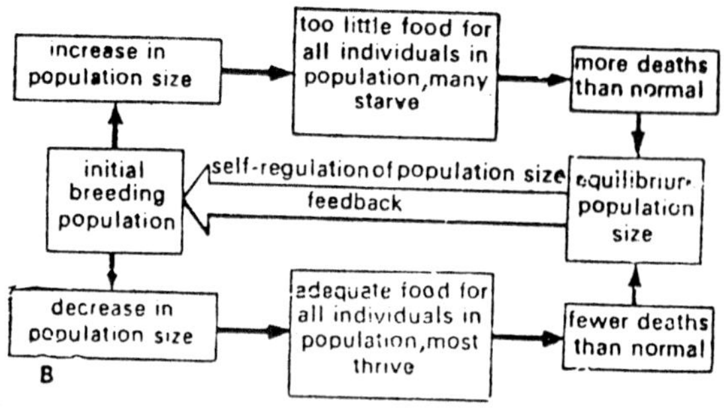

Fig. 5 : Generalized diagram of the operation of feedback mechanisms in natural ecosystems. A and B figures refer to the operation of these mechanisms in the regulation of population size in a hypothetical population of animals ; positive feedback leading to self-sustained change (A), and negative feedback leading to self-regulated homoeostasis (B) (after. Clapham, Jr., 1973).

Ecosystems have a rather delicate balance of inputs and outputs, but this balance is often not sufficient to avoid instability. Ecosystem diversity generally accompanies its physical stability, and ecosystems seem to have some regulatory or homoeostatic mechanisms. An essential feature of such regulatory mechanisms is the process of feedback operating both at the level of the individual and the entire system. Homoeostasis actually means constancy maintained by negative feedback, and the indidviduals of any community seem inter-related by means of homoeostatic mechanisms. The best illustration of this mechanisms is probably constituted by predatorprary interactions. The members of any climax community are involved in the cycling of materials and flow of energy. These members are interlocked by foodback loops and are hence adapted to the prevailing conditions. Each species has multiple relations with other species in the community. It has been commonly believed that a causal relationship exists between the members of prey and their food. Although it is agreed that oscillations do exist, it is now questionable whether in all cases the numbers of predators merely go up and down with fluctuations in the numbers of the prey.

Figure 5 shows an ecological example of positive and negative feedbacks. In both we start with a given number of animals. In the positive feedback loop, production of a large number of offspring leads to a large population than the original. This then becomes the new reproducing population, which can produce even more offsprings. The result is a self-sustained population increase, driven by the offspring feeding back into the breeding population. In the same way, if more breeders die than are replaced each year (which happens in some populations), this loss will be reflected in a smaller breeding population and will result in self-sustaining decline in population number by the same type of the positive feedback loop. In the negative feedback loop, the population is controlled by external factors, such as food supply. If the population increase in size, there will not be sufficient food to go around, and more individuals than usual will die from starvation. Conversely, if the population decrease in size, there will be more than enough food to go around, and the animals will be well- nourished, leading to smaller than normal die off. In either case, the equilibrium population size is regained. Most interactions between different species of organisms, and many interactions between a given species and its physical environment, depend on either positive or negative feedback. All organisms in an ecosystem are part of several different feedback loops at any point in time. Some of these relationships are negative, while others are positive.

The consequences of human activities, for example, through energy dissipation, air, water and land pollution is caused in the ecosystem. They tend to shift towards an increased entropy status, remote from complex groups of specialized species towards the generalisers, from species diversity to monotype, and from light and stable nutrient cycle towards loose and unstable nutrient cycles. Increased production of industrial materials frequently leads to acceleration of hydro-geochemical cycles, disturbance of input - output balance, accumulation of toxic substances such as hydrocarbons, metals, gases, over production or depletion of certain essential substances. This simplification of ecosystems involves or results in disorder and disintegration of the biosphere, shortening of food webs, decreases in species diversity and counteraction of forces of natural selection and evolution. Ecologists feel that our future concern should be the fact that most of the energy dissipation by the industrial society eventually causes an undersirable over simplification of the ecosystem, especially a curtailment of the food web. Our objective should be to prevent this and to strive for the maintenance of more highly complex ecosystem which are not so easily upset by perturbations and man-made catastrophes.

Principle of Limiting Factors

The principle of limiting factors has been far more widely used that that of the holocoenotic environment. The German biochemist Justus Liebig (1840) first drew scientific attention to the subject of mineral nutrition and formulated what has subsequently been called as the 'Law of the minimum'; by the deficiency or absence of one necessary constituent, all the others being present, the soil is rendered barren for all those crops to the life of which that one constituent is 'indispensable'. It is now usually incorporated with a 'Law of limiting factors' developed by a British physiologist F.F. Blackman (1905), who at the beginnning of this century investigated all physiological factors, photosynthesis in particular, with the generalized Vant' Hoff Chemical Rule that for every 10°C rise in temperature the rate of the reaction is about doubled or trebled. He found that this was not true for organisms as the rate approaches a maximum and then either remains constant or falls with increasing temperature. To explain this relationship he propounded the 'Law of limiting factors,' which can be defined as 'when a process is conditioned as to its rapidity by a number of separate factors, the rate of the process is limited by the rate of the slowest factor'. He illustrated this relationship by considering the limiting effect of carbon dioxide availability on photosynthesis and plotted hypothetical curves for different light intensities and carbon dioxide concentration (Figure 6). The sharp inflictions shown by these curves do not occur in reality, a more characteristic form being that of the pecked emperical curves : the transition zones of changing gradient of the curve being related to the exponential nature of the limiting process. Blackman's consideration of high

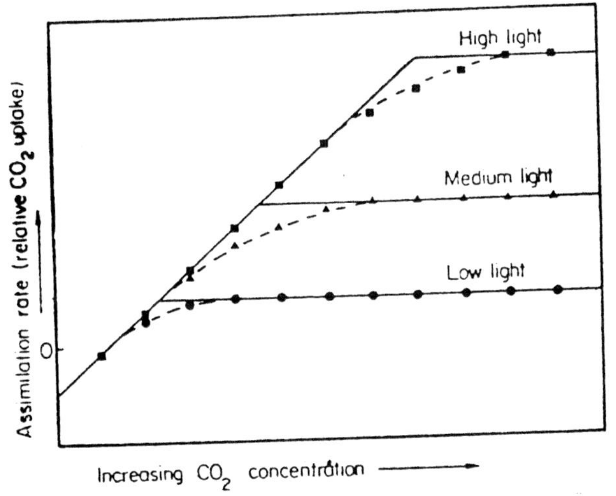

Fig. 6 : The relationship of photosynthetic carbon assimilation to ambient carbon dioxide concentration. The unbroken lines represent the hypothetical Blackman relationship and the pecked lines are based on experimental results. Rising carbon dioxide concentration increases the photosynthetic rate until the light intensity becomes limiting : at this point the curve inflects and further increase of carbon dioxide concentration has not effect unless the light intensity is raised. (after Ethrington, 1975)

temperature, supra-optimal conditions, showed that the law of limiting factors is valid not only for minimal supply conditions but also in circumstances where physiological tolerance of, or capacity for, a high level of some external factor is exceeded.

Cain (1944) has stated that physiological processes are multiconditioned. He says that, it is impossible to speak of a single factor as being the cause of an observed effect in an organism. He believed that it is erroneous to speak of a single condition of a single factor as being limiting.

Actually, it seems perfectly reasonable to assume that the holocoenotic environment and the principle of limiting factors are compatible. For example, water, or rather lack of water is a limiting factor in semiaried and arid regions. If water is added to the desert, the native desert plants are soon replaced by adventive weeds and other plants from the nearby irrigated areas. Certainly, the addition of water has far reaching effects in the desert environment because this environment is holocoenotic, but lack of water is the limiting factor, and no addition of any other factor to the desert environment will result in the same change in vegetation. If water is added permanently, the whole desert environment goes through a change - a change which results in an entirely different vegetation. When a limiting factor, then, is changed in nature and sets off a chain reaction in the ecosystem, it might well be termed a "trigger" factor. When a change in trigger factor upsets the delicate balance in an ecosystem, it is usually not possible to tell, when and where the chain reaction will end.

Numerous examples can be found to illustrate the effect of trigger factors in vegetation. One of the best example is provided in arid and semiarid zones of India by overgrazing in grassland which results in slow but inexerable deleterious change in the botanical composition. It triggers off succession and invariably *Dicanthium annulatum* and *Cenchrus ciliaris* which are highly palatable species are soon replaced by less palatable perennials and annuals such as *Oropetium themeaum*, *Aristida adscensionis* and *Eragrostis unioloides*. The importance of browsing is nowhere more pronounced than in the top feed trees such as *Zizyphus nummularia* and *Salvadora oleoides* which assume "bush form", where as *Prosopis cineraria* and *Anogeissus pendula* become "pillow cushion" form (Shankarnarayan, 1977).

Organisms themselves constitute a limiting factor in several cases. The predators, parasites and decomposers may also limit the population of an organism. The gases like carbon monoxide, sulphur dioxide, nitrogen oxides, hydrogen sulphide etc., are toxic in very small amounts and some organisms like lichens are so sensitive that even the traces of these substances are sufficient to kill them.

Principle of Tolerance

The principle of tolerance concerns with the limits of tolerance. V.E. Shelford of the University of Illinois in 1913, actually extended the concept of limiting factors so as to include the limiting effect of the maximum level of the factor on the organisms. According to the Shelford's law of tolerance "any environmental factor which is below the critical minimum or well above the critical maximum requirements of the organisms, would certainly limit the growth of these organisms in the given environmental areas". Naturally, organisms may be limited in their growth and occurrence not only by too little of an element or too low an intensity of a factor but also by too much of the element or too high intensity of the factor. For example, carbon dioxide is necessary for the growth of all green plants, small increase in concentration of carbon dioxide in the atmosphere will, under certain circumstances, increase the rate of plant growth, but very considerable increases becomes toxic. Likewise, small additions of arsenic to the human diet actually have a toxic effect, further increase in the dosage, however, soon proves fatal.

There occurs a range of gradient between the maximum and minimum limiting effects. Between the lower and upper limits of tolerance lies a broad middle sector of a gradient which is called the zone of compatibility, the zone of tolerance, the biokinetic zone or the zone of capacity adaptation. The region at either end of the zone of compatibility is called the lethal zone or the zone of resistance or zone of intolerance. The zone of compatibility too includes a broad range of optimum and narrow zone of physiological stress in between the range of optimum and lethal zones.

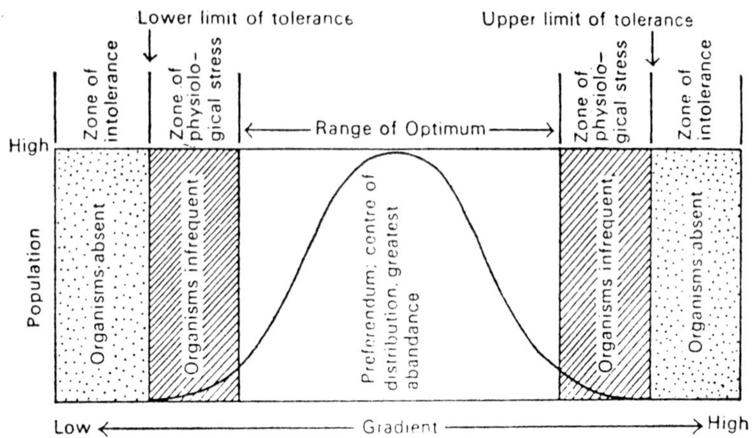

The upper and lower limits of tolerance are intensity of a factor at which only half of the organisms can survive. These limits are sometimes difficult to determine, as for example, with low temperature, organisms may pass into an inactive, dormant, or hibernating state from which they may again become functional when the temperature rises above a threshold. Similarly, at high temperature, there may be inactivation before the lethal level is attained. Even without dormancy occurring, there are normally zones of physiological stresses before the limits of tolerance are reached (Figure 7).

The species as a whole is limited in its inactivities more by conditions that produce physiological discomforts or stresses than it is by the limits of tolerance themselves. Death merges on the limits of tolerance, and the existence of the species would be seriously jeopardised if it was frequently exposed to these extreme conditions. Therefore, in retreat before conditions of physiological stress there is a margin of safety, and the species adjusts its activities so that limits of tolerance are avoided. There is a variation in hardiness of individuals within a species, so that some hardy individual find existence possible under conditions that disrupt other individuals. The population level of a species becomes reduced before the limits of its range are actually reached.

Principle of Dynamism

The environment and the organism are dynamic, that is, ever changing. The change may be short-term (cyclic), or long-term (non cyclic). Germination, flowering, fruiting phases are short term ones, and the changes in the vegetation structure and function in an ecosystem are long term ones. The long term change is usually cumulative and are called succession. The animals associated with the vegetation also change accordingly. Succession always leads to the climax, which is characterised by greater stability (more in harmony with the prevailing environment) and greater diversity (richness of species). Any attempt to keep the species diversity low (as in the crop fields) requires great efforts. As soon as the control is removed the vegetation progresses towards climax.

Ecosystem dynamics expresses its tendency to maintain itself in its characteristic dynamic state, through the processes of production, consumption and decomposition. Dynamics often implies the stability of the whole ecosystem, including its potentiality to resist exogenous disturbance and to replace itself in the event of catastrophe.

The effect of stress on ecosystem is best exemplified by eutrophication which is caused by forced input of nutrients into water. Increased primary production causes over saturation of water with oxygen but even then the oxygen content is often insufficient for complete decomposition of the organic matter produced and a part of the latter is deposited on the bottom sediment. Some oxygen also escapes into the air. In the hypolimion, denitrification results in escape of some nitrogen to the atmosphere, where as the increase in water pH, resulting from the assimilatory activity of phytoplankton, precipitates inorganic phosphate. Thus, the consequence of ecosystem fertilization are partly counteracted by some of the available elements being temperaily driven out of the cycle (Margalef, 1978).

McCormick (1978) made a comprehensive study of tropical and temperate ecosystems with regard to their response to stressors, and proposed three ways in which these systems tend to recover following perturbations. The first pattern involves the temporal and spatial replacement of species. The second pattern involves persistance of some species by altering rates of physiological processes in keeping with changes in environment. The third stretegy is resistance by certain individuals or even species.

Principle of Biomagnification

The process of accumulation of various elements and compounds along the food chain is called biomagnification. These substances enter a biological system aminly by three routes (*i*) aerial, (*ii*) terrestrial, and (*iii*) aquatic. Mercury, lead, cadmium, calcium, arsenic, chromium, copper, zinc, sodium, manganese, iron, potassium and tin increase as it passes from environment to the plant body. They further increase as they pass from plants to herbivorous animals and again increase in carnivorous animals and man. All elements are not metabolized by the organisms in similar amounts, for example, calcium is accumulated in wood of trees and bones of animals. It is of greater significance with respect to those elements and substances, which are harmful even in small quantities and tend to become accumulated at higher levels of the food chain, thereby causing greater damage. Thus, while the pollutant discharge may be at very low concentration or highly diluted form, quite safe for direct human consumption, but in course of time, the accumulated concentration in human body may create severe and danagerous effects.

Organic chemicals such as DDT and other pesticides also accumulate as they proceed along the food fhain (Figure 8). These chemicals persist in the environment for years, are transported through air and water to all parts of the world and get concentrated in eveyr food chain in every ecosystem on the earth. Aquatic micro-organisms absorb them in fat and oils, where they accumulate to concentrations many times greater than in water. Zooplankton that feed on countless contaminated phytoplankton cells contrates the pesticides still further in their tissues. Fish feeding on zooplankton further in their tissues. Fish feeding on zooplankton further concentrates the pesticides. Birds feeding on fish concentrates further. Many species of predatory birds - eagles, hawks, pelicans and the like - have shown serious adverse effects from this accumulation.

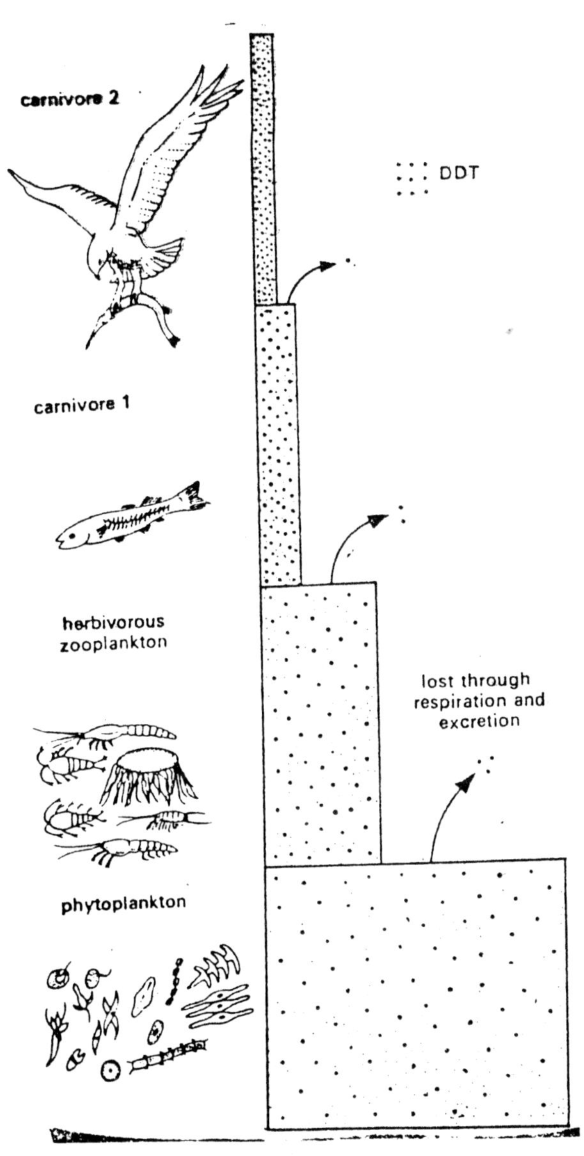

Fig. 8 : Biomagnification of pesticides in the food chain.

The rate of accumulation of pesticides is higher through aquatic route than through aerial and terrestrial routes. The cause of higher accumulation in aquatic environment is attributed to the chemical nature of the pesticides which have high liposolubility and lower water solubility. When the pesticides enters the aquatic environment, its movement is facilitated by water. It is then picked up by organic lipid containing particles which remain suspended in water. From this stage the pesticide enter the food chain and get accumulated in the biomass. Since, the pesticide is distributed throughout the water medium, the direct pickup of insecticides by the organism also results in the higher accumulation. In terrestrial environment the insecticide is picked up by the plants and soil invertibrates from the soil, but the system of bioaccumulation is less efficient than aquatic environment. In aerial environment the pesticides are picked up by plants and terrestrial animals from the air. This route is least effective among the three routes (Srivastava and Saxena, 1989).

Principle of Thermodynamics

These fundamental concepts of physics are related to ecology. The birth, growth and reproduction of an organism or an ecosystem are all accompanied by energy changes. Without energy transfer, which accompany all such changes, there would be no life. Energy is defined as the ability to do work. The beahviour of energy has been described by two laws of thrmodynamics :

The first law of thermodynamics called 'Law of conservation of energy' states that energy is neither created nor destroyed'. It may change forms, pass from one place to another or act upon matter in various ways, but regardless of what transfers and transformations take place, no gain or loss in total energy occurs. For example, when wood is burned, the potential energy present in the molecules of wood equals the kinetic energy released, and heat is evolved to the surroundings. This is an exothermic reaction. Whereas in an edothermic reaction, energy from the surrounding may be used into a reaction. For example, in photosynthesis, the molecules of the products store more energy than the reactants. The extra energy is acquired from the sunlight, but even than there is no gain or loss in total energy. Further, the total amount of energy involved in any chemical reaction such as burning of wood, does not increase or decrease, much of the potential energy stored in the substance undergoing reaction is degraded during the reaction into a form incapable of doing any further work. This energy ends up as heat, serving to disorganise or randomly disperse the molecules involved, thus making them useless for further work. The measure of this relative disorder is named entropy.

The first law can be seen operational in living systems. The radiant energy of the sun is absorbed by green plants. This radiant energy is first converted into electrical energy - energy of agitated electrons in the chlorophyll molecules. This electrical energy is further converted into chemical energy by the synthesis of complex molecules and also by the synthesis of ATP molecules. Hetrotrophs feed on the plants and thus the chemical energy flows into the heterotrophs. The heterotrophs recover the energy stored into chemical compounds during the final stages of the respiratory process and this recovered energy is stored in the terminal bond of ATP molecules. When the terminal bond of ATP is broken down, the chemical energy will be converted into kinetic energy (or energy of action) which is necessary for all the life activities. Thus, a series of energy changes take place in the living systems demonstrating the operation of the first law of thermodynamics in living things.

The second law of thermodynamics states that whenever energy is transformed from one kind to another, there is an increase in entrophy and decrease in the amount of useful energy. Thus, when coal is bruned in a boiler to produce steam, some of the energy creates steam that performs work, but a large part of the energy is dispersed as heat to the surrounding air. The same thing happens to the energy in the ecosystem. As energy is transformed from one organism to another in the form of food, a large part of that energy is degraded as heat and as a net increase in the disorder of energy. The remainder is stored in living tissues.

Laws of thermodynamics dictate that spontaneously occurring natural processes be accompanied by an increase of entropy or randomness. The change from an initial transitory state to a final stationary phase seem to involve a decrease in entropy and it may be inferred that a stationary phase is fairly stable against external disturbances since a low entropy system is unable to spontaneously move to a high entropy form. This results in succession of ecosystems with the ecosystem organisation increasing progressively (Figure 9).

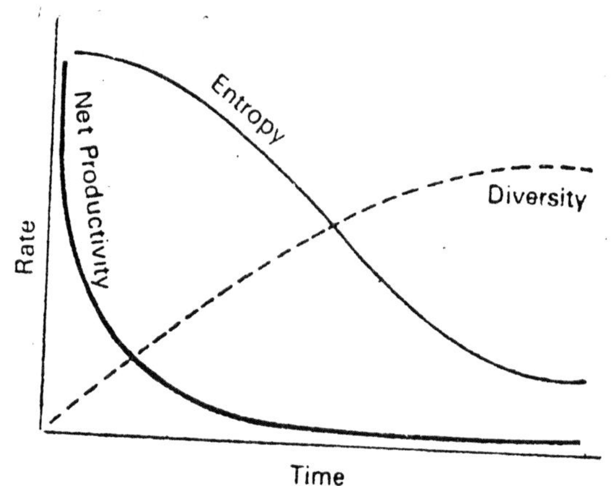

Fig. 9 : The general relations **among** net productivity, entropy and species diversity in **ecosystem** development (after Stumm and Stumm-Zollinger, 1972).

Generally, natural ecosystems evolve from unstable to stable condition. Such evolution involves decline in net productivity as well as entropy but increase in species diversity.

3. *Radiation*

The driving force in our ecosystem is radiant energy. The chief types of energy are sunlight, moonlight, starlight and the light produced by luminiscent organisms. The heated substratum in the day time release radiations in the night time, such radiations are known as thermal radiations. High energy radiations are also received in certain local areas due to the presence of radioactive rocks of fallout (Figure 10). Only sunlight has greatest ecological significance. However, the latter types of radiations are of considerable biological importance because of their potential and injurious effects on the organisms, yet they are only minor.

The sun is our parent star and is the fire that sustain all life. Its diameter is about 8,64,000 miles and contains 3,35,000 billion cubic miles of violent hot gases that weigh more than 2000 quadrillion tons. At the heart of this incandescent giat is the thermonuclear inferno which steadily feeds on its own matter—explosively fusing hydrogen atoms into lighter helium atoms and releasing stupendous floods of energy that raise the sun's internal temperature to a mind boggling 18 million degrees centigrate. Only a fraction of this prodigious output reaches our planet. Roughly 35 per cent of this radiation is absorbed : clouds, water vapour, dust, salt and smoke particles. Another 18 per cent is absorbed by the atmosphere and about 47 per cent reaches the earth. Thus, the total radiation received at the earth's surface consists of direct and diffuse radiation.

The solar energy is received in the form of electro magnetic waves. These are of different wavelengths between 2900 Å to 50000 Å or 290 mu to 5000 mu. The wavelength of maximum emission from the sun is 0.5 mu, which is in the visible portion of the spectrum (Sellers, 1965). 99 percent of the solar radiation is in the wavelength range of 0.15 u to 4.0 u, including 9 percent in the ultraviolet (0.4 u), 45 percent in the visible (0.4—0.74 u) and 46 percent in the infrared (0.74 u). The frequency of wavelength in visible spectrum ranges from 390 nm to 760 nm. Wavelength longer

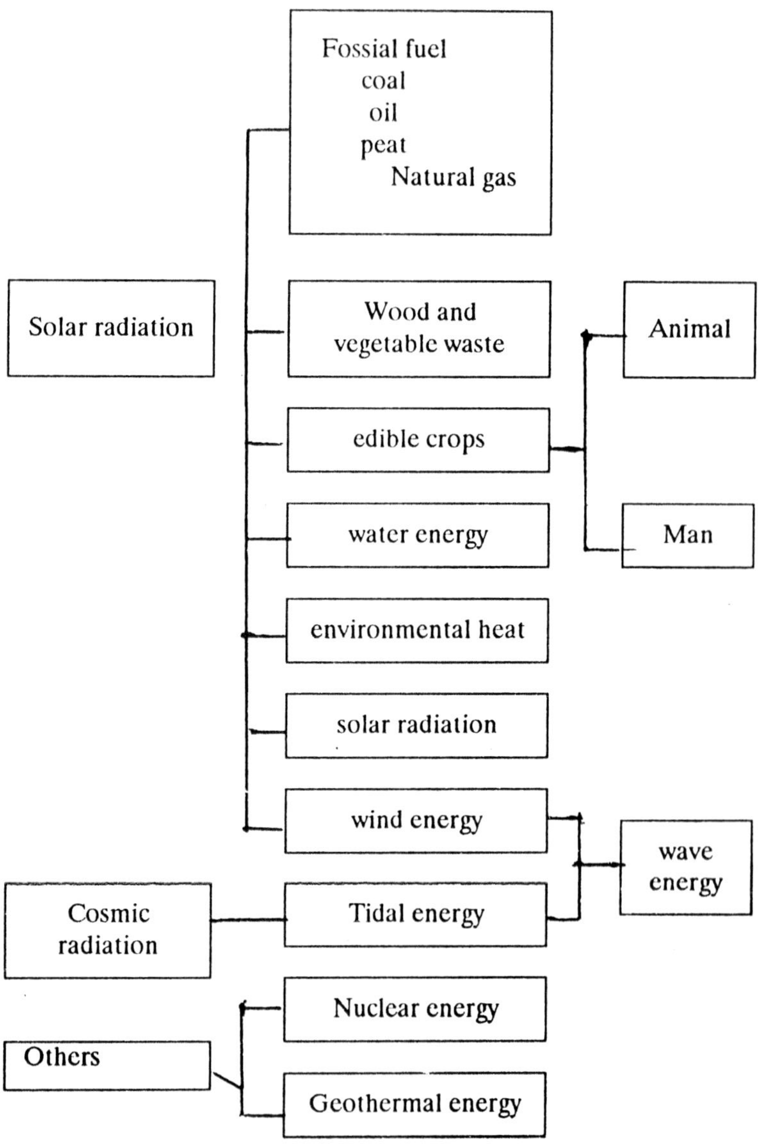

Fig. 10 : Types of radiant energy on the earth.

than 760 nm are considered to be infrared, and wavelength shorter than 390 nm are designated as ultraviolet (Figure 11).

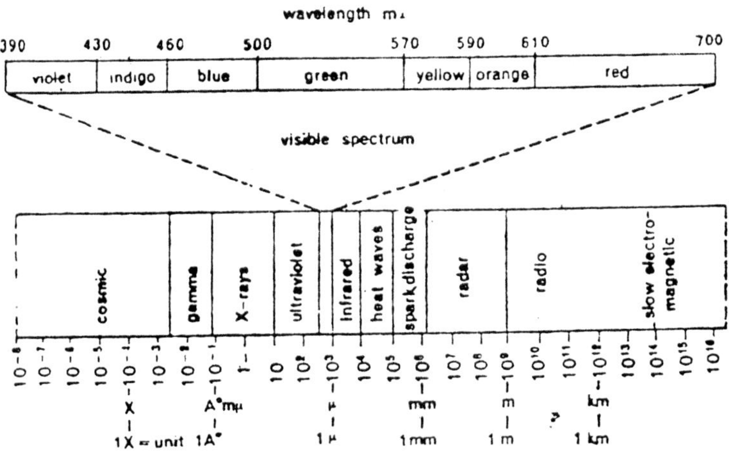

Fig. 11 : The electromagnetic spectrum of radiant energy (after Vernberg and Vernberg, 1970).

According to Geiger (1961) 86 percent of the net balance is ulilised in evaporation of water and the remaining 14 percent in atmospheric heating. A small fraction of the solar radiation which falls on the foliage of green plants is synthesized and stored as potential food for the heterotrophs. The very efficient utilize 2-5 percent incident solar radiation. However, on the global basis, only a fraction of 1 percent incident is used.

The intensity of light reaching the earth's surface varies with the angle of incidence, degrees of latitude and altitude, season, time of the day, amount absorbed and dispersed by the atmosphere and a number of climatic and topographical features such as fog, clouds, suspended water drops, dust particles, etc. When the angle of incidence is smaller, light have to travel a longer distance through the atmosphere, that is, why in equatorial regions light is most intense and decreases as we move towards the poles and at higher altitude above the mean sea level. The humidity and cloudiness of the atmosphere are most important factors responsible for variations in light intensity due to this reason the intensity of light is much greater in dry than in humid climates where cloud and fog are abundant. The direction and slope also affect light intensity. There will be no light on one side of the slope. In general, light intensity is weaker at dawn and sunset, while it is strongest when sun remains nearly overhead.

Vegetation brings about variation in light intensity reaching the ground. In tropical rain forests, major proportion of light intensity is absorbed by tree canopy and the light reaching the lower part of the ground vegetation is considerably reduced by 90-98 percent of that in the exposed areas. The amount of light reaching the forest floor depends upon the height of canopy, crown development of trees, age of trees and phenological characteristics of the species. Thus, in a forest, the mature tallest trees receive full insolation, undershrubs receive subdued illumination, and herbs grow in still feeble light conditions. However, in nature sun-flecks and movement of leaf canopy of trees play an important role in compensating for the reduced light and in making it available to the ground flora.

The light which enters in the aquatic media, comes from sun by passing through the atmosphere existing above the water surface and hence, it is subjected to all kinds of atmospheric factors like that of terrestrial environments. About 10 percent of the sunlight which falls over the water surface, is reflected back and rest 90 percent pass downwards in the water and is modified in respect of intensity, spectral composition, angular distribution and time distribution. The phytoplankton, zooplankton, suspended organic and inorganic particles either reflect or absorb the light rays. Further, in water, there is a selective absorption of light at various depths. The longer wavelengths are absorbed near the surface while the shorter wavelengths penetrate deepest. Thus, infra-red rays are absorbed in the upper layers of water (about 4 metres), red and orange rays are absorbed upto the depth of 20 meters, yellow rays penetrate upto 50 meters, green and blue rays penetrate upto 80—100 meters deep. Violet and ultraviolet rays penetrate beyond 200 meters depth. Depending upon the penetration of light, oceans are divided into 'euphotic zone' (upto 50 meter depth), 'disphotic zone' (upto 80—200 meter depth), and 'aphotic zone' (below 200 meter depth).

The solar radiation, as it reaches the earth, produces 'green house' effect in the atmosphere (Figure 12). The thick atmospheric layers over the earth behaves as a glass surface, as it permits short wave radiation from coming in, but checks the outgoing long wave ones. As a result gradually the atmosphere gets heated up during the day as well as night. If such an effect were not there in the atmosphere the ultraviolet, infrared and other ionising radiations would have also entered out atmosphere and the very existance of life would have endangered.

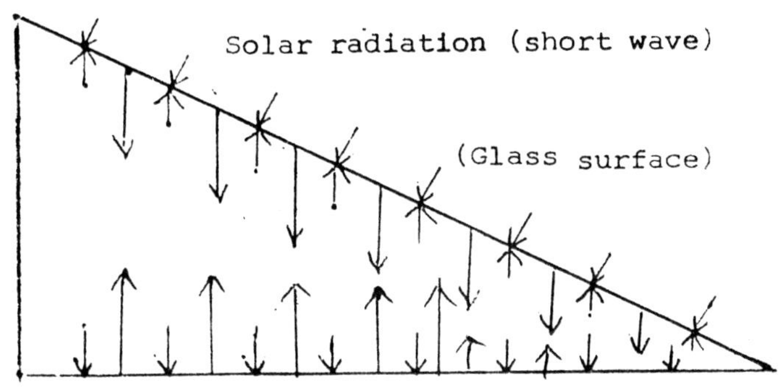

Fig. 12 : Green house effect in the atmosphere.

We might consider why the earth does not get warmer and warmer the years go by. The sun radiates continuously, yet the average temperature of the earth does not change to any real extent. This means that the earth must give as much heat energy to the outer space as it receives. The incoming solar radiation is during the day and the outgoing radiation are during the night (Figures 13 and 14). There is always a tendency of balance between the two, but differences occur, and the amount and nature of these differences generally contribute to the differences in the major climtes of the world. The way this balance is maintained, is a rather complicated process. This heat exchange is controlled by the atmosphere.

Depending upon the actual colour of the ground surface, the visible radiation from the sun is partly reflected and partly absorbed. Table 1 gives the percentage absorption of the visible solar radiation by some typical surfaces.

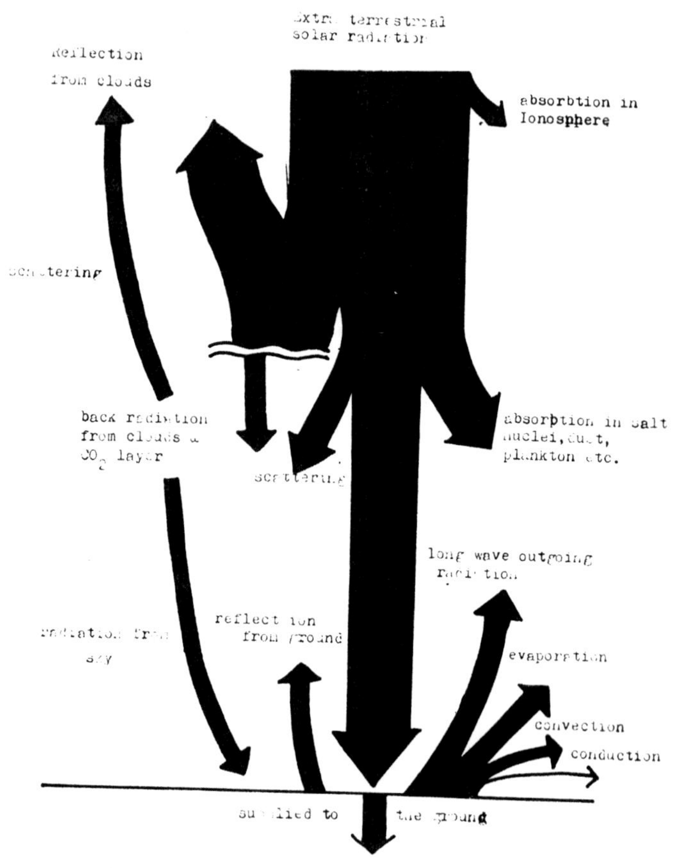

Fig. 13 : Energy exchange at about noon on a sunny day (After Geiger, R. 1961.

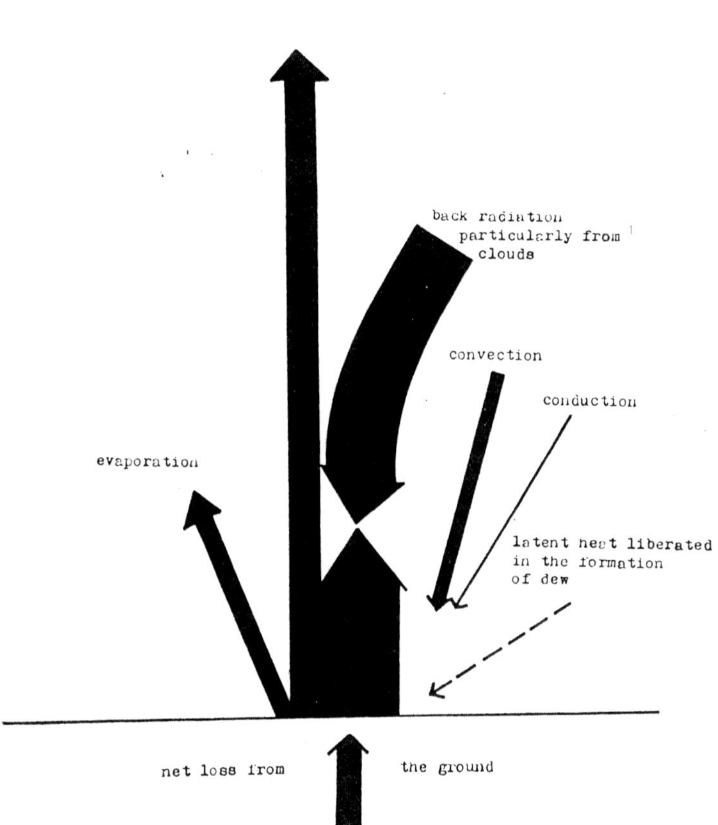

Fig. 14 : Energy exchange at night (After Geiger, R. 1961).

Table 1 : Percentage absorption of solar radiation by some common surfaces (Ramadas, 1974).

Surface	Absorption (%)
French chalk	0.0
White paint	20.0
Aluminium foil	15.0
Aluminium paint	20.0
Quartz powder (white sand)	28.0
Grey Alluvial soil	59.0
Brick	55.0
Concrete (cement)	60.0
Galvanised iron	65.0
Asbestos slate	81.0
Grass covered lawn (green)	68.0
Black paint	96.0
Charcoal powder	96.0
Black-cotton soil	84.0

The capacity of a surface to reflect radiation received from a source other than itself is called its *albedo* and is estimated in percentage of the incoming radiation. The albedo values of some typical surface are given in Table 2.

Table 2 : Albedo values of some common surfaces (McIntosh, 1963 ; Riehl, 1965).

Surfaces	Albedo (%)
Forest	5—10
Rock or buildings	10—15
Dry earth	10—25
Sand	20—30
Grass	25
Old snow	55
Clouds	50—65
Fresh snow	80
Water	2—40

Energy Exchange in Ecosystems

There are three possible pathways for radient energy to take once it reaches an organism. It may be reflected from the surface, it may be absorbed by the surface, or it may be transmitted through it.

Energy that is reflected from the surface is of no thermal benefit to an organism. Transmitted energy is of no value to an animal, plant leaves however transmit solar energy. Absorbed energy becomes a part of the thermal and physiological regime of an organism, and the quantity and distribution of absorbed radiant energy is of interest to the physiologist and the ecologists.

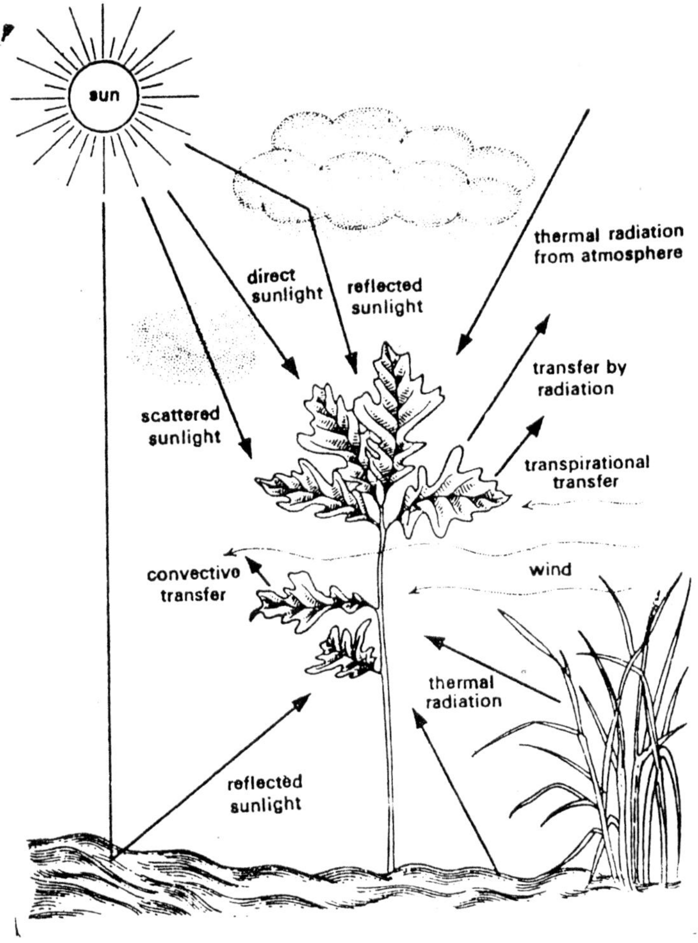

Fig. 15 : The various streams of energy between a plant and its environment (Gates, 1968).

Energy is transferred between the environment and the organism by radiation, convection, conduction and evaporation, sweating or transpiration. The various streams of energy flowing between a plant and its environment are shown in Figure 15. Not all these processes of energy exchange may operate at the same instant. An organism buried in the soil is without radiation and without convection but may have energy exchanged through the conduction of heat in the soil and mass transport of water vapour. An aquatic organism immersed in a lake may receive only a small amount of light, will receive and emit no infrared radiation, and will not exchange water vapour with the water, but will have strong energy exchange by means of convection and conduction. the terrestrial organise exposed to the atmosphere will have energy transferred by radiation, convection, conduction, and evaporation and is exposed to the greatest extremes of energy flow.

All energy absorbed by an organism must be accounted for. Conservation of energy exists here just as with any other system. An organism cannot store up energy indefinitely for it will get too hot, nor can it lose energy indefinitely for it will become too cold. There do exist short-term transients of warming or cooling. A leaf may warm or cool rapidly within a few seconds, a branch within a few minutes, and a massive tree trunk may take several hours to change its temperature.

Effects of Solar Radiation on Plants

Solar radiation either directly or indirectly affects almost all aspects as a plant's life. It controls the life-from, shape, tissue, differentiation, chlorophyll production, number and position of chloroplasts, leaf structure, stomatal movement, transpiration, pollination, flowering, seed and fruit maturation and dispersal etc. some of the well known effects are as follows :

Light is an essential factor in the formation of chlorophyll pigment in chlorophyllous plants. It also has a very strong influence on the number and position of chloroplasts. The upper part of the leaves which receive full sunlight has larger number of chloroplasts that are arranged in line with the direction of light. In leaves of plants which grow under shade, chloroplasts are very few in number and are arranged at right angle to the light rays, thus increasing the surface of light absorption.

The photosynthetic rate of a plant leaf increases linearly with increased light intensity upto the point of light saturation. The efficiency of the photosynthetic process, however, declines steadily with increasing light intensity. Thus, a leaf exposed to full sunlight is not very efficient in utilizing light energy ; at best it utilizes about 5 percent. At low intensity the photosynthetic rate is lower, but efficiency increases and may approach 20 percent. However, the plant yield increases under high light intensity, since more light reaches the lower leaves and even the lower layers to chloroplast within the leaf.

Plants have been grouped according to their tolerance to light intensities into shade loving, photophobic, sciophytes, and sun loving, photophilous, heliophytes. The shade plants maintain a high rate of photosynthesis in low light intensities, while the heliphytes are adversely affected by shade. There are some heliophytes which grow best in sun, but can grow fairly well under shade, such plants are called 'facultative sciophytes'. Similarly, there are facultative heliophytes which grow best at lower light intensity, but can grow well in sun-light. The morphological and physiological characteristics of heliophytes and sciophytes have been shown in Table 3.

Light affects opening and closing of stomata, influences the permeability of plasma membrane and has heating effect. All these in turn affects absorption of water. In many plants the respiratory rate increase with the increase in light intensity (*e.g., Canna, Nerium, Bougainvillea*). However, in certain plants respiration rate is decreased in intense light.

Stomata are the chief pathways through which the transpirational losses and exchange of carbon dioxide and oxygen between plants and atmosphere take place. The opening and closing of stomata is regulated by the presence or absence of light. The influence of light upon the transpiration is two fold. First, it causes greater stomatal opening and leads to greater transpiration. Second, incident light upon the leaf bring about an increase in its temperature and this leads to increased transpiration.

Table 3. Morphological and physiological characteristics of heliophytes and sciophytes.

	Characteristic	Heliophytes	Sciophytes
1.	Stem	Thick, short internodes, profusely branched ; well developed conducting tissues and mechanical tissues.	Thin, long internodes, sparsely branched ; poorly developed conducting and mechanical tissues.
2.	Leaves	Small, thicker, leaf blades not flat, compound, oriented obliquely to path of incident radiation ; more hairy ; thicker cuticle and cell walls, small-sized cello few small sized chloroplasts well developed pallisade, poorly developed spongy parenchyma, less chorolophyll content.	Large, thinner, oriented at nearly right angles to incident radiation ; less hairy or glabrous ; thin cuticle and cell walls, less developed pallisade, chloroplasts more and larger, well developed spongy parencyma, high chlorophyll content.
3.	Roots	Longer, numerous, more branched, high root/shoot ratio. More nodules in legumes.	Small, few, less branched, low root/shoot ratio. Less nodules.
4.	Physiology	Low photosynthetic rate per unit surface, high respiration rate, low water content, higher concertation of salts and sugar and high osmotic pressure, decrease in acidity in cell sap, low potassium content, early appearance of flowers, high resistance to temperature injury, drought and parasites.	

Light also affects the movement in many plants. The roots stems and leaves show variable responses to light. The effect of sunlight on the plant movement is called 'heliotropism' or 'phototropism'. The roots are negatively phototropic and the stem elongates towards light (positively phototropic).

Seeds of many plants are light sensitive, that is, either their germination requires light or is inhibited by light. Light has both promotory and inhibitory effect on the germination regulating mechanisms (Table 4). Some seeds prefer to germinate in light *e.g., Astercantha longifolia* and *Ruellia tuberosa*. Such type of seed germination can be termed as "photoblastic". While others show high germination in continuous darkness *e.g.*, *Cenchrus* and *Dectyloctenium* species (Sen, 1977). In most cases red light promotes germination and far red light inhibits germination. Investigations regarding the understanding of mechanism of seed germination in light have found the involvement of a pigment 'phytochrome'. The pigment occurs in two reversible forms (*i.e.*, Pr and Pfr) developed under red and far red light and the germination depends upon the balance between the two forms in favour of Pfr, which develops under red light :

$$\underset{\text{(inactive)}}{\text{Pr}} \underset{730\text{ nm}}{\overset{660\text{ nm}}{\rightleftharpoons}} \underset{\text{(active)}}{\text{Pfr}}$$

Table 4. List of light sensitive seeds.

Light necessary for germination	*Light inhibitory to germination*	*No effect of light*
Nicotiana tabacum	*Ailanthus glandulosus*	*Anemone mamarosa*
Alisma plantago	*Aloe variegata*	*Cystisus nigricans*
Capparis spinosa	*Ephedra helvetica*	*Datura stramonium*
Colchicum autumnale	*Lycopersium esculentum*	*Hyacinthus candidans*
Daucus carota	*Primula spectabilis*	*Linaria cymbalaria*
Ocimum americanum	*Tulipa gessenreiana*	*Pelargonium zonale*
Anagallis arvensis	*Nigella damacaena*	*Sorghum halapense*
Taraxacum officinale	*Cistus radiatus*	*Tragopogon pratensis*
Sueda maritima	*Mirabilis jalapa*	*Vasiccaria viscosa*
Nasturtium officinale	*Hedera helix*	
Lactuca sativa	*Ranunculus crenatus*	
Saliva pratense		
Phacelia sp.		
Lepidium virginicum		

Different wavelengths of light produce different growth effects. A plant grows better in the full spectrum of the visible light rather than in any one portion of the spectrum. Green light retards growth than either blue, violet or orange red ranges of spectrum, and this is not surprising because at green wavelength photosynthetic efficiency is lowest. The orange-red region of the spectrum results in lesser elongation of the stem and hypoctotyl than light from other parts of the specutrum. On the other hand, blue-violet light causes pronounced elongation. Ultraviolet light has a marked inhibitory effect upon the elongation of plant organ. Leaf expansion is greatest in full spectrum of light, slightly less in blue-violet and least in green light.

Table 5. Classification of plants according to their photoperiodic requirements (Salisbury, 1963).

Day-neutral plants

Cucumis sativus, Gossypium hirsutum, impatiens balsamina, Nicotiana tabacum, Phaseolus lunatus, Poa annua, Solanum tuberosum.

Quantitative short-day plants

Andropogon virginicus, Cannabis, sativa, Datura stramonium, Cosmos bipinnatus, Salvia splendens.

Quantitative long-day plants

Brassica rapa, Nigella avroensis, Secale cereale, Sonchus oleraceum, Sorghum vulgare.

Qualitative short-day plants

Bryophyllum pinnatus, Chenopodium rubrum, Ipomoea batatas, Lemna purpusilia, Coffea arabica, Calanchoe blosfeldiana.

Qualitative long-day plants

Agropyron smithii, Anagallis arvensis, Lolium temulentum, Avena sativa, Agrostis palustris.

Intermediates

Chenopodium album, Tephrosia candida, Saccharum officinarum.

The duration of light (photoperiod) is of considerable importance to most plants. Based on photoperiod responses, plants can be classified as short day plants, long day plants and day neutral plants (Table 5). Day neutral plants are those whose flowering is not affected by day length, but rather is controlled by age, number of nodes, previous cold treatment etc. Short day and long day plants are influenced by day length. When the photoperiod reaches a certain limit, it inhibits or promotes a photoperiodic response. The length of this period, so decisive to the response, is called the 'critical day length'. It varies among organisms but usually falls somewhere between 10—14 hours. Throughout the years plants "compare" this time scale with the actual length of day or night. As soon as the actual length of day or night is greater or smaller than the critical day length, the plant may flower of cease to flower whose flowering is stimulated by day length shorter than the critical day length. Long day plants are those whose flowering is stimulated by day length longer than a particular value. The latter usually bloom in late spring and summer. Long day plants and short day plants are further classified into subgroups on the basis of quantity and quality of light. Intermediate plants are those which flower at neither too short too long photoperiods. However, in a plant, *Melia elegans*, flowering occurs only at too small or too large photoperiods and the plant remains vegetative at intermediate day lengths. Such a condition is called 'amphiphotoperiodism'. The photoperiod is also effected by temperature and it governs the latitudinal distribution of plants.

Plants distributed in different geographical regions, in course of their long evolutionary history, have become ecologically adjusted to the light period of the place. Even within a species there are examples of photoperiodic ecological races as in *Xanthium strumarium* (Kaul, 1959).

The ecological understanding of responses of plants to different conditions of light is of great practical significance. For instance, the crop of betel leaves in India is grown in artificial shading. This results in greater expansion of leaf with less development of stiff or hard tissue and less of chlorophyll. The betel leaves that are soft, palatable and yellowish in colour fetch a higher price.

4. *Temperature*

Absorption of radiant energy raises the temperature of the absorbing substance. Since, substances very in their absorbing capacity, temperature differences exist our biosphere. The energy flows as heat from warmer to the cooler substances. The transport of energy in the biosphere is necessary to maintain the efficiency of life on earth, just as communication system maintains standards of living in a country.

Temperature is a measure of heat. Heat is a form of energy and is called as thermal energy. The same influx of energy on which photosynthesis depends is also the source of thermal energy that characterises the physical environment.

Thermal energy is exchanged between the organisms and the environment by movement of substantial volumes of the substance concerned, either in vertical direction (convection), or in horizontal direction (advection), or molecular impact without transfer of the matter itself (conduction), or by re-radiation at longer wavelengths.

Temperature of the surface soil fluctuates daily and seasonally. The temperature of the surface soil may be 30°C higher in the sunlight than in the shade and upto 17°C higher during the day than during the night. On the desert this spread may be as high as 40°C. The Thar Desert of Rajasthan shows a diurnal change of 20—30°C for all seasons. The lowest temperature recorded for any land mass is—70°C (Siberia in 1947). Higher temperature may likewise go often 85°C as in certain desert at noon.

Temperature fluctuations are comparatively less in the aquatic environments than in the terrestrial environments. The increase in depth of aquatic medium often increases the temperature fluctuations. the minimum temperature in the sea is—3°C, while in fresh water it never goes below 0°C. The maximum temperature in coeans generally goes upto 36°C, but in shallow pools of fresh water it may go higher. In deeper bodies of water, heating and cooling are restricted to the surface strata, but the deeper layers also get a lot of heat as a result of vertical circulation of water, where in, due to circulation of water, surface waters are brought to the deeper regions and *vice-versa*. Studies on the vertical stratification of temperature have led to the classification of water into three strata. The superficial layer is called 'epilimnion', which is warmer, and its temperature may rise upto 27°C during summers. The bottom layer is called the 'hypolimnion', which is coller, and its temperature may go upto 5°C. In between epilimnion and hypolimnion, occurs an intermediate zone called 'thermocline' or 'metalimnion'. The process of differentiation of fresh water habitat into these three strata is called 'thermal stratification'.

Atmospheric temperature varies from to night (Figure 16), from latitude to latitude, from altitude to altitude, and the season. A daily maximum of atmospheric temperature usually comes in the mid-afternoon, and minimum just before the sun rise. Since, the total insolation decreases with distance from the equator, obviously, the temperature values are maximum at equator, decreasing gradually towards the poles. Besides latitude and altitude, colour and composition of surface, plant cover, water content of soil, physiograhic factors such as steepness of slope, exposure of slope and direction of mountain chains greatly affect the temperature conditions. In nature, valleys and low lands are some times much cooler due to sinking in of the heavier cold air.

Fluctuations in temperature are brought about by varying intensity and quantity of solar radiation—*lateral variation* and also by certain physical phenomenon such as movement of air—*vertical variation*. As the air moves upwards away from the earth, it comes under low pressure which consequently results in the expansion of air. However, when the air moves downwards, it encounters high pressure and this results in the contraction of air masses. Due to these changes of expansion and contraction, the temperature of moving air changes, such changes are therefore due to internal forces and are called as *adiabatic changes*. In contrast to the adiabatic changes of temperature there is a natural vertical decrease in temperature as we proceed from lower to higher elevations, such changes are known as *lapse changes*. The rate of adiabatic and lapse heating and cooling of dry air is about 5.5°C and 3.5°C respectively for every 1000 meter change in altitude.

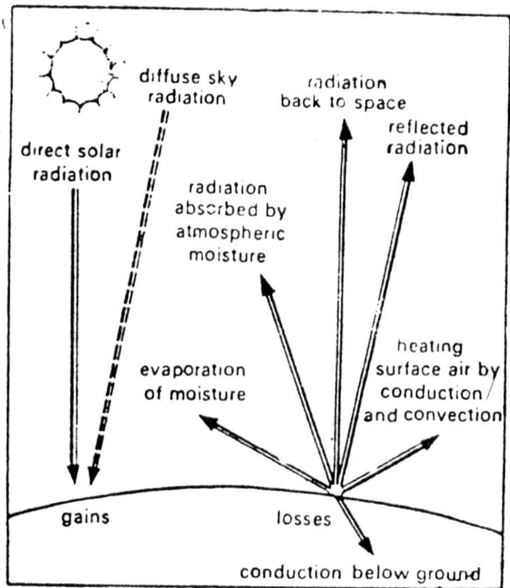

A Day-time surface heat exchange

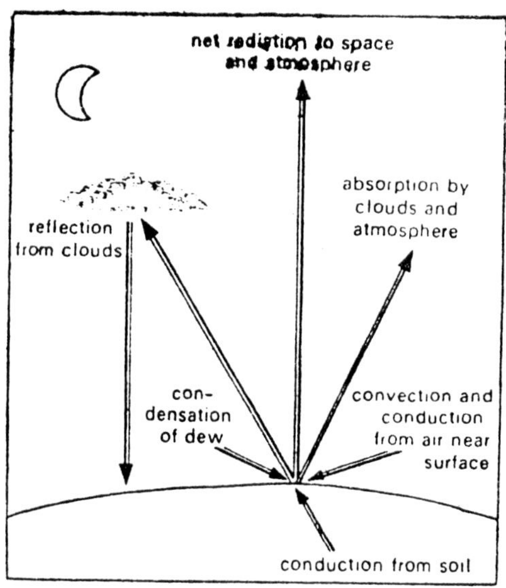

B Night-time surface heat exchange

A-Solar radiation that reaches the earth's surface in the daytime is dissipated in several ways, but heat gains exceed heat losses. B-At night there is a net cooling of the earth's surface, although some heat is returned by various processes (after Smith, 1974).

In general, the temperature of the air lowers with increasing elevation above the land surface. Occasionally, there will be a reversal of this trend, and an increase in temperature with height (a temperature inversion) will occur. This is often a night time phenomenon (Figure 17), although it is by no means confined to this period. Within a relatively clear atmosphere during the daytime solar radiation is absorbed by the earth's surface at a maximum rate. The earth in turn warm the lower atmosphere by conduction, by radiation, and by the convection currents. At night, rapid loss of heat from the earth's surface results in the eartn cooling more rapidly than the lower atmosphere. The heat exchange is then from the warmer atmosphere to the cooler earth, with a resultant decrease in temperature in the lower layers of the atmosphere. A typical night tine surface inversion will occur in the lower atmosphere. The inversion frequency, intensity and depth vary with weather and the topographic conditions (Basile, 1971).

Plants are at the mercy of the environment in regard to temperature. Except in sunlight of high intensity, temperature of leaves, stem, flowers and fruits are usually very close to ambient air temperature. Plants can be injured if they are overheated by the sun. Under intense irradiation the dark bark of a tree can become very warm—a danger both in winter, when a sharp temperature gradient is produced between the lighted and shaded sides of the tree, and in summer, when the heating can be sufficient to interrupt sap flow. Injury so caused is evident at the edges of clearing and along roads, especially in beeches. Green leaves are not heated very much by radiation ; furthermore, their temperature can be brought appreciably below that of the surroundings by transpiration. A cut off leaf is rapidly heated by sunlight, because of lack of water.

A plant can gain heat in several ways. Principally this gain is by absorption of radiation, conduction from air or water layer immediately over the surface of the leaf or stem. It can loose heat by conduction and convection to the surrounding medium, by radiation of long wavelengths, or as latent heat by the evaporation and transpiration of water.

Different species of plants vary greatly in the temperature that they can endure. Most of the plants grow only in a narrow range of temperature conditions *i.e.*, from 0°C to 45°C. However, some arctic

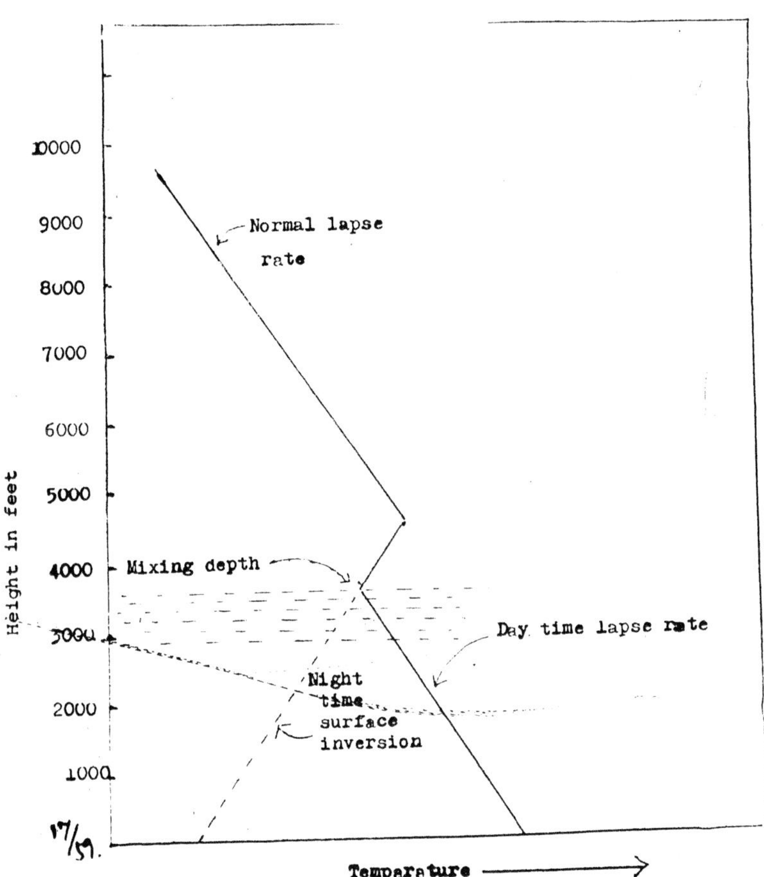

Fig. 17 : The temperature inversion conditions within the atmosphere.

and alpine plants do grow at sub-freezing temperatures, whereas several mosses and lichens grow at temperatures as high as 100°C.

Plants are subjected to a considerable range of temperature during the period of their growth. They grow only when the temperature remains within certain limits, mature and die when the temperature falls too low or become too high. The temperature at which a plant functions best, is called its *optimum temperature. Maximum temperature* is that above which the living activities are not detectable in a plant. *Minimum temperature* is that below which a plant cannot continue living activities. The three temperature limits are also called the *cardinal temperatures.* The cardinal temperatures vary with age of a plant, with its physiological conditions, and with changes in environmental factors (Figure 18).

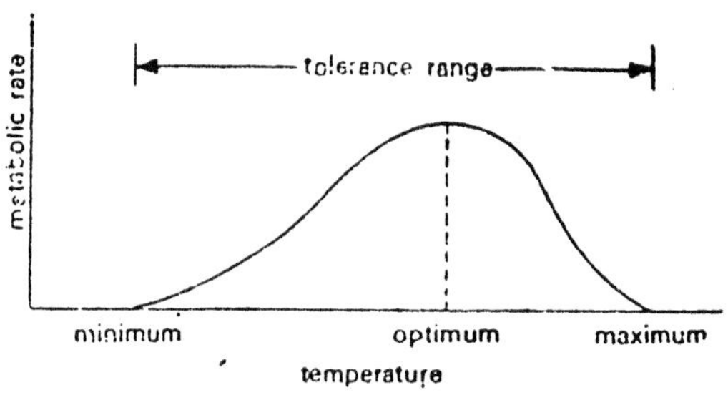

Fig. 18 : Effect of temperature on physiological activity of organisms.

When the temperature falls below the minimum for a plant, its growth activity ceases, due to non-availability of water, reduced cell wall permeability, freezing of cell-sap which leads to the rupture of the cell walls, freezing of water in inter-cellular spaces, conversion of starch into fat, sugars into oil and insoluble substances respectively. The *thermal death point* usually lies a few degrees above the optimum temperature for growth. High temperature can injure and kill the protoplasm, besides bringing about desication and disturbance between respiration and photosynthesis. The high temperature leads to ore rapid exhaustion of foof by respiration, stimulates the opening of stomata, even when the leaves are wilted.

The response of plants to rhythmic diurnal fluctuations in temperature is called *thermoperiodism.* Usually, the plants show better growth at higher day temperature and lower night temperature, than at constantly uniform temperature.

Temperature affects almost all the physiological processes of a plant such as the permeability of the cell wall, water absorption, photosynthesis, respiration, transpiration and through enzyme activity several other metabolic processes. Photosynthesis operates over a wide range of temperature. Most algae require lower temperature range for photosynthesis than the higher plants. The rate of respiration increases with the rise of temperature, but beyond the optimum limit high temperature decreases the respiration rate. The rate of respiration is doubled at the increase of 10°C above the optimum temperature, provided that other factors are favourable. However, optimum temperature for photosynthesis is lower than that for respiration. Lowering or increasing of temperature from the optimum limit soon brings about the minimum or maximum limit, any slight change in which leads to the stopping of the growth activity. This phase of an organism is called *dormancy*. There is again a definite limit of temperature at which an organism can remain dormant, any further change in temperature proves lethal and may even cause death. Even under dormant condition respiration and photosynthesis may continue, but at a comparatively slower rate. Low temperature further affect the plant by precipitating the protein in leaves and tender twings and by dehydrating the tissues. Extremely low and high temperature have adverse effect on the growth of plants Low temperature brings about cold injuries such as freezing or chilling injury. The plants which can tolerate very low (below 0°C) temperature are called frost resistant or cold resistant. Extremely high temperature cause stunting and final death of plants. This is called heat injury. Plants able to survive under high temperature are called heat resistant. Temperature effects the root growth. Large diurnal fluctuations in temperature particularly low day and high night temperature retard the leaf growth. In many temperate plants, low temperature, short duration treatment, result in rapid and more flowering when returned to normal temperature. Temperature affects the germination of seeds and sprouting of buds. Most seeds germinate at temperatures between 25 to 35°C, but the leguminous seeds require in general a higher temperature. In plants like *Oscimim sanctum* pretreatment of lower temperature is necessary to seeds before they can germinate at relatively high temperature. Similarly, the vegetative propogules of *Spirodella, Eleocharis, Cyprus* etc. and seeds of many plants require chilling under moist conditions for small duration. The seedlings are less tolerant to temperature fluctuations than the grown up plants and similarly the young growing parts are also susceptible to high or low temperatures.

The productivity of many plants is increased when the temperature increases. In the moist warm tropical areas the dead organic matter is rapidly decomposed. In contrast, forests with dead trees are characteristic of temperate and cool zones, because of low temperature the dead wood remains intact for a long time. In a tropical rain forest the fallen trees decay so rapidly that they are hardly noticeable (Figure, 19).

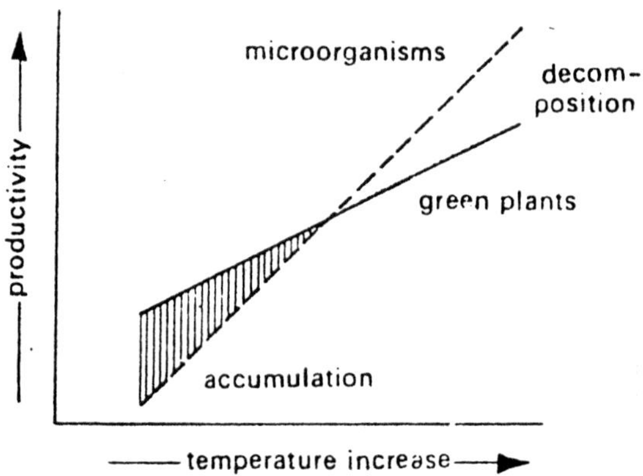

Fig. 19 : Effect of temperature on biomass accumulation and decompsition.

The temperature is of great significance among the physical factors of the environment in delimiting the distribution of various species of plants on earth. This is because various species have various temperature requirements and tolerance limits. On the basis of temperature world's vegetation has been divided into various classes as : (i) *Megatherms*—where high temperature prevail throughout the year and the dominant vegetation is tropical rain forest ; (ii) *Mesotherms*—where high temperature alternates with the low temperature, and the dominant vegetation is tropical deciduous forest ; (iii) *Microtherms*—where low temperature prevail and the vegetation is of mixed coniferous forest ; and (iv) *Hekistotherms*—where very low temperature prevails and alpine vegetation prevails.

5. *Atmosphere—Gases and Wind*

Our atmosphere is made up of a series of invisible shells of air covering the earth. Each new shell is just a little bigger and completely covers the previous shell of protecting air. The gaseous mantle forming the atmosphere extends into the outer space some 100 km or so above the earth's surface. it maintains its contact with all the major types of environment of the earth, interacting with them and greatly affecting their ability to support life. The atmosphere is a reservoir of several elements essential to life and it serves many functions including the filtering of radiant energy from the sun, insulation from heat loss at the earth's surface, and stabilization of weather and climate owing to the heat capacity of the air.

There are five concentric layers within the atmosphere which can be distinguished on the basis of temperature (Fig. 20). These are as follows :

1. Troposphere
2. Stratosphere
3. Mesosphere
4. Thermosphere
5. Exosphere

The earth's atmosphere from the surface upto about 20 km (8 km on poles) is called *troposphere.* The troposphere concerns us most because it is the only part that supports plants and animals. The important weather events, such as cloud formation, lightening, thundering etc take place in the troposphere. Air temperature in the zone gradually decreases with height at the rate of about 6.5°c per km. Towards the upper layers of the troposphere, the temperature might decrease upto −60°c. The water vapour in the atmosphere, and most of the air, is confined to the troposphere. Air in the troposphere moves side-ways (laterally) and up-down (vertically) keeping it well mixed. The troposphere gradually merges into the next zone, which is known as ***tropopause.***

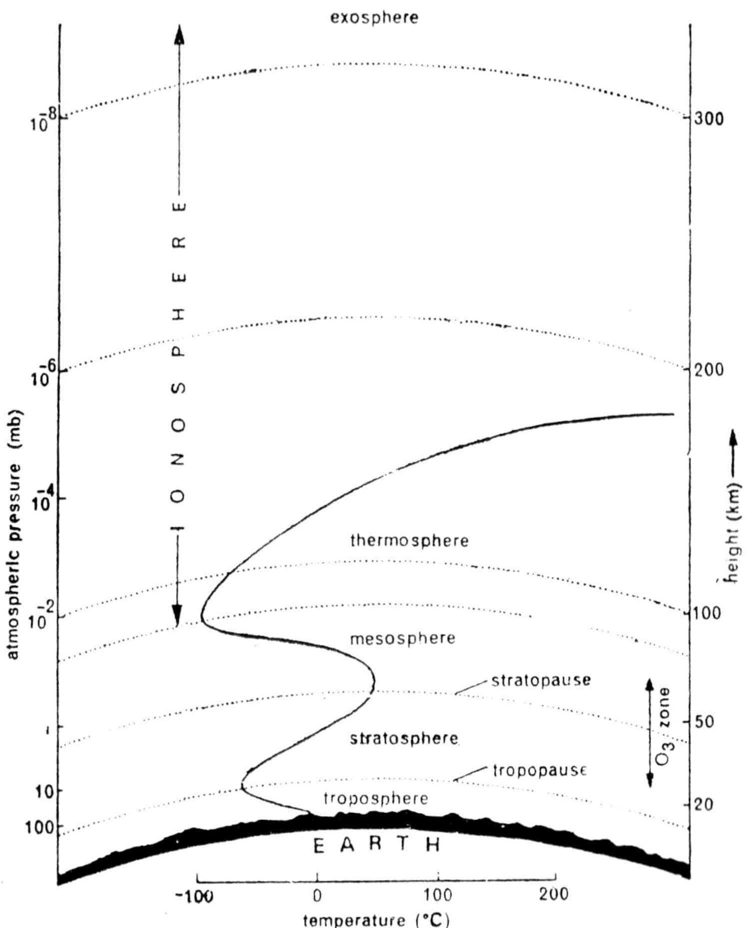

Fig. 20 : Diagrammatic sketch showing the principal zones of atmosphere along with variations in temperature.

Stratosphere is the second zone of about 30 km, where temperature are about 90°c. In this zone the temperature shows an increase from a minimum of about −60°c to a maximum of about 5°c. The increase in temperature is due to ozone formation under the influence of ultraviolet rays of solar radiation. Ozone is formed from oxygen by a photochemical reaction, in which solar energy splits the oxygen molecules to form atomic oxygen which then combines with oxygen molecules to form ozone. Stratosphere is the region of horizontal motion with no up-and-down wind motions. In the lower layers of the stratosphere, the horizontal winds reach the highest speeds of any wind in the entire atmosphere. The high speed winds are known as jet-streams. The stratosphere is above clouds, violent storms and precipitation. The upper layers of the stratosphere form the *stratopause.*

Mesosphere is the third layer of the atmosphere, about 40 km in height. In this zone, temperature shows a decrease upto −80°c. The upper limit of the mesosphere in termed as *mesopause.*

The region of atmosphere above the mesosphere is the *thermosphere,* which extends upto 500 km above the earth's surface, and is characterised by steady increase in temperature with height. In this region ultraviolet radiation and cosmic radiation cause ionisation of oxygen and nitric oxide. Hence, this region is also called as *ionosphere.*

Exosphere is the region above the thermosphere. It lacks atoms except that of hydrogen and helium and extends upto 32190 km from the earth. The earth's magnetic field becomes more important than gravity in the distribution of atomic particles in the exosphere. Exosphere has a very high temperature due to solar radiation.

All these layers of the atmosphere are of interest to the ecologists since together they form the total blanket of air which moderates the solar energy reaching the biosphere, and also serve as a blanket and regulates the earth's radiation escaping into the space.

Air Composition

The gaseous mixture of the troposphere is called air. It has a relatively constant gaseous composition. The composition of dry air, except water vapour and air-borne dust particles have been shown in Table 6. Both water vapour and dust particles occur in extremely variable concentration. The concentration of water vapour in air ranges from virtually zero percent to more than 4 percent. As a rule its abundance is a function of altitude and temperature. Dust is even more variable component of air.

Table 6 : Composition of dry air by volume (Clapham, Jr, 1973).

Gases	per cent (By volume)	Gases	Per cent (By volume)
Nitrogen	78.0841	Crypton	0.00011
Oxygen	20.9486	Xenon	0.00009
Argon	0.9340	Hydrogen	0.00006
Carbon dioxide	0.0318	Methane	0.0002
Neon	0.00182	Nitrous oxide	0.00005
Helium	0.00052	Ozone	0.000004

The four principal gaseous constituents of the air are nitrogen, carbon dioxide, nitrogen and water vapour, in varying quantities spatially and temporally.

Nitrogen forms the main bulk of the air (78%). It combines with very few things and so remains inactive. Koromondy (1969) stated that through electrochemical and photochemical fixation about 35 mg/m^2/yr of nitrate is formed and through biological fixation it ranges between 14-700 mg/m^2/yr. The biological nitrogen fixation is usually carried out by bacteria like *Azotobater* (aerobic), *Clostridium* (anaerobic), blue green algae like *Anabaena* and *Nostoc,* symbiotic bacteria like *Rhizobium.*

Oxygen is an indispensable gas for life, and constitutes about 20.94 per cent of the air. It is used during respiration by all living organisms and is also utilized during weathering of rocks in soil formation processes. If autotrophs of the biosphere were removed somehow, the entire oxygen of the atmosphere would be used up in the weathering process and life would disappear from our planet.

Carbon dioxide is the third important constituent of air. It is a the product of respiratory activities of the organisms and is also produced during the burning of fuels. Its concentration varies from hour to hour during the day. In the morning carbon dioxide concentration is on the high side because of accumulated respiration of the community in the absence of photosynthesis. The oxygen content gradually increase by noon and carbon dioxide content correspondingly decrease. In the thickly populated places and in the vicinity of forests and grasslands the carbon dioxide content of the atmosphere may be three times higher (0.09%) than the normal value (0.03%). If the carbon dioxide exceeds 1.0 percent in the atmosphere it becomes injurious for the organisms.

Water vapour is perhaps the most important gas in the atmosphere. Because of water vapour, we have weather as we know it. Without it we could not see our weather, although water vapour itself is an invisible gas. When water vapour condences, weather is made visible in the form of clouds, rain, fog, and snow. Water vapour also acts as a storer and giver of energy. The process of changing water and water vapour from one form to another is the basic way the energy of our atmosphere is balanced and distributed. A great amount of sun's energy required to change water into vapour. When vapour changes back into same form of water in the liquid or solid state (when it condenses) ti gives out heat.

Air As a Medium for Organisms

Air is not an easy and suitable medium to support life and actually no organism ever originated in air, though, certain aquatic and terrestrial organisms have become secondarily adapted for aerial existence. Air has lower buoyancy than water, which leads to stronger pull of gravity on the organisms towards the earth. In air organisms are exposed to evaporation of water from their bodies, which threatens with death due to desiccation. Changes in temperature are much more drastic in air and there is a greater danger of chilling or overheating. Further, light exposure are much longer and more intense in air. Even more miner supply becomes acute in air.

Density of Air

The density of air is greatest at the ground and sea level. This is because millions of air molecules above press down on those below them, packing the molecules close together. In fact, the weight of this enormous blanket of air presses against the entire earth's surface with an average pressure at sea level of 5.88 pounds for each square centimeter of surface area. The gas molecules get farther apart the higher we go. Some where beyond 643.72 to 804.65 km, above the earth's surface, they are bounced out into space by their last collisions with other molecules. This very thin fringe of gas molecules is where our atmosphere ends and space begins.

Air Cools as it Rises

When air is lifted, pressure decreases and air expands. Distance between molecules grow greater. The temperature drops as molecular speed slows down. The air's ability to hold water vapour becomes less. Depending on how much the air is cooled, and how saturated with water vapour it is, condensation of water vapour will occur and form clouds, rain or snow. Air rise in three ways :

(1) It can be lifted by the natural shape of the landscape. As air flows upslope on mountains it expands, cools and begins to condense its water vapour. If the air is very moist, the vapour forms rain. On the down slope, the reverse process takes place. Air becomes warm as it descends. In mountain areas where prevailing winds and air flow come from one direction only, the upslope side is green and heavily wooded. The downslope side is frequently dry and sandy, with very little growth except small shrubby bushes.

(2) Air rises when the land surface beneath it is heated. The air layer in contact with the land surface becomes heated. This causes increased molecular action forcing the air to rise, expand and cool at higher levels.

(3) Third type of lifting occurs when large masses of air come together. The different air masses have different temperature and density. The air mass having greatest density has the greatest weight. The lighter less dense air mass is forced to flow up over the heavier colder mass.

Origin of Air Masses

We have already seen that air warms up by coming in contact with heated surface of the earth, and cools by coming in contact with cold land or sea surface. Air that moves over large bodies of water absorbs large amount of water vapour. The air must remain there for a long time so that its new characters become firmly established throughout its vertical height. If the contact with the surface is brief, only a thin layer of air, nest to the earth's surface is changed. Time is needed to exchange that thin layer of water vapour with other layers of gases above it.

Some air masses are difficult to tell from other air masses. On the other hand, there are certain air masses whose characteristics are so definite that no one could mistake them. Chief among them are :

(1) *Cold and dry*—Some air masses that come from cold ground, far from water areas are very cold and dry. It becomes dense, ice-cold, and dry form the surface all the way upto the stratosphere.

Because it is very cold, it can hold very little moisture. With moisture lacking there can be no rain.

(2) *Warm moist*—It has spent few days or weeks moving slowly, or even resting, over large body of warm water.

(3) *Cool moist*—It originates over cold areas. The air is apt to produce frequent rain squalls as it moves over mountains. Heavy clouds develop in this air mass and severe icing conditions exist in the clouds.

(4) *Warm dry*—A fourth major air mass, having a personality all of its own is the combination of high temperature dryness.

Movement of Air Masses

Although an air mass often remains for many days at its source without much movement, giant forces and pressures continue to work. These forces cause pressures to built up behind the air mass so that it may be pushed from behind and moved from the place where it build up. They generally move from region of higher pressure to that of low pressure. These movements are also controlled by (a) *Coriolis force,* and (b) *Centrifugal force.*

Coriolis force is brought about by direction of the rotation of earth with a slight deflection. In the northern hemisphere the direction of coriolis force is towards north in relation to the earth's surface, and therefore air moves towards east. In the southern hemisphere the situation is reverse. The centrifugal force determines the speed of air mass, but not direction.

Due to greater heating at the equator, the warmed, rising air flows towards either pole. The direction in which the earth and the atmosphere rotate is from west to east. The warmed air moving towards the poles rotates with the same speed, west to east, as the equator. It is moving more rapidly towards the east than the rotating surface of the earth beneath it. It is called a west wind (blowing from the west). We also have a just opposite situation when cold polar air begins to move away from the poles towards the equator. The polar region rotate very slowly as compared to equatorial regions. They move with much greater speed than that of the cold air. An east wind is the result.

Winds Spin in Pressure Centres

In the northern hemisphere, cold polar air masses move towards the south and warm tropical air masses move towards the north. These two air masses intermingle the year round. When the cold air mass pushes under the warm air, a rotating motion develop. This is due to the spinning of the earth.

The rotating motion of these interlaced masses produces a counter clockwise spin where they meet. This counter clockwise spin of the air in the northern hemisphere is called *cyclonic circulation*. The centre of this spinning wheel of winds is of low pressure area. The spin of the winds around a high pressure area is clockwise—*anticyclonic circulation.* A low pressure centre causes winds to flow towards it. A high pressure centre pushes winds away from it. The whole circulation picture is completely reversed in the southern hemisphere.

Cyclones

Cyclones are caused by atmosphere disturbances and are accompanied by torrential rains and sometimes tidal waves. They travel in the form of a violent whirl from the high seas towards the coasts. They are usually a tropical phenomenon.

In the Atlantic and the Eastern Pacific, cyclones are called *Hurricane,* in the Western Pacific—*Typhoons*, in the Australia *Willy-Willies,* and in the Phillippines—***Bagul.***

A cyclone develops as a result of thunderstorm activity over several successive days. A lightening flash at this time releases a force of as much as 100,000,000 volts.

When lightening strikes, the energy that is liberated, heats the air through which it passes. The air suddenly expands. As the current of heavy lightening strikes build up, the heating of the air also intensifies. Over a period of time the heated air begins rising upwards, leaving a low pressure area behind.

It is then that cool air allround rushes in to fill the low pressure zone. As the process of heating up of the air goes on, more cool air which is denser penetrates, eventually building up a cycle of vertically rising and horizontally moving air—a whirlpool. When this intensifies, the earth's centrifugal force developed by air motion along a curved path forms a vigorous 'cyclonic system'. The extent of a cyclone may be between 150 km and 1000 km, and it can be as high as 6 km. It travels about 25 km per hour and may cover upto 500 km in a day.

An interesting feature of a tropical storm is the 'eye'. This is the core of the cyclone and it is completely calm. In this central region high velocity winds turn into a light breeze. The eye of the storm is about 25 km in diameter.

Among the storms that hit our coasts between 1891 and 1960, 314 developed in the Bay of Bengal and 82 in the Arabian Sea. Our storms occur mainly during two seasons—April-May and October-November-December.

Effect of Air Masses on the Environment

(*a*) *Expansion of deserts*—In the arid and semiarid parts of the world, the deserts have expanded, because of poor land management. In the absence of perennial vegetation, air masses moves unchecked and therefore lifts greater quantity of surface sand. Wind errosion naturally leads to deposition of sand on areas where the wind moves unchecked, including the adjoining fertile areas. In India, the Rajasthan desert has, in this way advanced towards western Utter Pradesh, Haryana and Punjab. The deposited sand makes the edaphic environment of cultivated fields unsuitable for the good growth of crop plants. In order to check this menace, wind breaks and shelter belts are being planted. Plant roots bind the soil and protect it from erosion. The shoot canopy acts as a physical barrier for dust storms. Land and plants on the leeward side of the shelter belt escape the ravages of wind, sand and dust.

(*b*) *Rainfall*—Air masses have many beneficial effects on the environment. Most of the rainfall at any given place is regulated by air movement. In Northern India, in early summer westerly winds are dry and desiccating because they originate on dry land. The easterly winds in the month of June, July onwards originate in the Bay of Bengal and move from sea to land. Therefore, they are wet and cause widespread monsoon rain in the country.

Effect of Air Masses on Plants

(*a*) *Dispersal*

Many micro-organisms, fungal spores and bacteria are freely transported in moderate wind to long distances. Similarly, the pollen grains of many species are wind transported from one flower to another (*anemophily*). Pollen grains in some genera like *Pinus* are specially adapted to float in air due to the presence of air bladders. Seeds and fruits of many plants are also wind dispersed. Wind dispersed seeds and propagules leads to better establishment and a more successful and healthy growth.

(*b*) *Dwarfing*

Plants growing under the influence of drying winds generally suffer from dehydration and consequent loss of turgidity. Under these conditions, their organs become dwarfed. This phenomenon is common in trees found on sea coasts, arctic or alpine areas.

(*c*) *Deformation*

Strong winds from a constant direction sometimes cause permanent alteration in the form and position of shoots. In open situations for example, seashores and high mountain tops where the strong winds blow all the year round in one direction, the trunk and branches of trees are twisted chiefly in the direction of prevailing wind. In such plants, generally, the growth of buds because checked on the windward side (Fig. 21).

Fig. 21 : **The effect** of abrasive ice **particles and strong winds on the branch growth.** The buds on the **windward side are either killed or they grow** opposite the **windward side.**

(*d*) ***Breakage and uprooting***

A wind of much high velocity may cause the breaking of living branches of trees and sometimes even the complete uprooting of trees. However, in forests where the canopies of individual trees at different heights reduce the velocity of wind to about 80 percent, such effects are uncommon.

(*e*) *Lodging*

In strong windsweak plants like wheat, maize, sugarcane, jawar etc. are bent against the ground. These lodged plants, if their stems are not too mature, may again become partially erect. This is due to differential growth of the meristematic lower nodes of these plants.

(*f*) *Transpiration*

Strong winds cause an increase in the rate of transpiration. Wind quickly removes the layer of humid air just above the leaf surface. The rate of transpiration depends upon the dryness or wetness of the atmosphere. Initial transpiration leads to the formation of a moist layer of air above the leaf surface. This moist layer reduces the transpiration rate but wind removes the transpired vapour quickly and therefore the overall rate of transpiration rises considerably. The mechanical to and for movement of twigs and leaves also helps in rapid moisture loss due to physical contraction and expansion of cells.

(*g*) *Anatomical effects*

In the areas subjected to strong winds, the leaves of plants become small and rolled. The transverse section of the stem shows eccentrically developed secondary wood, known as compression wood, on the compressed side. This eccentric growth of secondary xylem prevents the stem from getting easily broken. In some herbaceous plants collenchyma formation may be stimulated in the wind deformed organs to provide mechanical support and save the plants from wind injuries.

6. *Water*

Water is perhaps one of the most peculiar of our natural resources. Water is essential not only for the sustenance of human life and activities but for the 'quality of life' as well. It is the essence of life on earth and totally dominates the chemical composition of all organisms. The ubiquity of water in biota as the fulcrum of biochemical metabolism rests on its unique physical and chemical properties. It provides both food and drink and has been used for recreation, transport, cooling, waste disposal and more besides.

While we consider it a renewable resource, in the sense that it is possible for us to control it, regulate its use, improve its quality and ensure that we will have a continuing supply of it, nearly every community has a water problem. One-fourth of the population today is troubled with water shortage, poor water or both, and the prospects are for even more difficulty in the future. It is generally recognised that atleast in our country water is no longer available in unlimited quantity. It is already a scare commodity and in terms of its requirements, it is destined to become more and more scarce. On the other hand, it is a non-renewable resource, in the sense that for the entire earth there is fixed and relatively unchanging amount.

The amount of water which is present on the earth is vast. More than two-third of the earth's surface is covered with water. The water which is present today, was formed centuries ago, during a time when our earth was changing from a molten mass to essentially the form in which we know it today. During the cooling period, chemical changes occurred, gases were released and from these gases water was formed. This water has appeared at different times in different ways and has covered varying amount of the earth. However, overall of these changes and times and places the amount of water has remained the same.

Out of the total amount of water present on the earth, 97 per cent of it is salty, and of the bare 3 percent remainder, 4/5 is blocked in the form of polar ice and glaciers, and gets eventually lost to the sea. Another major fraction remains in the deep crevices of the land, leaving about 0.3 percent for use of plants, animals and man.

Forms of Water in Nature

For practical purposes we can classify our water resource as existing generally in four conditions, namely (1) Salt water, (2) Atmospheric water, (3) Fresh water, and (4) ice and snow. The most abundant is salt water which exists in the seas and oceans. Second is the water which exists in vapour form in the atmosphere and the soil. Third is the fresh water which exists in a liquid states in lakes, rivers and underground aquifers of the earth. Fourth is ice and snow which is a solid state found on high mountains and poles.

Hydrologic Cycle

Not only does water exist in these four conditions but it has a

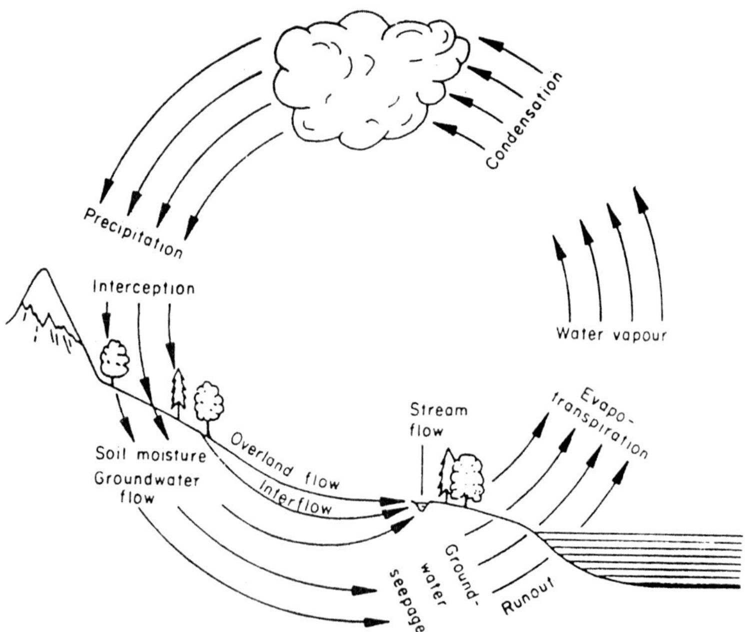

Fig. 22 : The hydrologic cycle (after Ward, 1973).

unique feature among most of the resources in that it is constantly in a cycle. The cycle of water begins with rainfall and ends in condensation of water vapour in the form of clouds. The clouds on cooling and condensation produce rain or ice. This circulation is called the *hydrologic cycle.* This cycle is in continuous flow (Fig. 22). There is probably no place on the earth's surface in which the water cycle does not occur.

Figure 23 represents the hydrologic cycle in terms of average global conditions on land, with the rate of low and amounts of storage indicated by the width of flow, lines and areas of circles

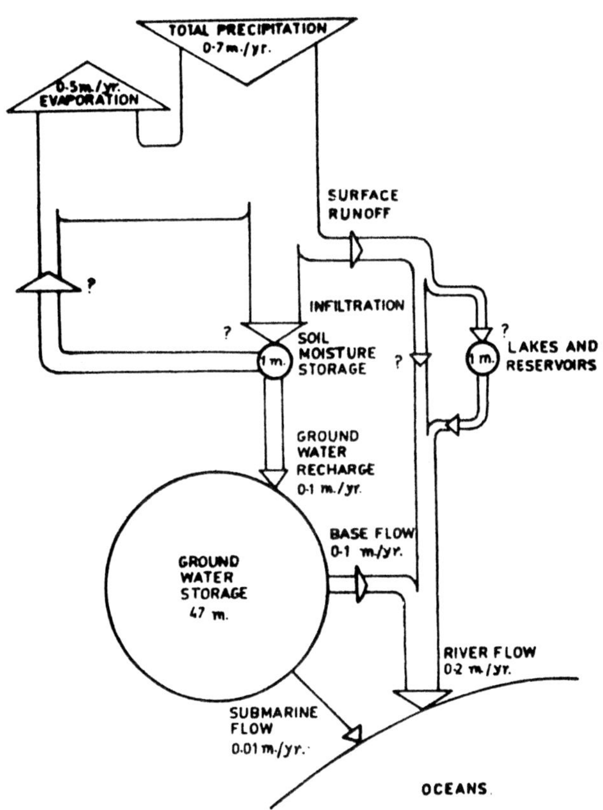

Fig. 23 : The hydrologic cycle (after Gehlar, 1972).

respectively. The implied average time for a drop of water to travel through the ground water system, from rainfall to the seas, is somewhat over 400 years. This is because ground water spends long time moving through very small passages in the soil.

The existence of water cycle is responsible for ecosystem water balance. Fig. 24 shows the path of water through the system when precipitation exceeds transpiration and evaporation. Rain may pass the canopy as thro-fall or be intercepted by foliage from which it may either evaporate directly or be channelled into stem-flow or leaf-drip and increase the input to the soil surface. Infiltration replanishes the soil storage reservoir or leads to drainage. If the infiltration capacity is exceeded, lateral run-off may occur, reducing the input to some parts of the soil, and increasing it to others. Lateral movement below the water table may also be a source of income or loss. Water uptake by plants and its loss, by transpiration, represents the main return pathway to the atmosphere, reinforced by some evaporation from the soil surface. Between periods of rain and soil storage capacity buffers these

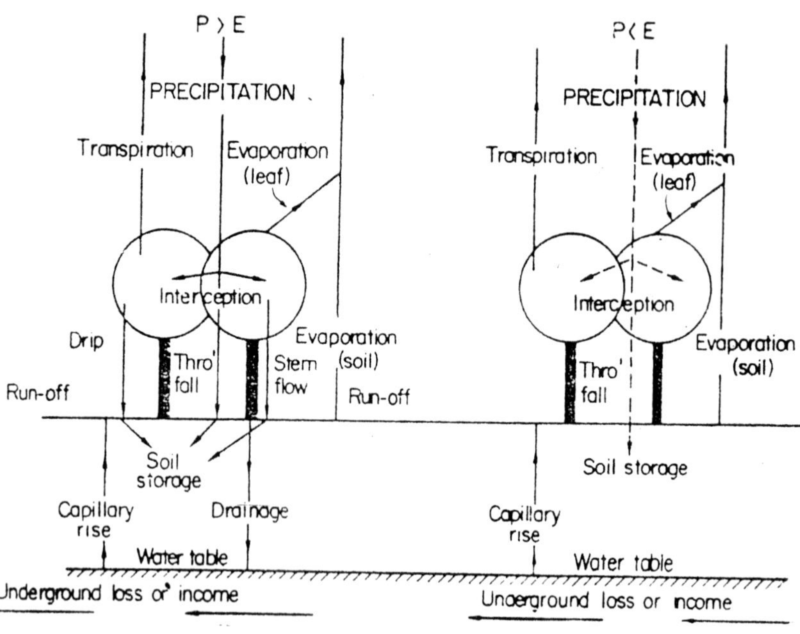

Fig. 24 : Ecosystem water balance (after Ethrington, 1974).

changes and may be replanished by capillary rise from the water table. Figure also shows the inversion of the dominant direction of water movement when evaporation exceeds precipitation. Much less water enters the soil and capillary rise may be all important both in supplying the plant with water and in determining the pedogenic processes. Surface enrichment with solutes may occur compared with the strong leaching processes (Ethrington, 1974).

Rainfall

Although the average precipitation over the country as a whole is about 1000 mm, this is very unevenly distributed in space and time. The west coast and the Assam region are areas of heavy rainfall, receiving 2500 mm and above annually. The eastern part of the peninsula and the northern plains receive moderate rainfall of 1000-2500 mm annually. The Punjab plains and upper western part of the Deccan plateau receive low rainfall of 250-1000 mm. While the Rajasthan desert and Ladakh plateau of Jammu and Kashmir are regions of very low precipitation of less than 250 mm. The bulk of the precipitation occurs in the south-west monsoon period covering 4 to 5 months of June to October. A large part of the country experience acute water shortage in the other months. It is only the south-eastern coasts of peninsular India that receives the major share of the precipitation in November and December from the north-east monsoon.

Surface Water Resources

As a rough estimate, the annual rainfall over the whole country would be equivalent to about 3700 billion cubic meters. Of this around 1250 billion cubic meters is lost by evaporation—transpiration, and another 790 billion cubic meter by seepage into the soil, thus leaving 1660 billion cubic meter as surface flow into the river system. Fourteen major river system share 83 percent of the drainage basin, accounts for 85 percent of the surface flow and serve 80 percent of the total population of the country. There are other 44 medium and 55 minor rivers which are mostly seasonal in nature. However, all the river water flow cannot be utilised because of the numerous limitations imposed by topography, climate, soil conditions etc. It has been estimated that only about 666 billion cubic meter of water can be utilized from various rivers without large inter-basin water transfer. Moreover, because of uneven distribution of rainfall over the year, it becomes necessary to store up the flows in the monsoon period for regulated releases during the non-monsoon months.

Ground Water Resources

It has been estimated that out of about 790 billion cubic meter of water that seeps into the soil, about 430 billion cubic meter remains in the top soil layers and produces soil moisture which is essential for growth of vegetation. The remaining 360 billion cubic meter percolates into the porous strata and represents the actual enrichment of underground water. Out of this, the water that can be extracted economically is only about 255 billion cubic meter.

Clouds

Often the first sign of a coming change in the weather is a band of this, wispy clouds very high in the sky. Clouds are not only an indication of this kind of weather to come. They are an essential part of the weather itself.

Clouds are what we see when invisible water vapour, present in our atmosphere, condenses and becomes visible in the form of water droplets. Clouds are made up of these small drops of water. if the temperature is below freezing, they contain small particles of ice. Very often the same cloud contains both water drops and ice.

Three Levels of Clouds

Clouds are fairly easy to classify. They are present in only three different levels of the troposphere. Families of clouds are found at high levels (usually above 18000 feet in temperate latitudes), an intermediate level (6000-18000 feet), and at low levels (below 6000 feet). The one exception is cloud that extend continuously from the lowest to the highest levels. The clouds of one level are different enough in appearance from those of the other levels to be easily recognised.

Three Basic Cloud Forms

There are only three basic cloud types, though there are various combination of these :

The first type occurs only at very high levels near the stratosphere. Here the temperature is always below freezing. *Cirrus* clouds are composed of ice crystals. These high level cirrus clouds are easily recognised by all of us because of their delicate silky appearance.

The second basic cloud type is called *cumulus*. They always have delicate form, with the larger cumulus clouds giving an appearance of a head of cauliflower or piled-up balls of cotton. Cumulus type of clouds are usually the result of raising currents of air. The base of cumulus clouds is the level where the water vapour in the lifted air is cooled to its dew point. This type of cloud is usually having currents of air. A growing cumulus is apt to have strong updraft and downdraft and considerable agitation within it.

The third basic cloud type is called *Stratus*. These clouds do not have definite contous, but are quite smooth and without shape. Usually grey in colour, they may appear like one smooth sheet over the sky. Fog is simply a stratus cloud resting on the ground. There is very little up-and-down motion in these clouds.

Table 7 describes ten different cloud types. Although there are many more individual types, these ten are the most important. They provide all the kinds of weather we have. The many additional types of clouds are variations of one or more of these ten.

Fog

Fog is another visible form of weather which can change our view of the world around us. Thick fog puts us in a world all our own, limiting our view to the few meter near us. Fog may be only a few meter deep of it may extend upto several meters. It consists of tiny particles of moisture suspended in air.

Fog occurs mainly within an air mass, although they can be part of a front. Cold-front fogs are rare and do not last very long. Warm-front fogs are more extensive and are usually associated with a slow moving warm-front. Cooling and condensing of water vapours occurs when the air is gently lifted by the gradual upslope of the land. This cooling effect is added to the condensation already taking place within the warm-front itself.

Fog is formed by the same process of cooling and condensing that causes clouds to form. Air at certain temperatures can hold only so much water vapour. The temperature at which these droplets begin forming is called the dew-point.

Table 7 : Summary of Cloud Types

Height	Cloud Name and Symbol	Description	Remarks
1	2	3	4
High Cloud 18,000 to 40,000 feet	1. Cirrus-Ci	White, wispy with silky edges. Semi-transparent.	Made up of ice crystals. May appear in bands or separate long wisps.
	2. Cirrocumulus-Cc	Fibrous lumpy small bands called "Mackerel Sky." Ripple appearance.	Seldom appears unless Cirrus also present. Real Cc is seldom seen.
	3. Cirrostratus-Cs	Thin milky veil. Does not obscure sun or moon.	Responsible for "Halo Rings" seen around sun or moon. If Cs thickens, strong indication rain or snow coming in other, denser clouds.
Middle Clouds 6,000 to 18,000 feet	4. Altocumulus-Ac	Layer of patches of flattened round clouds usually arranged in rows.	Several varieties of Ac exist. May indicate frontal clouds approaching.
	5. Altostratus-As	Gray fibrous film almost obscuring sun or moon, or heavy gray sheet.	Rain or snow may fall from As. No halos are visible in As.
Low Clouds Below 6,000 feet	6. Stratus-St	Smooth layer of foglike cloud. Lowest of clouds, without shape.	Drizzle or rain may fall. Frequently found in valleys and coastal areas.
	7. Stratocumulus-Sc	Layer or series of patches of roll-like clouds. Considerable shadow effect —fluffy and way surfaces.	Often associated with strong surface winds. Frequently follow rains.

(*Contd.*)

1	2	3	4
	8. Nimbostratus-Ns	Low, formless dark cloud of bad weather.	Usually gives continuous rain or snow. May develop from a thickening and lowering layer of altostratus.
Low to the Highest Level	9. Cumulus-Cu	Billowy, thick clouds with definite contours. Flat bases, rounded tops with knobs. Strong light shadows.	May develop to great heights with strong surface heating, rain showers may occur. Small, flatter Cu indicate fair weather.
	10. Cumulonimbus-Cb	Towering Cu usually with anvil-shaped top. Usually a roll round at forward lower edge.	Typical weather: heavy showers, frequent hail. lightening, thunder, gusty winds, and turbulence.

How Fog Forms ?

One type of fog forms when the dew-point is raised to the surface air temperature. It is rather uncommon, but occasionally occurs when water evaporates from the ground after rains. This increases the water vapour content of the air. It may also occur when air passing over water surface adds more moisture to its contents and raises the dew point. In other cases, fog may result rain falls through lower layers of air, increasing the moisture content of the lower layers. The most common type of fog, however, are those that formed when the temperature of the air which is at or near the ground surface is cooled to its dew point. Such cooling if surface air is brought about in two ways (1) by radiation of surface heat into space on clear nights, and (2) by a process called advection (transfer of heat through horizontal motion of air).

Radiation Fogs

Ground fog is a good example of radiation fog. The earth's surface cools during night. If there is enough moisture in the air, and the wind is not strong, ground fog will occur as the surface air temperature cools to dew point. Ground fogs usually forms a little before sunrise and disappears as the sun's heat begins to raise the surface temperature.

Another type of radiation fogs occur in enclosed valleys where the coldest, densest air is to be found as it drains into the valley bottoms. Ground fog will be most dense in such cold air 'traps'. Radiation fogs do not form over snow surfaces or water surfaces.

Advection Fogs

The best known fogs are advection fogs. The most common of this is sea-fog. A sea-fog is formed when air moves from warmer areas of the water to colder areas. Fog can also form as air moves from warmer land surfaces to colder water surfaces. This frequently happens near coastal areas. During the day, on shore movement of sea fog is burned off by the sun's heating of the land surface. At night, as the ground cools off, sea-for frequently extends inland.

Another type of advection for occurs when very warm tropical air moves towards higher, cooler latitudes. Such warm moist air requires only slight cooling to reach the dew point temperature.

Steam fogs occur when cold air moves over warm water surface. This type of fog forms over inland lakes, rivers and over water areas of the Arctic. On clear autumn nights when the air is moist and the wind light, dense steam fog frequently appears over portions of large rivers.

Precipitation

The term precipitation generally means a process of seperating some substance from its solution. This applies to weather, too, because water is separated from the air when precipitation occurs.

In order to produce some kind of precipitation, the small suspended water or ice particle in clouds must grow to a size large enough and heavy enough to fall from the cloud to the ground. If air currents are rising into the clouds, the precipitating particles must be heavy enough to overcome the upward push of the rising air.

In order to form the small droplets which become clouds, small dust, smoke or other tiny particles of floating matter in the atmosphere must be present. Water vapour condenses on these particles. A rather complicated process, not completely understood, causes some of these droplets to grow in size until they are large enough to fall as precipitation.

To better understand the great variety of precipitation patterns weather produces, brief descriptions of the different kinds are given below :

Rain

When precipitation occurs above 0°c, drops of liquid water usually falls to the earth. These drops fall as rain if they are atleast 0.52 mm. It is usually associated with stratus clouds. It often accompanies fog.

Dew

Water drops condensed directly from the air surfaces that were cooled during the night by radiation.

Frost

Thin ice crystals which are deposited directly out of the air on surfaces cooled by radiation. Air temperature must be below 0°c.

Sleet

It occurs when rain drops fall through a very cold layer of air which freezes them. Sleet is usually transparent and bounces when it hits the ground. It usually occurs during the winter and means that the air near the ground is well below freezing. Sleet falls out of a solid overcast sky.

Halt

It occurs as pellets of ice ranging in size from 5.2 mm to more than 52 mm in diameter. Not usually transparent, a hailstone is made up of several layers of ice piled one on top of the other. Hail occurs with towering cumulus and cumulonimbus clouds. These clouds contain strong currents of air and shower like precipitation. The lower levels of these clouds have temperature above freezing while the upper layers are considerably below freezing. As rain drops form and begin to fall, many are caught by the strong updrift. These are carried into the freezing zone where they become ice and fall down. As the frozen drops fall, their surfaces pick up more water in the warmer zones. Often the same drops are again caught in the updrift. The cycle is then repeated. Hail is the result of several trips of a water drop through this cycle. Large hailstones are the result of many such trips and of every strong raising currents of air.

Snow

In temperature below freezing, water vapour crystallizes directly to a solid form or water instead of condensing into liquid water. Snow is such precipitation in the form of hexagonal ice crystals. Usually many of these crystals are melted together resulting in large snow flakes. The size and consistency of snow depends on the temperature at formation and the warmer temperature (just below freezing). The more granular smaller type results from much colder temperature.

Rime Ice

It forms when small drops of fog or drizzle strike a surface and immediately freeze. Rime ice is opaque and usually forms in white layers. This kind of ice is easy to remove and seldom causes much damage.

Glaze Ice

Clear hard ice results when larger raindrops spread out as they hit a surface. Clear ice is difficult to remove or break up.

7. *Soil*

From the time immemorial, the soil or the earth has been recognised as the mother of us all—plants, animals and man. Soil receives at last their discarded forms. It is imperishable store house of eternity.

Soil is the nation's permanent asset and its most precious resource. Almost every one, from the farmer to the window box gardner, is aware of the importance of soil and the necessity for having soil of certain quality to support plant growth (Narayana and Shah, 1966).

Dokuchayev (1900) a Russian worker writes, 'by soil one should mean the day or external horizons of rocks (whatever rocks) modified naturally by the combined influence of water, air and various kinds of organisms, living and dead.

The soil is the proportion of the solid substratum in immediate contact with the roots of the plants growing in it and also containing algae, fungi, bacteria, insects, worms, snakes and mammels, often called soil flora and fauna, surrounded by minerals and dead organic matter of the soil and by air and water contents. According to Charles E. Kellong, Chief of the soil survey, U.S.A., soil is that thin film between earth and sky that supports all living things. Beneath lie the sterile rocks above it are the air and sunshine. From it, all plants and animals, and man himself draw their nourishment, either directly or indirectly from other living things that live in soil. To it their bodies return. There is no life without soil, and no soil without life (Narayana and Shah, 1966).

Development of Soil

Today, almost all areas of the earth's land surface are covered with a thin mantle of soil that varies in thickness from a few centimeters to several meters. But this condition was not always so. At one time in the history of the earth, there was no soil. To gain a better appreciation of soil and its importance to the terrestrial environment it would be worthwhile to discuss here the process through which soil was formed and is still being formed.

To tell the story of how soil first appeared we must go back in time a few billion years—to that point in the development of the earth when most volcanic activity has ceased and there was a gradual cooling of the surface. During this period the surface of the earth was a consolidated mass of volcanic rock. This rock took various forms such as lava, basalt and granite. As fare as can be determined there was no life present on the earth.

Through millions upon millions of years the exposed rock surfaces were acted upon by various weathering agents. The action of water and wind upon the rock caused it slowly to disintegrate. Chemical reactions occurred to form acids and alkalies. These agents began to eat away the surface of the rocks. Millions of years later, after the earth had cooled and the weathering process had proceeded, small fissures and crannies appeared in the surface of the rocks that held small amounts of water. The tiny openings provided footholds for the earliest and most primitive of plants—bacteria, algae, fungi and lichens. This early plant life aided in the breakdown of the original rock material through further chemical action that disintegrate the rock and released chemical nutrients which were, in turn, required by the plants. Secondly, dead plant material combined with the fragmented rocks to form the first primitive soil.

During this period another process was taking place in the vast seas that covered most of the earth's surface. Aquatic plants, especially algae, were extracting calcium compounds from water. Over millions of years the insoluble calcium, precipitated by plant processes, settled slowly to the bottom of the seas, layer upon layer, until the weight of the upper layer so compacted the lower layer that beds of lime-stone were formed. While all this was going on, the earth's surface was buckling and heaving, and portions once below the seas rose above them to form dry lands. The combination of physical, chemical and biological activity through the years served to disintegrate the surface of the original rock material and produce the beginning of that thin unconsolidated mass of material that we know as soil.

As soil layers developed and became deeper, they formed a base for more advanced kinds of plants life. Early relatives of such plants as the ferns and mosses formed a suitable environment in which to develop. These plants continued the process of biological weathering, as their fibrous roots penetrated into the surface of the soil and into the many porous openings into the rocks. Expansion and contraction water within the fissures developed down of rock surfaces. As the plants went through their yearly and life cycle of growth, death and decay, organic material was added to the primitive soil and thus paved the way for the coming of the higher plants—grasses, herbs, shrubs and trees—the forms dominant in our landscape today.

All these processes that have proceeded for such a long time in the history of the earth are still going on. Visitors to the high peaks in the Chambla Command Area, and other igneous rocks can observe lichens clinging tanaciously to exposed rock surfaces. These simple plants today are duplicating the activities which took place hundreds of millions of years ago in forming soil.

Once the higher plants became established on the land, soil formation became more rapid. The larger plants are very productive, producing countless tons or organic matter that added to the soil, has built up what we recognise as top soil.

The breakdown of parent rock material has proceeded at a more rapid pace in the recent history of the earth's surface. As the earth's crust became more stable, definite water courses were established and the erosive action of water and wind, freezing and thawing, the advance and receding action of glaciers have all combined to speed up the process through which the surface crust of the earth was broken down. Water, wind and glaciers have also acted as transporting forces, bringing bits and pieces of rocks, sand, silt and clay from long distances to deposit it in beds and layers at different points on the earth's surface. The long history of floods and the development of flood plains along the lower reaches of our streams has provided an ideal place for silt-laden flood waters to deposit their load of soil materials.

While we consider customarily of erosion as detrimental process with respect to soil, but we should keep in mind that weathering and erosion have been major forces in the development of soil—including some of the world's most productive agricultural regions.

The soil formed through all these processes varies markedly from place to place. These differences are apparent not only in their ability to support different kinds of plant life, but also in their general makeup and texture (Carlozzi, 1965).

Maturation of Soil

Maturation of soil is result of interaction of many factors over a long period of time. The controlling factors assigned are climate, vegetation and parent rock. Soil workers have often used such terms as old and young soils, mature and immature soil or stabilized and fresh soil. In their definitions they have evidently concentrated on the chemical stages in leaching. Thus, a mature or stabilized soil has been defined as having full development of A and B horizons in equilibrium with the prevailing weathering forces. Four major maturation processes may be recognised :

(1) *Melanisation*—The humus derived from the dead organic matter gets mixed in the upper layers of the soil which become dark coloured. It occurs mostly in the regions with low humidity.

(2) *Podzolisation*—In regions with high rainfall or high humidity and low temperature, the minerals in the humus become leached from the upper horizon (eluvial) and get precipitated in the middle of B horizon (illuvial) forming a hard pan. This leaves an ash coloured surface layer of the soil from which the soil derives its name Podzol.

(3) *Gleization*—In very cold climates, the underground water lying above the rock layer continuously reacts with the partly weathered mineral matter. The hydrolysis and reduction of the minerals result in the formation of a hard gley horizon.

(4) *Laterization*—In very hot and humid climate, the rapid decay of organic matter and release of bases from organic combination result in the solubility of silica and formation of oxides of iron, aluminium and manganese etc. This results in a red coloured soil usually rich in iron and deficient in bases and organic matter.

Vertical zonality bringing about maturation of soil has not been observed in India so far. Pandeya (1959) claims to observe that at least some Madhya Pradesh soils have attained certain degree of maturity. However, they do not show a marked vertical colour zonality. They are also not acid ones. Having known this interest arises to evaluate the maturation process at work in the soils.

Maturation of the type of temperate zone may not be found in India since the climate is so markedly different. In the temperate climate due to low temperature and restricted precipitation soil profiles show a clear zonation. Both physical and chemical changes occur in such soils. However, under tropical climates of India physical changes like effect of frost etc., usually do not occur, atleast in Madhya Pradesh. Except in the initial fragmentation of sun exposed rocks the main weathering agent appears to be rain water and substances dissolved in it. At higher tropical temperature and abundant water hydrolysis would be very rapid and the only physical weathering would be transportation. Indian soils are very ancient ones in comparision to European soils which date back only as far as last glacial age. It may be added that maturation will proceed only at places where precipitation is more than evaporation and where the soil will not get desiccated to start back accumulation of bases by capillary rise.

Maturation under grasslands is probably like 'calcification'. Under forests, atleast in Madhya Pradesh, maturation will probably be 'laterization'. Laterization essentially brings about leaching of bases like calcium and making the upper soil neutral to acidic. This favours removal of silica and iron, while aluminium is left behind. Sal soil appears to be of this nature. It has been pointed out that maturation in Indian soils should be judged by its base status and not by visual zonal colour distinction. Absence of colour distinction in Indian soil profiles is probably due to heavy precipitation resulting in infiltration of leachate to greater depths. Concentration zones are usually deep seated in Indian mature soils.

Factors Affecting Soil Formation

The effect of various factors which operate in the formation of soil may briefly be understood by consideration of each one of them :

(1) *Parent material*

The nature of the parent material generally effects the rate of soil profile development as well as the texture and nutrient content. The parent material is the product of the bed-rock and if it consists of a mixture of clay, silt and gravel, the soil type that develops from such a parent material will be fine textured and fertile loam. If the parent material consists of quartz and sand the soil type will be coarce textured infertile sand. The parent material influences broadly physical and chemical properties of the soil.

(2) *Climate*

Soils are the residual products of the action of meteorological agencies upon rocks, and so there should exist an intimate relation between the soil of a region and the climatic conditions that prevail. The effect of climate on soil formation are direct because the climate effects the weathering of rocks, and also indirect because the climate effects plant and animal life. The moisture relations of the soil are partly a function of the climate, for example, soil developed in humid climates are subjected to intense leaching and will have profiles different from those developed in arid zones. The influence of temperature upon soil formation are primarily on the rate of weathering of rocks and soil minerals. Higher temperature result in more weathering and therefore increase of clay content in the soil, but the content of nitrogen and organic matter in the soil decreases in regions of high temperature, because the organic matter is burnt and decomposed at such temperatures.

(3) *Living organisms*

The effect of living organisms on soil formation is remarkable. The plant alter the micro-climate, their residue are incorporated into the mineral body. They transfer the elements from the lower to the upper horizons. The vegetation favours soil development by diminishing erosion. Different types of vegetation effect soil development differently. In view of the differences in their size or life-form, depth of rooting and physiological behaviour. They contribute the organic remains of the soil, for example, the roots of annuals which remain n the soil, or the organic matter supplied by the forest trees in the form of falling leaf on the surface of the soil. Deep rooted species contribute more to the soil development while the shallow rooted ones less because the deep roots help in weathering of the parent material in the lower layers and absorb nutrient elements which are subsequently retained to the soil surface in leaf fall. The micro-organisms also help in rock weathering and in the decomposition of organic matter. Many of the chemical processes in the soil are directly or indirectly due to the activity of fungi and bacteria.

(4) *Topography*

The effect of topography on soil development is mainly through its effect on water relations, erosion, temperature relations and vegetation cover. The rainfall that falls on the soil always tends to run along the slopes and collect in depressions, as a result soil on steep slopes receive more water than those in the flat areas. The soil that are developed from the slopes are therefore characterised by reduced amount of water entering the soil. The process of erosion is also maximum on the slopes and therefore the soil remains in the youthful state of development.

(5) *Time*

The length of time required for soil development depends upon the different factors which contributes to the soil formation.

Physical Properties of Soil

The proper use of soil is largely determined by its physical properties, which determine the flow and storage of water, the movement of air and the ability of the soil to supply nutrients to plants. The common physical properties of the soil are profile, skeleton, porocity, colour, permeability, moisture holding capacity, surface drainage, erosion, etc.

Soil profile

A vertical section through soil body usually shows a number of layers of horizons of varying thickness. These horizons in natural sequence are collectively called as 'soil profile'. Unfortunately, there is no universally accepted system of horizon designation. The upper part of the profile may be designated as 'A' horizon' the middle part 'B' horizon, and the consolidated material that has been affected very little by the plant root as 'C' horizon. Subscript such as 0, 1, 2, 3, 4 are used to denote the sub-divisions of A, B, and C horizons. Figure 25 indicate a hypothetical soil having the principal horizons. No one should be expected to have all these horizons well developed, but every soil has some of them.

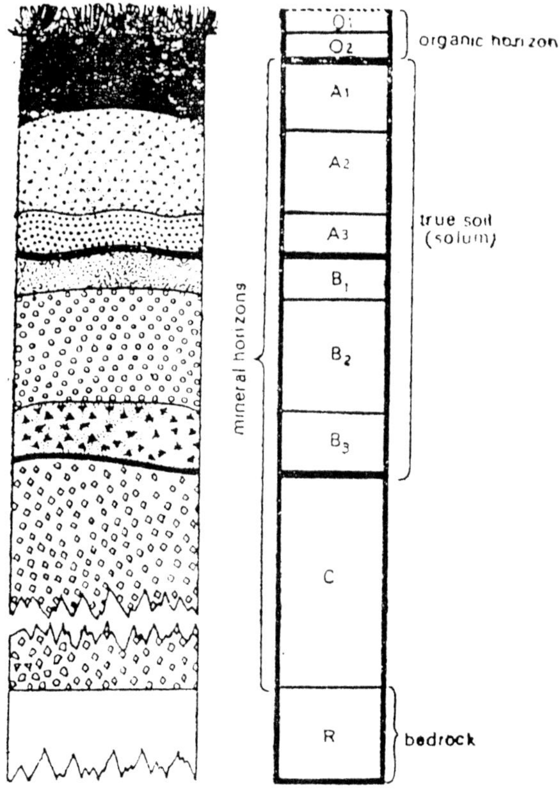

Fig. 25 : A generalized profile of soil. O_1 : Loose leaves and organic debris; O_2 : organic debris partly decomposed or matted; A_1 : A dark coloured horizon with a high content of organic matter mixed with mineral matter; A_2 : A light coloured horizon of maximum leaching; A_3 : Transitional to B but more like A than B; B_1 : Transitional to B but more like A than B; B_2 : A deeper coloured horizon of maximum accumulation of clay minerals or of iron and organic matter; B_3 : Transitional to C; C : Weathered material (regolith); R : Consolidated bedrock (Smith, 1974).

The 'A' horizon is called the zone of leaching, as the minerals and organic solutes are washed down this layer by the precipitation. It is also called as eluvial zone. All the materials get accumulated in 'B' horizon, therefore it is called the zone of accumulation. The 'C' horizon is mainly comprised of weathering matter and is sometimes called as illuvial zone. Soils which show profile development are known as zonal soils. Sometimes the soils have no profile and are known as azonal soil. Azonal soils develop due to short period of development, wind drifted nature of soil, as in sand dunes and continuously disturbed by biotic agencies.

Soil skeleton

The soil horizons are heterogeneous systems, made up of materials in three different states—solid, liquid and gases. Proper proportion of these constituents are necessary for the soil to be a good medium for plant growth. The solid part of the soil consisting of both organic and inorganic matter forms the skeleton. The air and water fill the intervening spaces.

The inorganic part of the soil is made up of particles ranging in size from that of fine clay to large rocks. The Table 8 gives the appropriate sizes of fractions that we commonly recognise.

Table 8 : Soil skeleton

Soil type	Particle size (mm)
Very coarse sand	2.0 - 1.0
Coarse sand	1.0 - 0.5
Medium sand	0.5 - 0.25
Find sand	0.25 - 0.10
Very fine sand	0.10 - 0.05
Silt	0.05 - 0.002
Clay	< 0.002

All these fractions of the soil can be broadly classified into two categories (1) coarse fraction, and (2) fine fraction. Coarse fraction consists of large particals which are generally more than 0.05 mm in diameter. It includes mostly sand. Fine fraction consists of small particals which are less than 0.05 mm in diameter. It generally consists of silt and clay particals. The coarse raction comprise the skeleton of soil, while fine fraction generally gives a mechanical support. The fine fraction is the active fraction or protoplasm of the soil. It posses the property of absorption, particularly of mineral ions, water molecules etc.

Soil Texture

Texture has been fieldman's term for how a soil feels or how it behaves under cultivation. Soil texture relate to the relative proportion of sand, silt and clay that are present in the soil (Table 9). A large amount of sand in the soil make it coarce and light. If silt is present in large amount, the soil is medium textured. While a large amount of clay in the soil makes it sticky when wet and hard when dry, such a soil is heavy. The United States Department of Agriculture has devised a method for determining the textural classes of soil from its mechanical analysis. The method is clear from the Figure 26.

Table 9 : Textural classes of soils.

Soil classes or Textural names	Range in relative percentage of mineral particles		
	Sand	Silt	Clay
1. Sandy soil	85-100	0-15	0-10
2. Loamy sand	70-90	0-30	0-15
3. Sandy loam	43-80	0-50	0-20
4. Loam	23-52	28-50	7-27
5. Silt loam	0-50	50-88	0-27
6. Silt	0-20	8-10	0-12
7. Sandy clay loam	45-80	0-28	20-35
8. Clay loam	20-45	15-53	27-40
9. Silty clay loam	0-20	40-73	27-40
10. Sandy clay	45-65	0-20	35-45
11. Silt clay	0-20	40-60	40-60
12. Clay	0-45	0-40	40-100

Soil Classification

There are several systems of soil classification. On the basis of the mode of their formation, particularly the nature of the origin of mineral matter, soils are sometimes classified as (i) Residual soils, and (ii) Transported soils.

Residual soils are those in which the whole process of soil formation, that is, weathering and pedogenesis occurs at the same place. Thus, in these soils the soil formation occurs at a place where their parent matter is present.

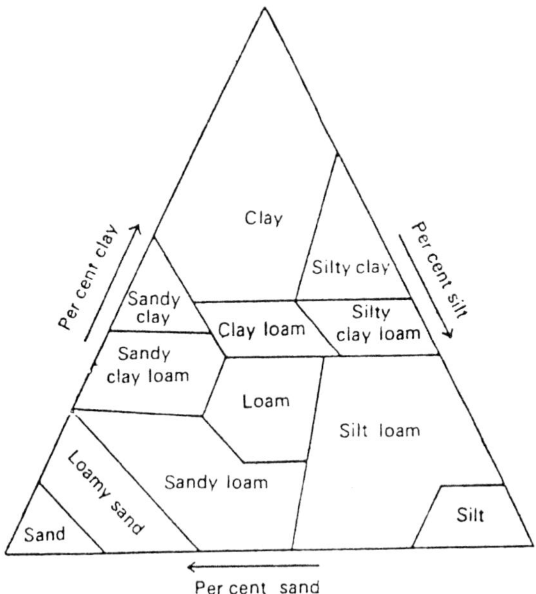

Fig. 26 : Classification of soil texture.

Transported soils are those where the weathered material is taken away at other places. This is at these different places, where, through pedogenesis, soil formation in completed. The weathered matter may be taken away to other places by several agents. Depending upon the nature of these transporting agents, the transported soils mat be :

1. Celluvial (by gravity)
2. Alluvial (by running waters)
3. Glacial (by glaciers)
4. Eolin (by wind)

The earliest attempts of a detailed classification of soils were made in Russia. This classification is based upon a combination of climatic/vegetation data and soil morphology. Dokuchayev's (1900) final classification of Russian soils in given in Table 10.

Table 10 : Classification of soils (Dokuchayev, 1900).

Class A. Normal or Zonal soils

Zones	1. Boreal	2. Taiga	3. Forest-steppe	4. Steppe
Soil types	Tundra (dark-brown soils)	Light grey podsolized soils	Grey and dark grey soils	Chernozem
Zones	5. Desert Steppe	6. Desert		7. Subtropical and tropical forest
Soil types	Chestnut and brown soils	Aerial soils Yellow soils White soils		Laterite or red soil
		Class B. Transitional soils		
8.	Dry land moor soils or moor-meadow soils	9. Carbonate containing soils (rendzinas)		10. Secondary alkali soils
		Class C. Abnormal soils		
11. Moor soils		12. Alluvial soils		13. Aeolian soils

The soils are also classified on the basis of the profile development into zonal, interzonal and azonal soils. This classification is chiefly based upon the various factors, such as nature of parent matter and climatic factors etc., which influence the soil development. Each type has been further sub-divided into suborders (Table 11).

Table 11 : U.S.D.A. revised system of classification of soils (Thorp and Smith, 1949)

Order	Suborder	Great soil groups
Zonal soils	1. Soils of the cold zone.	Tundra soils.
	2. Light-coloured soils of arid regions.	Desert soils Red desert soils. Sierozem. Brown soils. Reddish-brown soils.
	3. Dark-coloured soils of semi-arid, subhumid, and humid grasslands.	Chestnur soils. Reddish chestnut soils. Chernozem soils. Prairie soils. Reddish Prairie soils.
	4. Soils of the forest-grassland transition.	Degraded Chernozem. Noncalcic Brown or Shantung Brown soils.

(Contd.)

1	2	3
	5. Light-coloured podsolized soils of the timbered regions.	Podsol soils. Grey wooded, or Grey Podsolic soils. Brown Podsolic soils. Grey-brown Podsolic soils. Red-yellow Podsolic soils.
	6. Lateritic soils of forested warm-temperate and tropical regions.	Reddish-brown lateritic soils. Yellowish-brown Lateritic soils. Laterite soils.
Intrazonal soils	1. Halomorphic (saline and alkali) soils of imperfectly drained arid regions and littoral deposits.	Solonchak, or Saline soils. Solonetz soils. Soloth soils.
	2. Hydromorphic soils of marshes, swamps, seep areas, and flats.	Humic Gley soils (includes Wiesenboden). Alpine Meadow soils. Bog soils. Half-Bog soils. Low-Humic Gley soils. Planosols. Ground-Water Podsol soils. Ground-Water Laterite soils.
	3. Calcimorphic soils.	Brown forest soils. (Braunerde). Rendzina soils.
Azonal soils		Lithosols. Regosols (includes Dry Sands). Alluvial soils.

The most recent classification has been given by U.S.D.A., 7th Approximation, which recognised ten orders. Each order is identified on the basis of characters of its horizons, that is, sequence, morphology, colour, constitution, etc. Further subdivisions of each order is made into sub-order, great groups, subgroups, families and series (Table 12).

Table 12 : Various orders of soils in the U.S.D.A. 7th Approximation (based on Soil Survey Staff, 1960).

Order	Approximate equivalents in Thorp and Smith (1949)
1. Entisols	Azonal soils, and some Low Humic Glay soils.
2. Vertisols	Grumusols.
3. Inceptisols.	Ando, Sol Brun Acide, Some Brown Forest, Low-Humic Gley, and Humic Gley soils.
4. Aridisols	Desert, Reddish Desert, Sierosem, Solonchak, some Brown and Reddish Brown soils, and associated Solonetz.

(Contd.)

1	2
5. Mollisols	Chestnut, Chernozem, Brunizem (Prairie), Rendzinas, some Brown, Brown Forest, and associated Solonetz and Humic Gley soils.
6. Spodosols	Podsols, Brown Podsolic soils, and Ground-Water Podsols.
7. Alfisols	Grey-Brown Podsolic Grey Wooded soils, Noncalcic Brown soils. Degraded Chernozem, and associated Planosols and some Half-Bog soils.
8. Ultisols	Red-Yellow Podsolic soils, Reddish-Brown Lateritic soils of the U.S., the associated Planosols and Half-Bog soils.
9. Oxisols	Laterite soils, Latosols.
10. Histosols	Bog soils.

Porosity

The total porosity or pore space, of a soil is dependent upon both its size distribution and structure. In terms of total porosity clay soil are more porous than sandy soil. The pores are, however, so small that in the absence of structural development or aggregation, there is essentially no movement of water or air into or through the soil. If a soil is too porous water moves through too freely. On the other hand, heavy clay soils may hold much water that is not available for plant growth.

Colour

Soil colour is mainly due to the presence of iron and manganese compounds and the organic matter in soil. Organic matter makes soil black, brown on grey. Iron and manganese oxides impart red, brown and yellow colours to soil depending on the oxidation status and extent of hydration of these compounds. Red, yellow or reddish brown colour are produced by well oxidised salts of the above metals. Sometimes the colour is used as an indication of soil drainage. Well and moderately well drained soils have generally a uniform brown colour when moist but in some cases it may be various shades of red or yellow. A pale or grey soil denotes poor drainage for long periods. A black soil usually indicates that the land is either rich in organic matter or has remained wet for long periods. A preponderance of quartz particles makes soil grey or whitish. Mica particles produce glittering appearance.

Permeability

Permeability refers to the ease of movement of water and air through soil. Sandy or gravelly soils have rapid permeability; loam soils have moderate permeability and the clay soils are slowly permeable.

Moisture Holding Capacity

This refers to the amount of water that can be stored by a soil for use by plants. Water, the earths most abundant compound is a vital constituent in all living matter. It is highly mobile liquid and can exist in three physical states-solid, liquid and gas. Hence, it may be considered as the life blood of the earth.

When water reaches a soil profile, it percolates down to the bed rock as *gravitational water*. Some water is retained by the soil particles due to adhesion - *hygroscopic water*. Some amount of water is retained by the soil particles due to cohesive forces - *capillary water*. Capillary water is an important source of water for plants. This water is held by a force of 1/3 atmosphere. Water retaining capacity of soil at 1/3 atmosphere is known as the field capacity of soil. In general the coarser the soil, the lower is the moisture holding capacity. Black soils in India have higher moisture holding capacity than the red soils. The comparative water holding capacity of important classes have been shown in Figure 27 and Table 13.

Table 13 : Comparative water holding capacities of soils.

Kind of soil	Water held by 100 kg of soil when saturated. (kg).
Sand	25
Sandy clay	40
Cultivated soil	52
Garden soil	81

Surface Drainage

Surface drainage is the relative rate of removal of water which is in excess of the amount that can be absorbed by the soil. If water is removed so slowly that the soil remains wet for a long time, the surface drainage is poor. It is fair if water is removed at such a rate that soils do not remain wet for a long period.

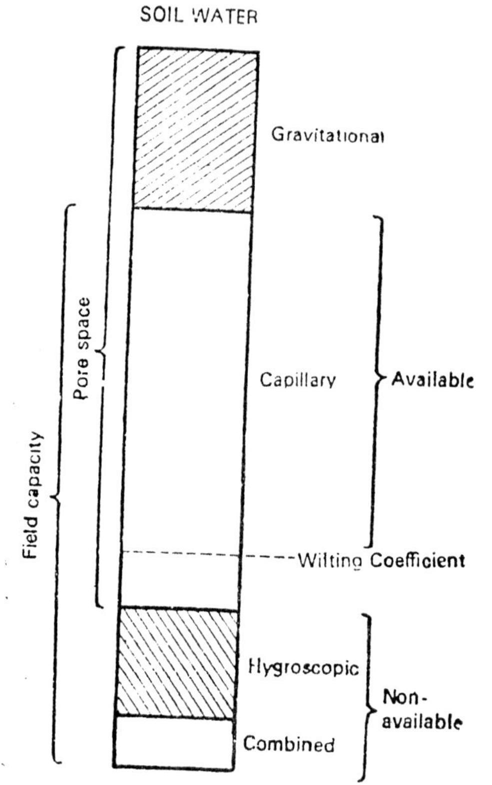

Fig. 27 : Generalised diagram showing the different forms of soil water and their possible relationship with soil and plant water status.

Chemical Properties of Soils

Soil is a mixture of various inorganic and organic chemical compounds and exhibits certain significant chemical properties :

Inorganic Components of Soil

The chief inorganic components of soil are the compounds of Aluminium, Silica, Calcium, Magnesium, Iron, Potassium and Sodium. Soil also contains smaller amounts of compounds of Boron, Manganese, Copper, Zink, Molybdenum, Cobalt, Iodine, etc. Most of these inorganic salts exists in soil in the form of weak solution. Soil solution may contain complex mixtucres of minerals as carbohydrates, sulphates, nitrates, chlorodes and also organic salts. The chemical nature of the nutrients solution depends upon the nature of the parent matter through which water has percolated and climatic conditions of the region. For example, temperate soils with high rainfall has hydrogen ions in abundance, due to which leaching of basic nutrients like calcium, Magnesium and Potassium occurs and fertility of soil is greatly reduced. The fertility of a soil depends upon the amount of plant nutrients that it possesses and their availability to the plant.

Nitrogen is the essence of life. Nitrogen in the soil is mainly derived from the organic matter, and to a small extent from rain and air. It stimulates the vegetative growth of the plants and keep them healthy.

Phosphorus is the vehicle of life. It is derived from rocks, leaf and other organic deposits. If deficient, it effects germination, but when in excess it hastens maturity.

Potassium is found combined with silicates, chlorides, nitrates etc. It prolongs the growing season of the plant and stimulates starch formation in crops like potato.

Calcium is normally found in the form of calcium carbonate and as oxide and hydroxide. It neutralizes the soil and makes plant food readily available to the plant. It also helps in building up the structure of the soil.

Copper, zink, molybdenum etc. are present in the soils in micro quantities, and are known as trace elements. In the absence of them, plants suffer from deficiency diseases. Application of even very small dose of these micronutrients cures the disease and promotes a healthy plant growth.

The macronutrients and the micronutrients have varied functions in plant growth. They contribute to the material from which protoplasm and cell wall are constructed. They increase the hydration of the cell colloids, permeability of membrances and osmotic pressure of plant cells. They provide substances to the cell sap and regulates its pH. They contribute or counteract the effect of toxins and act as catalyst or enzymes. Of the different minerals the most important are Nitrogen, Phosphorus and Potassium. Most of the chemical fertilizers that are in use are composed of NPK.

Organic Components of Soil

The chief organic component of soil is *humus*, which chemically contains amino acids, proteins, purines, pyrimidines, aromatic compounds, hexose sugars, sugar alcohols, methyl sugars, fats, oils, waxes, resins, tannins, lignins and some pigments. Further humus is a dark coloured, odourless, homogeneous complex substance. The chemical composition of well digested humus is given in Table 14.

Table 14 : Chemical composition of humus.

Constituent	Composition by weight (%)
Carbon	55 - 60
Hydrogen	4 - 5
Oxygen	35 - 40
Nitrogen	5 - 6

Formation of Humus

Humus consists mainly of an organic layer derived initially from plants and subsequently modified by physical and chemical processes under the influence of climate. The biological and biochemical changes that leads to humus formation are complex and not clearly understood, but briefly freshly fallen litter decomposes as follows : during the first few days after litter fall, the water soluble substances are usually leached out by rainfall. These substances consists partly of sugars, amino acids and aliphatic acids. The constitution of the remaining is still unknown. Various anions and cations are simultaneously released and percolated down through the soil profile. This stage of litter decomposition is not usually much influenced by microbial activity, but subsequently soil micro-organisms like bacteria, actinomycetes, fungi, algae and protozoa may degrade the orginal components of leaf tissue into simpler ones by a process called mineralization.

Some of the organic constituents of the litter are decomposed very rapidly, others less rapidly. Sugars and starches are the first to be degraded, then the proteins, the hemicelluloses and finally the lignins. Sugars and starches are changed to carbon dioxide and water, proteins to carbon dioxide, water and ammonia, the latter being rapidly converted to nitrate. The simple organic substances formed as the result of microbial activity are the substrate from which are formed more complex humic compounds. Less than one quarter of the total organic material produced by plants is converted into humus, the rest is lost through oxidation, leaching and erosion. The process of leaf litter disappearance therefore, illustrates the close and complex relationship between the living soil community and its environment. This process is far from straight forward and moreover, the problem to which solutions have not been found far outnumbers those which have been solved (Madge, 1966).

Functions of Humus

The important function of humus have been enumerated as follows :

1. Humus gives darker colour to soils which therefore absorb more light and hence are warmer.

2. Humus tends to elleviate the undesirable conditions of very sandy soils on the one extreme and very heavy clayey soil on the other.

3. Humus improves soil tilth of heavy soil by reducing their cohesion, stiffness, shrinkage and cracking. It makes the soil scour easier on the plough.

4. Humus not only encourage granulation, but also more important, stabilizes the granular structure.

5. Through better granulation, humus improves the water holding capacity, drainage and aeration of soils. Similarly, it increases the resistance of soils of erosion.

6. Humus acts as a sponge and increases the water absorption of soils.

7. Humus promotes natural mulching in soils which helps in retaining moisture by reducing evaporation.

8. Owing to its high absorptive power for nutrients, even small quantities of humus greatly enhance the base exchange capacity of soils.

9. The acid humus is capable of reacting with complex soil minerals and mobilise their plant nutrients. Thus, organic matter in the soil not only serves by itself as a source of plant nutrient, but also releases the nutrients from the minerals of the soil.

10. It is estimated that nearly 98 percent of the nitrogen, 33 percent of the phosphorus and appreciable quantities of sulphur, potassium and other plant nutrients are held in their humus. Although, humus is highly resistant to microbial attack, it is not immune to mineralization. Humus is also oxidised though very slowly and the nutrients present in humus are made available to plants.

11. Since almost whole of the soil nitrogen is in its humus, it is evident that only means of storing nitrogen in soils is by maintaining the humus content of soils.

12. Soil humus is also the source of micronutrients and hormones to plants.

Thus, plants and animals reduces, during their passage from the fresh state, through soil humus, to carbon dioxide, water and a small quantity of inorganic salts not only provide a complete and self regulating supply of plant nutrients, but more important, they also promote the essential physical conditions of soil most favourable for plant growth. Humus is beneficial for all types of soils - sandy or clayey.

Recently, synthetic substance like 'krillium' which are equally effective in granulation of soil, have been put on the market, but they are too costly and beyond the earth of the farmers.

Soil Reaction

Many chemical properties of soils centre around soil reaction. as regards their nature, some soils are neutral, some are acidic and some basic. The acidity, alkalinity and neutrality of soils are described in terms of hydrogen ion concentration or pH values. A pH value of 7.0 indicates neutrality, a pH value above this figures (7.1 - 14.0) indicates alkaline condition and a pH value below (0.0 - 6.9) indicates acid conditions. Normally the pH value of soils lies between 2.2 and 9.6. It was Sorenson (1909) who introduced the concept of pH and defined pH as "log of the reciprocal of hydrogen ion concentration present in the medium". If the hydrogen ion concentration in the medium is 1/1000 gm/litre, the log of this is 3 (pH). This has given a very convenient scale to express the hydrogen ion concentration of the medium. Smaller the value of the pH, the greater will be the hydrogen ion concentration. The pH value and its interpretation in terms of soil reaction are given in Table 15.

Table 15 : The pH value and its soil reaction.

Soil reaction	pH
Strongly acidic	less than 5.5
Medium acidic	5.5 - 6.0
Slightly acidic	6.0 - 6.5
Very slightly acidic	6.5 - 7.0
Neutral	7.0
Very slightly alkaline	7.0 - 7.5
Slightly alkaline	7.5 - 8.0
Medium alkaline	8.0 - 8.5
Strongly alkaline	More than 8.5

The pH value of soils normally varies between 4 and 10. Soils having high amount of calcium may not have pH more than 7.5, where as at pH value above 8.5 it is quite likely that the sodium ions are in excess in it. Soils developed in humid regions normally have a pH of less than 7.0 as these are highly leached, while those developed from lime deposits have a pH in the alkaline range. Warm, dry climate soils are strongly basic. In India, acidic soils (pH below 5.5 to 5.6) occur in the high rainfall areas of Western Ghats, Kerala, Eastern Orissa, West Bengal, Tripura, Manipur and Assam. The saline, alkali or basic soils or India occur in Uttar Pradesh, West Bengal, Punjab, Bihar, Orissa, Maharashtra, Madras, Madhya Pradesh, Andhra Pradesh, Gujarat, Delhi and Rajasthan.

Soil reaction is of considerable importance in a variety of processes within the soil and also within the plants. The pH value and its significance in plant growth has been studied by various workers and certain general conclusions have been arrived at : (1) The pH value of the soil and the distribution of plants are controlled by certain variables such as length of the growing season.

The length of the growing season if it is high result in increased tolerance of plant species to soil acidity. (2) Total moisture available in the soil. If there is high humidity, there is likely to be more alkalinity in the soil, for example, some of the earlier classifications of vegetation as one suggested by Unger (1936) classifies the vegetation into (a) lime loving, (b) lime tolerant, and (c) lime avoiding species.

Highly acidic and highly alkaline soils are often injurious for plant growth. Neutral or slightly acidic soil, however, are best for the growth of majority of plants.

Soil Colloids

These are the soil particles which are infinetely small and are electrically charged. The average size of the colloidal particles is little more than 1/254000 inch in diameter. The particles however, occupy an enormous surface area. It has been estimated that the combined surface area of the colloidal particles in a cubic foot would be somewhere about 150-200 acres. This helps to explain the high water holding capacity of certain soils.

Soil colloids can be inorganic or organic. Both together form what is called the soil complex. This colloidal complex has an important bearing on soil fertility. Due to its absorptive properties, it acts as a reservoir of plant material. Inorganic and organic soil colloids also modify physical properties of soil such as structure to a great extent. Colloidal organic and mineral matter are essential for the formation of the granular structure of the soil. Soil with such a structure can be more easily worked and easily drained and are rapidly and completely aerated for which reason they are able to produce better crops.

Cation Exchange Capacity

Clay have negetive charge and attract positive ions such as Calcium, Magnesium, Potassium, Sodium etc. If an acid solution is treated with potassium chloride solution, the potassium displaces the hydrogen ion in the soil :

$$\text{(Colloid)}H^{+} + Kcl \longrightarrow \text{(Colloid)}K^{+} + HCL$$

The different cations differ in the forces with which they are held. These differences in force depends upon (a) valence that is, the amount of positive charge that the ion carries, (b) of size of the cation, (c) the degree of hydration of the cation, (d) kind and percentage of clay, and (e) the percentage of humus in the soil. Based upon these differences the cations can be arranged in a series :

H > C > Mg > NH_4 > K > Na > Li

The least amount of force is with hydrogen ions, and the fertility of soil which will help plants to grow depends upon the degree with which they would be available for plants. This degree is known as Cation exchange capacity.

The colloidal complex that we get in the soil generally behave as salts of weak acid like silicic acid and Humic acid. They also form bases with aluminium and iron, these bases react with colloidal micelle giving a weak salt. Both acidic as well as basic salts are left in the soil. Acid salts combine with cations and give rise to a soil rich in cations, this is metwith in humid climates (pH low-acidic). Basic salts combines with anions generally in hot humid climates (pH high - alkaline).

The nature and extent of exchangeable bases in the soil influence its properties to a great extent. A soil with high base status with calcium as the principle cation easily acquires a granular structure. Such a structure minimises the bad effects of the high clay content and ensures good aeration and percolation of water, thus preventing them from getting eroded. On the other hand, a medium saturated soil has inferior structure and is badly drained.

Anion Adsorption Capacity

Comparable to cations there is another phenomenon known as anion absorption capacity. This is a condition developed in certain soils where the colloidal particles carry positive charge and absorb the negative charged anions like NO_3, Cl_2, PO_4 etc. This kind of soil develops into categories known as amphoteric soil, which are quite common in tropical zones. They are amphoteric in the sence that they are capable of changing their pH either on acidic or on basic side. This is done in the soil by counter balance of positive charge which is developed by the hydrated iron and aluminium compounds which are present at the edge of the clay particles. The most important significance of this anion exchange in the soil is in respect of PO_4 ions, because it enters into many inorganic exchanges in the plant roots. Therefore, the anion exchange capacity can also be defined as the amount of bound phosphate at any given pH of the soil.

There is another phenomenon, that is, the occurrence of the 'chelating agents' in the soil. This phenomenon was discovered by Martell and Calvin (1952). The chellating agents are organic anions which are more complex. They form stable complex compounds with calcium, Iron, Aluminium, Manganese, Copper and Zinc. As a result these positive ions are not available in the soil freely for plant and the plant develops deficiency diseases.

Major Soil Types of India

Geographically soils of India are divided into the following three groups : (1) Mature soils of peninsular India, (2) Alluvial soils of Indogangetic plain, (3) Scanty soils of Himalayas (Figure 28).

The Deccan peninsula is characterised by soils of Vindhyan origin. During the Cretaceous and Tertiary period basaltic periods basaltic lava flows formes the parental material for the soils of Deccan peninsula. This region is characterised by the presence of 'laterite' which caps many of the hills and plateaus of Deccan India. Laterite is a surface formation caused by weathering in regions where temperature and humidity are high. Laterites are soft when first quarreid but it hardens on exposure and after hardening it resists erosion more strongly than the rocks on which it rests. The black cotton soils occur in West Madhya Pradesh, Maharashtra, Parts of Andhra Pradesh, Bihar, West Bengal, Assam, Tamil Nadu, Orissa etc. Red soils occur in large areas of Madras, Mysore, North East Andhra Pradesh and other regions. Laterites soils occur on the summit hills of Deccan, Madhya Pradesh, Eastern Ghats, parts of Orissa, Assam, Karnataka and Kerala.

Alluvial deposits are of Himalayan origin and found in the Indo-Gangetic plains. These deposits in depression extends to a depth of approximately 13000 meters. The soils are fertile and support three-fourth of the Indian population. In India, alluvial soils occur in the belt of the submountain regions running along the base of the Himalayas (in Punjab, Uttar Pradesh, Bihar, West Bengal) and velleys of Narmada, Tapti (Madhya Pradesh), Krishna and Cauvery.

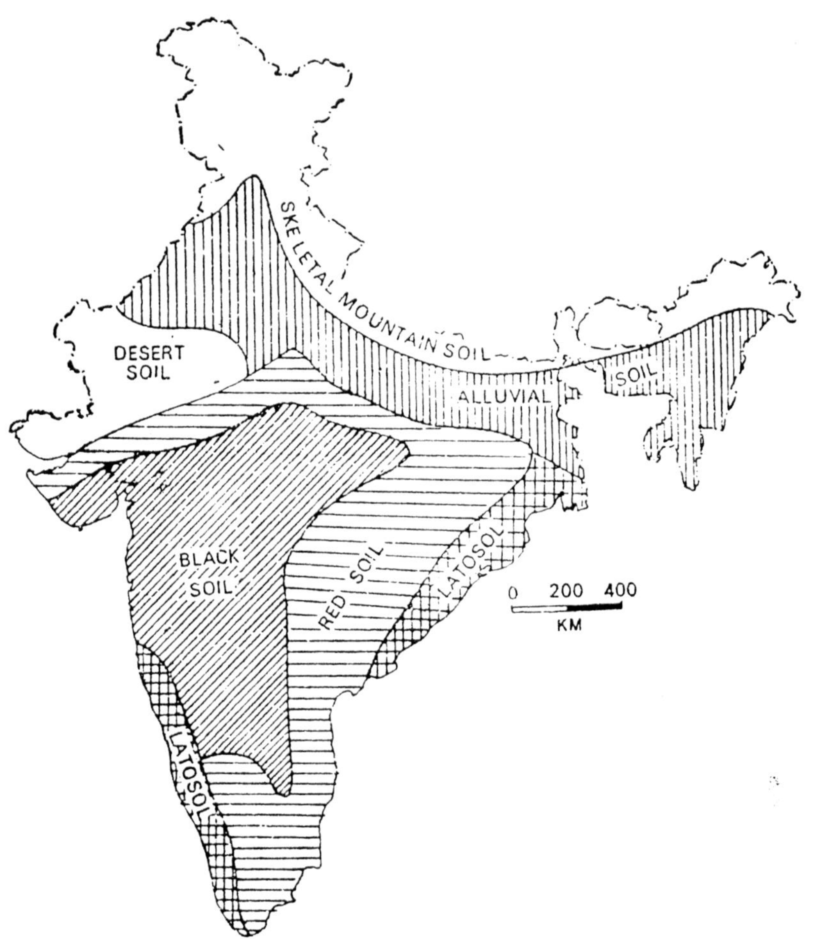

Fig. 28 : Outline map of India showing soil types.

The Himalaya area contains immature soils of fresh origin.

Besides these there are mountain soils, desert soils, peaty and other organic soils. Peaty soils are mostly found in Kerala. They are black and acidic (pH 3.9).

8. *Topographic Factors*

The topographic factors of the habitat include geological strata, altitude, slope, exposure, erosion, silting and blowing up of sand. Earth's surface is not similar in different regions and may show several irregularities. Such topographic features influence vegetation by producing variations in climate. Such variations in climate due to these factors give rise to characteristic local or even microclimates, or may even modify soil conditions. The chief topographic factors are (1) height of mountains, (2) direction of mountain chains, (3) steepness of slopes influence the terrestrial vegetation and animals.

Effect of different altitudes can be better seen on high mountains. With an increase in altitude above mean sea level, there are changes in the values of temperature, pressure, wind velocity, humidy, intensity of solar radiation. Due to these changes, vegetation at different altitudes differ much (Figure 29). In Himalayas, temperature variations are quite evident, and there is a general zonation of vegetation from lower to higher altitudes. The successive zones of vegetation from base upwards are tropical and subtropical, temperate and alpine.

The direction of mountains steer or deflect wind into different directions and capture moisture from wind on certain sides. It may also effect insolation, temperature and humidity. In the northern hemisphere the south facing slopes are warmer and xeric than the north facing slopes, and reverse in the case in the southern hemisphere. Water vapours may accumulate only in some preferred directions. This may be the reason that on certain sides of the high mountain, one can see luxurient forests, where as on the other side, there occurs scanty vegetation.

The steepness of the slope plays an important role in determining the climatic and edaphic characters. It affects the amount of solar radiation received during the day. The steeconess of the slopes increases the exposure to sun. The slopes receive more

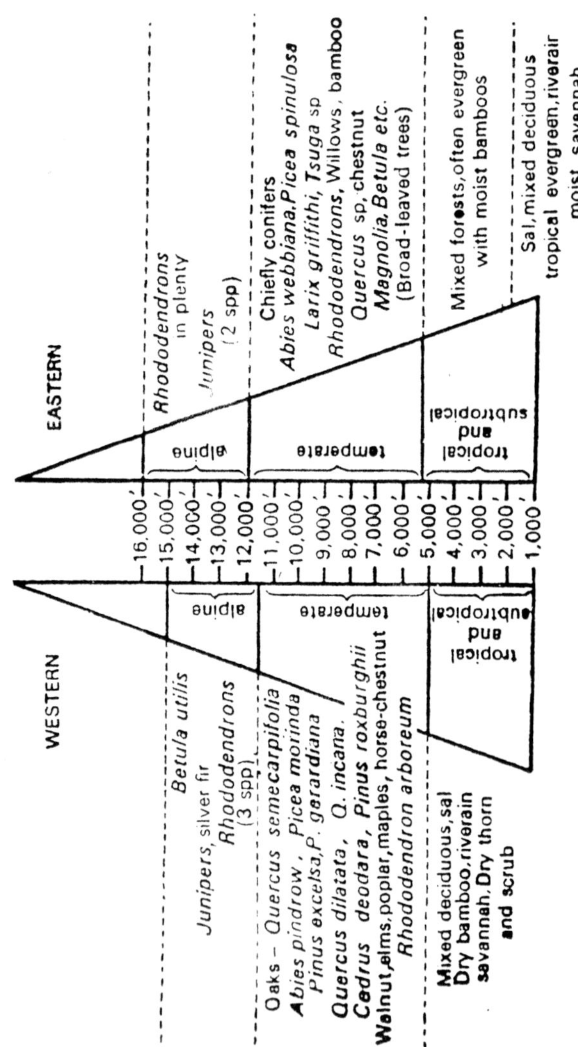

Fig. 29 : Altitudinal zonation and distribution of various species of plants on the Western and Eastern Himalayas.

rainfall facing the windward side while the leeward side is generally drier. Rainfall lost by runoff increases with the angle of the slope, and water absorbed correspondingly decreases. The slope plays an important part in determining the character of the soil. The downward movement of rain water removes soil from a slope and carries it down and deposits usually dry and its surface unstable on which plants are available to establish themselves firmly. The angle of a slope determines the amount and type of soil accumulated. It is only on nearly ground or gentle slopes that considerable amount of soil accumulates. The gentle slope support vergin forest with hygrophilous ground vegetation, whereas the steep slopes have sparce vegetation.

Thus, topography exerts indirect influence on biota by influencing the abiotic environmental factors like wind, energy, and moisture regimes of the habitat.

9. *Biotic Factor*

The association of organisms in any community is a potent biotic factor. Our biota represents an arrey of life-forms which differ in shape, size, behaviour and number. They differ in their environmental requirements and life activities. The organisms, whether plants or animals, are interdependent and derive their nourishment from the common pool of materials, that is, the environment. The biotic factors include the influence of living organisms, both plants and animals, upon the vegetation. The biotic effect may be both direct and indirect. It may be beneficial or detrimental.

All the biotic components of a community interact with each other in various ways. These interactions are directed to procurement of food, space and leaving behind their progeny for survival of the species (Misra, 1974). In the forest trees, shrubs, herbs, mosses and lichens interact with one another and adjust according to environmental conditions. Trees cast their shadow on many shade-loving plants which grow around or beneath them. The micro-organisms such as bacteria, algae, fungi and virus affect the life of plants in many ways. Decomposition of dead parts of organisms causes significant addition of organic compounds and humus to soil. In this way vegetation modifies the habitat to a considerable extent. Similarly, animals affect the plant life in several ways. Many animals use plants as their food and for shelter as well. Pollination, fruit and seed dispersal, parasitism, symbiosis are the common examples of such interactions.

The interaction between the living and the non-living environment forms the so-called eternal triangle of life. Hence, every organism must share its environment either with members of its own species or with members of other species. The relationship between organisms of the same species is oftern referred to as the *intraspecific*, while that between two different species is called as the *interspecific*. Such relationships occur in all grades among living organisms. It may be either beneficial to both partners or harmful to both, or beneficial to one and harmful to the other, or its may be neutral for the other.

Relationships among Organisms

Haskell's elaborate classification of 'coactions' between species has been adopted by Burkholder (1952), who on the basis of several combinations of 0 (no significant interaction), + (growth, survival, or other population attribute benefited), and – (population growth or other attribute inhibitted) between two species, have been grouped into nine types of interactions. The possible combinations are OO, --, ++, +O, –O, +–, three of which (++, --, and +–) have in turn been commonly subdivided, and the whole scheme resulted into nine types of possible interactions, which have been explained in Table 16. Malcolm (1966) recognised ten familiar interactions (Table 17). The column marked 'on' lists the effects of the interaction when it is going on, and the column marked 'off' lists the effects when it is not going on - the numbers 1 and 2 refer to the two organisms or populations that are interacting.

Neutralism

Neutralism is the most common type of interspecific interaction. In neutrality neither of the partners is affected. That would be saying that the effect of the interaction is the same when it is going on as when it is not going on. If there are interactions, they are of a more subtle and indirect type.

Table 16 : Possible biological interactions among populations (Haskell, 1949; Bruckholder, 1952)

	Type of interaction	*Species 1*	*Species 2*	*General nature of interaction*
1.	Neutralism	0	0	None of the populations is affected.
2.	Competition : Direct interference type	-	-	Direct inhibition of each species by the other.
3.	Competition : Resource use type	-	-	Indirect inhibition when common resource is in short supply.
4.	Amensalism	-	0	One population is inhibited, the other is unaffected.
5.	Parasitism	+	-	One population, the parasite, generally smaller is benefitted at the expense of the other, host.

(Contd.)

1	2	3	4	5
6.	Predation	+	−	Predator, generally larger, destroys the smaller prey population.
7.	Commensalism	+	0	The commensal benefits while the host is unaffected.
8.	Protocooperation	+	+	Interaction is favourable to both but it is not obligatory.
9.	Mutualism	+	+	Obligatory interaction affecting favourably both populations.

0 = no significant interaction.
+ = growth, survival or otherswise the population is benefitted.
− = population is inhibited.

Table 17 : Biological interactions (Malcolm, 1966).

Interaction	Effects			
	on		off	
	1	2	1	2
Neutralism	0	0	0	0
Competition	-	-	0	0
Mutualism	+	+	-	-
Un-named # 1	+	+	0	-
Protocooperation	+	+	0	0
Commensalism	+	0	-	0
Un-named # 2	+	0	0	0
Ammensalims	0	-	0	0
Parasitism	+	-	-	0
Un-named # 3	+	-	0	0

Competition

Competition refers to the type of interaction in which the two individuals or species compete for a limited resource. Competition dominate community organisation when the physical environment is uniform. Under uniform conditions the organisms do need a place to live and extract the energy and materials, thus they compete for a suitable place, even though the requirements vary. When competition takes place between individuals of the same species, we call it 'intrespecific competiton', while in the other case, where two organisms, belonging to different species compete, the situation is known as 'interspecific competition'. In a community, both these types of competition are possible. As a result of interspecific competition the weaker individuals are eliminated and the stronger ones remain to reproduce. While with interspecific competition the weaker species are aliminated from the community. This characteristic of organisms to compete with their own progeny and members of the other species enables some of them to dominate within the community.

There is no doubt that species evolution in a community is driven in part under the stress of competition. We can assume that at the start, all the species in a community do less well than they could because their niches overlap (by niche is ment any and all resources used by members of the species). If they do not compete for energy and materials, they would all grow faster or reproduce more vigrously or live longer and so on - they would be more successful. The observed persistence of many species in nature is really due to environmental heterogeneity (Kumar, 1981).

Mutualism

We know more about mutualism, because it has been given more attention. It is a relationship between two dissimilar organisms that are mutually beneficial. The organisms involved in mutual relationship cover a wider range, namely (1) plant-plant, (2) animal-animal, or (3) plant-animal associations.

It is an obligatory interaction that is strongly beneficial to both species. The term 'symbiosis' has often been applies to this relationship, but symbiosis properly refers to intimate association of two or more dissimilar organisms, regardless of benefits or the lack of them and hence includes mutualism, commensalism and parasitism (Smith, 1974). Mutualism may be 'facultative', when the species involved are capable of existence independent of one another, or 'obligate', when the relationship is imperative to the existence of one or both species.

The most common type of mutualism is met within organisms which are in close contact and in which there is physiological interdependence. The classical example of mutualism is the interaction between an algae and a fungus in the lichen (Hale, 1967). The lichen are composed of fungi enclosing algae (Figure 30). The term consortium has been proposed by the students of lichens for this type of association to emphasize the intimacy of their association. The fungus part of the lichen contributes moisture, nitrogenous waste, respiratory carbon dioxide as well as shelter for the algae. The algae in turn produce carbohydrates for themselves and for the fungal partner. Lichens may consequently survive in relatively dry terrestrial environments. The fungal species of the lichens cannot survive without the algal species and *vice-versa*.

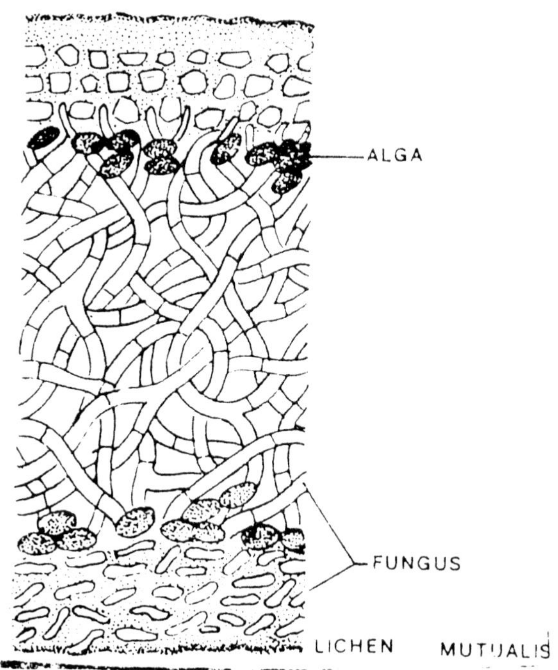

Fig. 30 : Obligatory mutualism in lichens between algae and fungus.

The most intimate form of mutualism involve organisms which live directly *within* other organisms. For example, the root of certain leguminous plants form root nodules as a result of the invasion of nitrogen faxing bacterium - *Rhizobium*. Bacterial root nodules have also been reported from 10 genera of eight other families of angiosperms (stewart, 1966). In them the host provides nutrients for the bacteria, and the bacteria in turn fix atmospheric nitrogen, making it available to the host plant (Nutman, 1976).

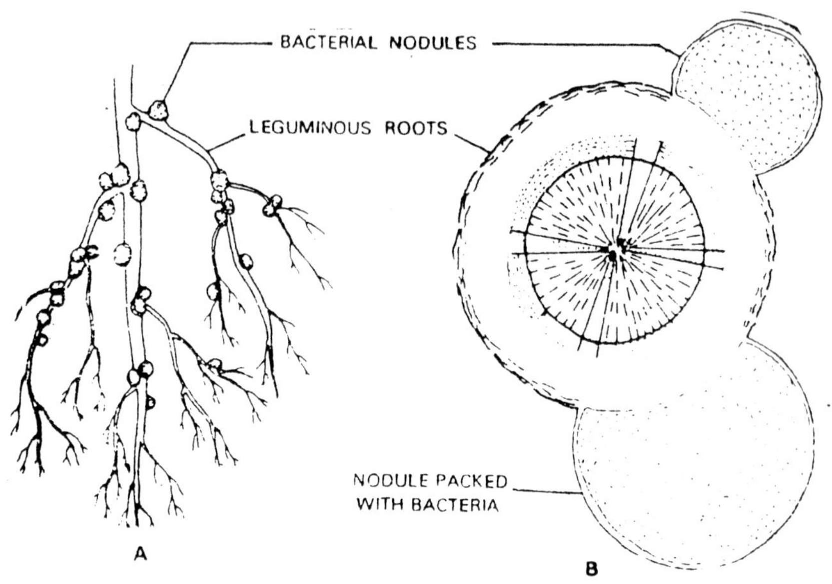

Fig. 31 : (a) Leguminous root with bacterial nodules.
(b) T.S. Leguminous root passing through bacterial nodules.

Bacterial nodules in leaves are less common, but have been reported from approximately 370 species of 5 genera in the families Myrsinaceae and Rubiaceae. According to Humm (1944), "Leaf nodules are typically small elevations of less than 2 mm diameter. They may be irregularly scattered over the surface of the leaf, located along the leaf margin only or present in two rows along each side of the mifrib". Bacterial leaf nodules of *Psychotria bacteriophila* Val has a symbiotic association with *Klebsiella rubiacearum*. Centifanto and Silver (1964) have shown that the bacteria can fix nitrogen, but that it is only of secondary importance. Bacteria free seedlings supplied with nitrogen do not survive, which has led to speculation that hormones or other growth substances secreted by the bacteria are of primary importance. Lersten and Horner Jr (1967) proposed that the constantly renewed cell wall of the mesophyll cells bordering, and within, the nodules provides a convenient carbohydrate substrate for the bacteria. The membranes found between bacteria could be sites for enzymes capable of digesting the carbohydrates of the mesophyll cell wall. These membranes, furthermore, could also be involved in the transfer of whatever substance or substances the bacteria must contribute to the plant.

Certain algae and plants, and algae and animals form mutualistic associations (Figure 32). The coralloid root of *Cycas* contains inside, nitrogen fixing blue green algae and bacteria which fix atmospheric nitrogen for *Cycas*. In return they get protection and nutrients. Intracellular association between unicellular algae and invertibrates represent the coexistence of unlike genomes residing in the same cytoplasmic environment. Table 18 illustrates the wide range of taxa of algae and animals involved in such a mutualistic association. In general, mutualistic algae resist host destruction by avoiding digestion either by counteracting the function of lysosomes with vacuoles containing the algae, or by possessing cell walls which resist hydrolysis by the host (Trench, 1979).

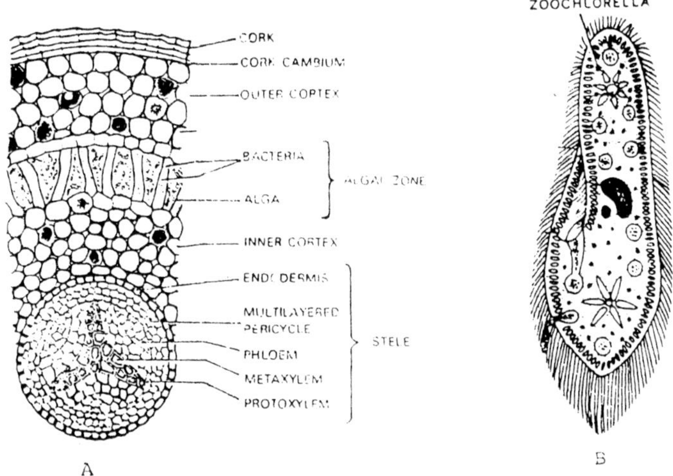

Fig. 32 : Mutualistic association involving algae :
(A) Blue green algae and Cycas root.
(B) algae (*Zoochlorella*) and *Paramecium*.

Table 18 : Some algae-animal symbiotic association (Kumar, 1981).

Algae	Host	Habitat
Aphanocapsa	*Ircinia*	Marine
Cyanocyta korschifoffiana	*Cyanophora paradoxa*	Feshwater
Prochloron	*Didesmid ascidians*	Marine
Chlorella	*Hydra*	Marine
Chlorella	*Paramecium*	Freshwater
Platymonas convolulae	*Convoluta roscoffensis*	Marine
Gymnodinium microadriaticum	*Calms, Cnidarians*	Marine
Licmophora	*Convoluta convoluta*	Marine

The mycelium of fungi is found either attached to roots of certain plant, or grows inside the roots to form a mutualistic association known as mycorrhizae (Figure 33). The main effect of mycorrhizal infection on plant growth is the stimulation of phosphorus uptake due to exploration by the external hyphae of the fungus, beyond the root hair. The fungi in turn gets food from the host plant.

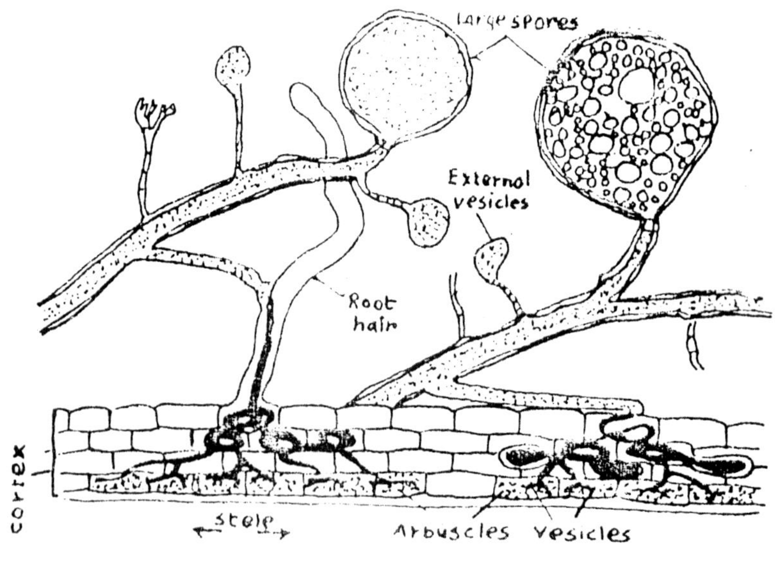

Fig. 33 : Diagramatic representation of Vasicular-Arbuscular Mycorrhiza showing vesicles and arbuscles.

Mutualistic associations are very common in terrestrial plants and animals of the tropical regions. In mutualism animals are benefited by plants and *vice-versa*. Bees, moths, butterflies, etc. derive food from the nectar, or other plant product, and in return bring about pollination. Seeds and fruits are commonly dispersed and transported by animals. The fruits are eaten by birds, mammals, etc. and seeds contained in them are dropped in the excreta at various places. Foliage and other plant parts are eaten by herbivorous animals, and their dung is manure (Sharma, 1978). an obligate coupling exists between *Ficus* and its insect pollinators and between swollen thorn *Acacia* with its associated Pseudomyrmecine ants (Ramirez, 1969).

The birds are often found riding on the back of the grazing animals, where they feed on the ectoparasites such as ticks and mites. The grazing animals are helped in more than one way for they are freed from these pests and they also serve to give the signal of the approaching enemy. The association of the ostriches and the zebras is said to be for the mutual benefit of each other. The keen sight of ostriches and the keen sense of smell of the zebras are of great mutual help in guarding against the attack of the predators.

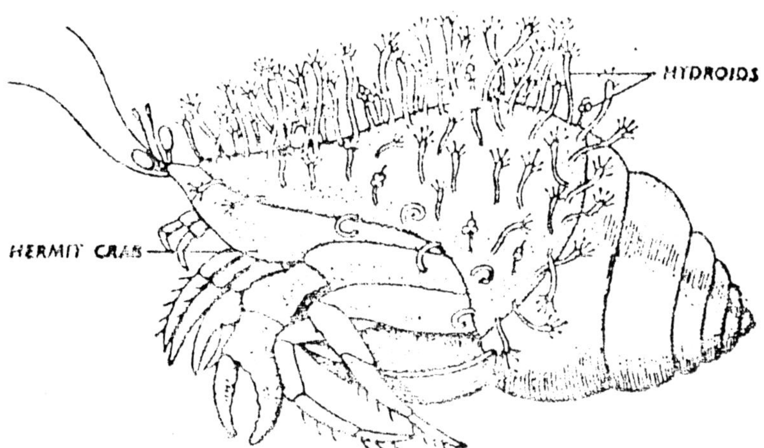

Fig. 34 : Gastropou shell occupied by a hermit crab with attached hydroids.

The most remarkable example of mutualism is found between the intestinal flagellates and the wood eating termite and cokroaches. These hosts are not capable of digesting wood on their own accord and depend upon the protozoans present in the intestine. The protozoans digest cellulose for termites and in return obtain food and shelter from the termites and cockroaches.

Among the marine invertibrates numerous examples of symbiotic relationship between two animals are met with. The association between sea-anemone *Adamsia palliata* and the hermit crab *Eupahurus prideauxi* is a classical example of mutualism found in animals. The hermit crab lives inside the shell of the gastropod. In certain instances outer surface of the gastropod shell is completely colonized by the hydroids (Figure 34).

Proto-Cooperation

Proto-cooperation is less extreme types of interaction than mutualism in which the interaction is clearly beneficial to both species, allowing the equilibrium population levels of both to be higher than they otherwise would be. It is not obligatory for either species. Examples of proto-cooperation are many, including the clearing of fishes by other fishes. The cleaners of course get food from the fishes they clean, and in turn the cleaned fish get rid of parasites and organic debris. Similarly, large birds such as herons and ibises nest in the lower branches of relatively unprotected trees, while the snakes congregate around the bases of the trees. This protects the bird from tree climbing predators such as racoons. In turn the snake feed in part, on fish dropped by the birds and the occasional baby birds that fall out of the nest (Ehrenfield, 1970).

Commensalism

In this type of association, one species does better, and the other does neither better nor worse. The one that does better is the commensal, and the unaffected species in the host. A commensal as an organism requires shelter, support, locomotion or food supply from the host, without any harm to the host. Among plants the lianas are common examples, which are rooted in the ground and maintain erectness of their stems by making use of other trees (objects) for support. They maintain no direct nutritional relationship with the trees upon which they grow. Epiphytes (Figure 35) such as Vanda are the best examples of commensalism. These plants survive better by being attached to other plants. They differ from lianas in that they are not rooted into the soil. They use other plants as support and not for water or food supply. They neither harm nor benefit the obliging host. In epiphytes there is a special layer - velamen over the root surface. The cells of the velamen can take up abundant water rapidly from the atmosphere.

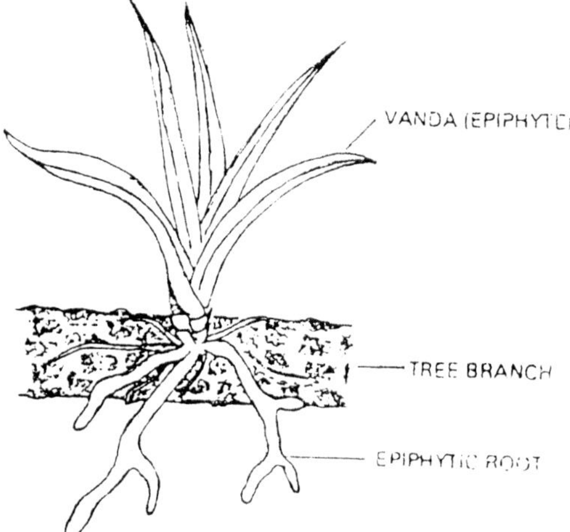

Fig. 35 : Epiphyte Vanda on a tree branch.

Some plants grow on the surface of animals. For example, *Basicladia* (Cladophoraceae) grows on the backs of freshwater turtles as commensals.

In oceans there are many commensals. *Eurythoe* is a common commensal that is seen in *Suberites* and other sponges. the decapod crustacean *Polyonx* lives inside the 'U' shaped tube of *Chaetopterus*. Inside the tube it is protected against the enemies and at the same time it derives its food as well as oxygen from the water that is forced out of the tube by the movements of the parapodia of the *Chaetopterus*. The polynoids are likewise found inside and chitons, as well as in the ambulacral grooves of starfishes. The oyster crab, *Pinnothers ostreum* is found in the mantle cavity of the oyster. In addition to shelter, it also gets food from the host mollusc, oyster, without causing any harm.

The staphylinid beetle lives in the termite colony as a scavanger. The beetle is seen even riding on the head of termite. The beetle gets living space, food and transportation (Figure 36)

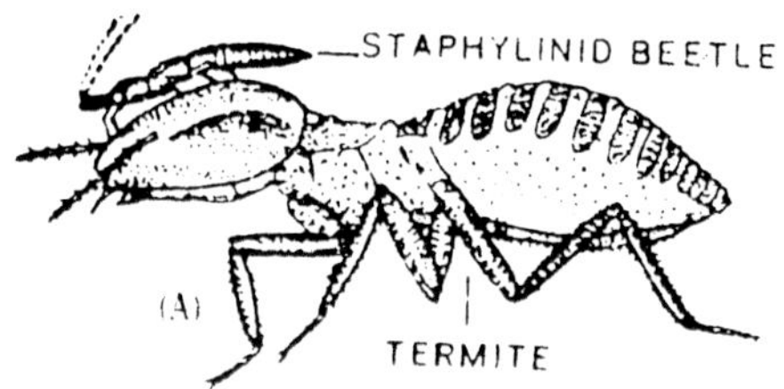

Fig. 36 : Commensalism for food - termite and staphylinid beetle.

The sucker fish has the dorsal fin modified as a sucker with the help of which it is attached to the body of shark, so that the sucker fish gets transportation. The attachment is not permanent, the sucker fish releases the attachment after some time and swims in search of food (Figure 37). This situation is which the small organism gets free transportation is called 'phoresis'.

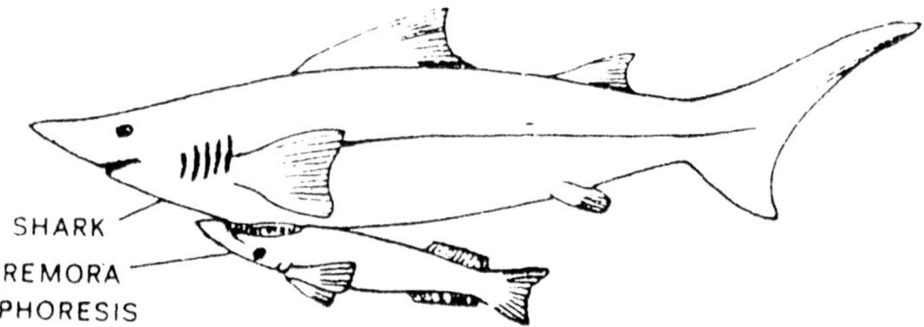

Fig. 37 : Commensalism for transport - *Phoresis* (shark) and *Remora* (sucker fish).

Ammensalism

It is the counterpart of commensalism. Where as in commensalism one population benefits from the interaction and the other remains unaffected, in ammensalism one population becomes less successful during the interaction and the other is unaffected. In many cases the harmful effects are due to certain chemical substances secreted by one population as specific toxins into the environment. These chemicals are known as allelochemicals. They are of three types : (1) allomones, (2) kairomones, and (3) depressants.

Allomones are chemicals which give adaptive advantage to the organisms that produce the chemical. Examples include repellents, which provide a defence against attack, such as the odour of a shunk : esçape substances; the ink that is used by an octopus to confuse prepators; venoms and toxins found in snakes and scorpions; and atractants secreted by carnivorous plants and flowers scent secreted by flowers to attract pollen carrier insects.

Kairomones are chemicals that give adaptive advantage to the receiving organism. For example, the chemicals released by nematodes stimulates certain fungi to develop traps for nematode worms are used to protect the nematodes from predators. Similarly, the odour of a cat is smalled by mouse, and function as danger signal of the predator to the receiver.

Depressants released by certain organisms inhibit or poison the receiver without benefit to the releasing organism. Several algae like *Anabaena, Aphinizomenon* and *Gymnodium* cause death of fishes, ducks and other animals. *Hydrodictyon* and *Scenedesmus* are known to produce antibiotics causing death of several bacteria. The grass *Aristida oliganlha* inhibits nitrogen fixation by bectria and blue-green algae (Rice, 1974). In *Partheneium argentatum* grown in gravel culture, trans-cinnamic acid has been found to accumulate in the culture, which interferes with the growth of plants. Similarly, bitter almonds are known to form HCN which inhibits poppy seed germination. In *Eucelia farinosa*, the leachate from the leaves contains 3-acetyl-6-methyl benzaldehyde, which causes reduction in the growth of tomatoes (McLarsen and Peterson, 1967).

Majority of the inhibiting chemicals are produced as secondary substances by plants and released into environment through roots, leaves, fruits and other tissues. The suppression of growth through the release of chemicals by a higher plant is known as allelopathy.

Antibiotics produced by bacteria, algae, fungi, actinomycetes and lichens are widespread in nature and may be one of the reasons why bacteria pathogenic to man cannot multiply well in the environment. A number of antibiotics such as penicillin have been used extensively in human welfare.

Parasitism

It is a kind of harmful interaction between two species. One individual which receives benefit at the expense of the other individual is called *Parasite*. The suffer is called the *host*. The parasite may live inside (endoparasite) or outside (ectoparasite) the host. Parasitism is mianly for food, but at the same time the parasite derives shelter and protection from the host. A parasite usually parasitisis a host which is larger in body size than it. Further a parasite does not ordinarily kill its host, atleast not until the parasite has completed its reproductive cycle. The organisms which derive their food partly and remain in contact with their host only for a short period of their life-cycle, are not true parasites. A wide variety of plants and animals are parasites.

Species of *Cuscuta* are total stem parasite on other plants. it is an ectoparasite (Figure 38). Their young stem twines around the host stem, from which adventitious roots (haustoria) develops that finally penetrate the ste of the host. *Orobanche*, *Striga*, anc *Balanophora* (Figure 39) are found as total root parasites on the roots of higher plants. *Dandrophthoe falcata* and *Viscum album* (Figure 40) are partial stem parasites, they grow rooted in branches of the host tree. *Santalum album* and Thesium are partial root parasites.

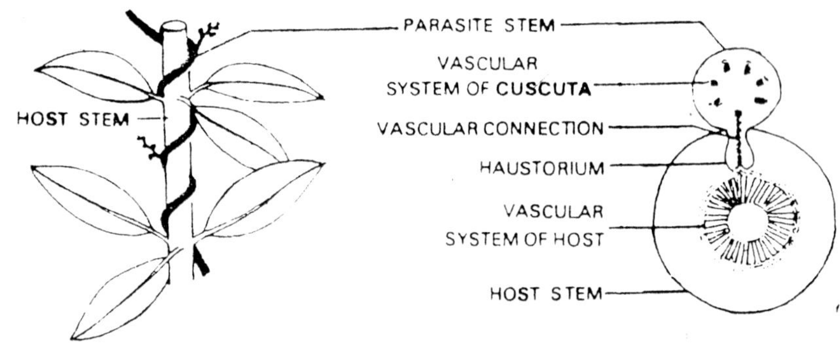

Fig. 38 : Total stem parasitic *Cuscuta* on host plant, (B) Section through host and parasite.

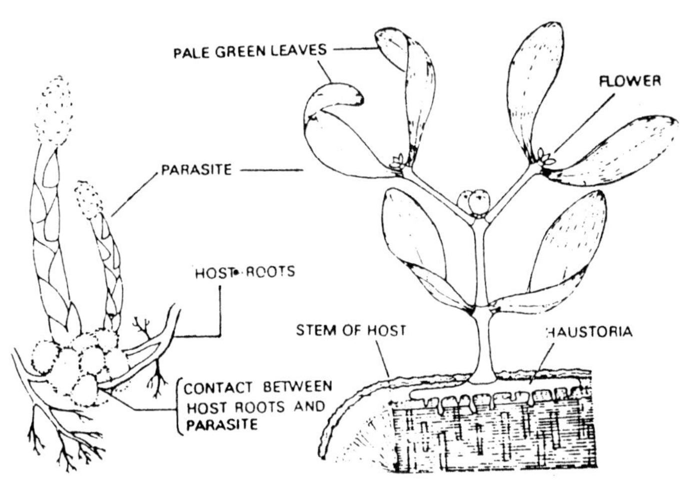

Fig. 39 : Balanophora on host roots. **Figure** 40. Viscum album host branch.

Majority of parasites are micro-organisms, of which fungi (*Sclerospora graminnicola, Albugo candida, Puccinia, Ustilago, Clavinceps*), bacteria (*Xanthomonas citri, Phytomonas malvaearum, Xanthomonas oryzae, Pseudomonas solanacearum, Pseodomonas magniferae indicae*), and viruses (Potato virus I, Nicotiana virus 10, Papaya Mosaic virus, Sugaracane mosaic virus, Banana virus I) parasitise plants. Besides these Mycoplasmas, rickettsias also parasitise a number of plants.

Animals parasise a number of plants. *Meloidogyne arenaria thamesi* nematodoe have been reported to develop root knot.

Animals parasitic on animals belong to protozoa, various invertibrates and a few vertibrates. Parasites which live inside the cell are termed as intracellular parasites *e.g.*, malerial parasites. Parasites which live in between the cells or in cavities are termed intercellular or extracellular parasites for example, tapeworm, Ascaris etc.

Predation

In predation the predator kills the prey and directly feeds on it. The eating and being eaten up relationship among the organisms lays the foundation of the ecosystem structure and function. Most of the predatory organisms are animals, but there are some plants also, for example, *Nepanthus, Darlingtonia, Sarracennia, Drosera, Utricularia, Dioneae* etc., which consume insects and other small animals for their food. They are also known as carnivorous plants. The leaves or foilar appendages of these plants produce proteolytic enzymes for digestion of the insects.

Animals like ducks, fishes. etc. eat aquatic plants frequently. Insects, squirrels, mice, rodents. etc. consume seeds of terrestrial plants as food. Moreover, they brouse seedlings of plants. Herbivorous animals kill plants and use unharvested herbs, shrubs or even trees as their food. Different plants receive varying degree of predation as a result of browsing and grazing. Generally annuals suffer more due to grazing than the perennials. Selective grazing results in growth of unpalatable species and decrease in species diversity (Harper, 1969). Tramling by hooves of animals are far more damaging than browsing. The first casulity in trampling is the disappearing of sensitive species to leave a sparce cover of hardier and usually less palatable species. The second casuality is pounding and disaggregation of organic matter and soil humus, so that it can be flown away (Shankarnarayan, 1977).

The various interactions described above exert a stress to drive genetic changes in the tolerances of the community organisms. We expect that - if organisms in a population pick up genetic known-how for coping with an environmental stress like an interaction, the known-how shows up as adaptive genetic changes the stress triggers selection that minimise the stress. Well, then it would seem that living things eventually would pick-up enough traits to avoid all interactions.

10. *Pyric Factor*

The pyric (fire) factor in terrestrial ecosystem especially in temperate and tropical regions assumed a position of great impact before man came to this planet. Man did not invent fire, they were a part of our natural environment. With the spread of civilization man-caused fires have become far more numerous than those caused by nature.

Kinds of Fire

Whenever, the soil is overlaid with thick accumulation of litter, it may catch fire. Fires of this type are flameless and subterranean, and are called *ground fires*.

Fire often sweeps over the ground surface rapidly, the flames consuming litter, living harbs and shrubs and scorching the bases of any tree it may encounter. These are called *surface fires*.

In dense woody vegetation fire may travel from the canopy of one plant to another. Such fires are known as *crown fires*.

When burning conditions change, one of the above types of fire gets frequently converted into another.

Meteorological Basis for Fire

Thunderstorms with their lightening and thunder excited the curiocity of man from his earliest beginning. The thunderstorm is a vital part of the world around us, but in many ways underreamed of by us the ancients and the casual observers of weather today. There are three types of thunderstomrs which cause lightening fires (1) Local, (2) General, and (3) Intermediate. Nearly all of the peak-loads of lightening fires occurs from the general type of storms.

To have thunderstorms and consequently lightening, there must be air masses in which certain conditions of temperature and moisture are present. These masses of hot and cold, and moist air must meet. Although, there are several theories in regard to the development of lighening, they are all concerned with the friction of rain drops or ice crystals within this electrical generator we call the thunderstorm.

Although thunderstorms and lightening fires may occur at all hours of the day and night, they predominantly occur in the afternoon hours. Summer months are obviously the months of highest lightening frequency.

The upper third of the mountain slope apparently receive more lightening ingition strikes and remaining areas receive the rest rather evenly (Figure 41). Kourtz (1967) has reported that 60 percent of the strikes were within the top third of the hills, 19 percent were within the middle third, and 21 percent within the lower third. Most of the time ignition from a lightening strike occurs in the litter, rotten-wood and duff; less frequently in snags or spiketop trees and occasionally in green leaves.

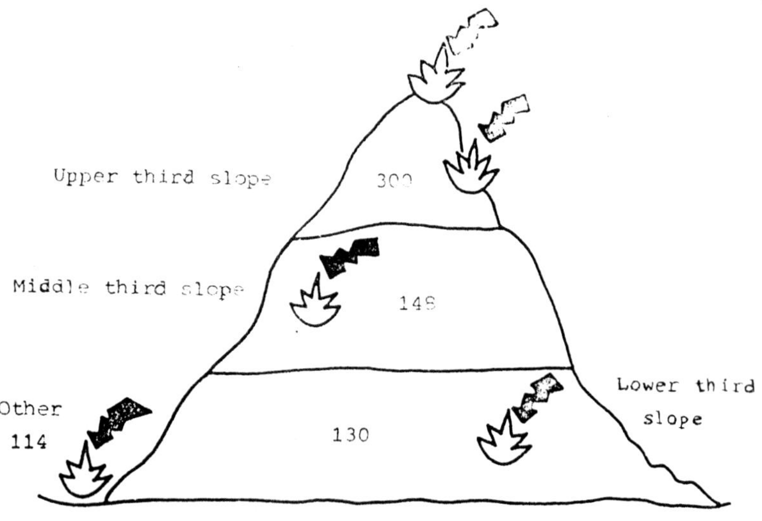

Fig. 41 : Ignition point by lightening in relation to topography (after Momareck Sr, 1967).

Effects of Fire

Each fire produces some effects we consider good and some we consider bad, with the balance depending on both the nature of the fire and - point often over looked- the nature of our objective. It was Komarek (1963) who as an ecologist realized that there is in nature no bad or no good fire.

On the one hand, fire produces a wide complex of changes in the total environment - in the soil, in the vegetation, and in the animal life. The soil laid bare by burning is liable to erosion by heavy rain or by wind. After burning, the relative frequency of different species may be altered; species may be eliminated or new species may invade; the growth characteristics of certain species may be changed and the biomass of the community may be altered atleast temporarily (Lloyd, 1968).

On the otherhand, fire is a versatile tool in the management of our forests. Komarek (1966) has rightly said that, fire is the only natural agency widespread enough to hold plant succession on a vast scale in the absence of man. The first fire thins by killing many of the smaller trees, but the larger ones survive to form another even aged group. New seedlings continually spring up under the older trees, but as fast as they appear, they are killed by the next surface fire. In this way, forests are kept open park like, and the trees are free from understory compition. These beneficial effects of fire raise a natural question in our mind - why it is not used more by the foresters ? The reason for this may be of a psychological one than anything else. We are taught from early childhood that "fire is bad, don't play with the mathces, don't touch the heater etc". As we grow older we are brain washed by propoganda of fire killing our wild animals, destroying the beautiful forests and laying vast areas subject to erosion, and all this is true to a certain extent, but "I dare say that this damage is a direct result of man's exclusion of fire to such a degree that fuel accumulated to a very dangerous point. Over protection in thereal villian and not the fire" (Perkins, 1967).

Thus, we know fire as a master, we have learned something about fire as a servant - but we still have much to do before we can direct this servant so as to win its most effective service to our own advantage (Phillips, 1965). We know of course that wild fire prevention and suppression are essential in the management of today's forests, with their many uses.

11. *Autecology*

Ecological study of individual species is called 'autecology' Ecological study includes environmental inter-relationships of a species in a population or set of populations at all stages of life-cycle.

Studies on individual species first started, when man adopted agricultural practices. This was followed by the emphasis on the studies of economically important plants. Thus, Misra and Puri (1954) very rightly stated that agriculture and silviculture are extensivions of autecology. A detailed study of the ecology of a species throws light upon the facts of considerable importance, as it furnishes data on dynamic inter-relationships, populations and their distribution. Although, autecological work has been done extensively, still only a few species have been worked out in detail. The detailed study of the individual plant and its relationship with the environment has got more importance in the applied fields of Botany such as forestry, range management, plant pathology, agriculture, soil conservation, weed control, etc. Misra (1957) in his Presidental address to the National Academy of Sciences, India expressed the view that a correct assessment of the ecological niches of any plant can be gained by a study of its autecology.

Aims of Autecology

In autecological studies, an assessment of potential and aggresiveness and susceptibility of a species under varying environmental conditions are studied. Studies on the life-histories greately contribute to the understanding of the distribution, environmental responses, adaptation and speciation. It also helps in explaining the structure and dynamics of the communities. The essential purpose of such studies is to uncover the responses of various forms of stimuli and compulsions of the environment on life. These studies are used to arrive at inferences with respect to mode and magnitude of growth and dispersal in relation to the physical and the chemical environments (Sen, 1988).

Historical Review

In India, the beginning of the autecological studies of herbs was made by Mukerji (1932) studying the genus *Artemisia*, its species, varieties and ecads occurring in Kashmir. Prof. R. Misra and his students have given much attention to this type of study. He realised that for an understanding of biological requirements of plant species to cope with the environmental flux, a much greater net of experimental results are necessary. Misra (1944) worked on *Potamogeton perfoliatus*, leaf-form varieties in this plant were observed to be preconditioned by carbohydrate supply to the growing shoots. Misra and Rao (1948) showed that *Lindenbergia polyantha* Royle and *L. urtaecifolia* Link & Otto are two ecotypes of *L. polyantha* Royle dependent upon available calcium in the soil. Srivasta and Tandon (1951) studied the autecology of *Trapa bispinosa*. Bakshi (1952 a, b) in a series of papers have shown that *Anisochilus eriocephalus* Benth growing on house tops roofed with country tiles has very high seed out put and seed dispersal, but poor germination and 92.3 percent mortality among the seedlings, a high susceptibility to interspecific competition and water logging with poor aeration have driven the species to the peculiar habitat. Furthermore, he has shown that, *A. carnosus* is an ecotype of *A. eriocephalus*. Bakshi and Kapil (1954) worked on the autecology of *Mullugo nudicaulis* and *M. cerviana*. Pandeya (1953) observed that the two fodder grasses *Dicanthium annulatum* and *D. caricosum*, although similar in appearance are variable in shape, size and morphological attributes, depending upon the environment. The transplant technique has shown that *D. caricosum* with one spike per raceme and growing in a dry and sandy place may grow into a form resembling *D. caricosum* var *mollicomus* (sys *D. nodosum*), if water supply is made favourable. Sharma (1953) discovered that *Sida acuta* is eliminated by *Cassia tora* and other plants as its natural habitat is open areas. Kaul (1959), and Sharma (1955) worked on *Xanthium strumarium* and showed that the root-soil relations, interspecific competition and photoperiodic behaviour as relevant points in its distribution. Mall (1955) has shown that the dry bed of pools acquires a highly specialized plant community in which prostrate and mat forming habit is common. They bear strong tap root system, which in many cases becomes fleshy. Mall (1956) studied the autecology of *Chrozophora rottleri*. A Juss, particularly the germination behaviour and variation in osmotic concentration of the plant in relation to soil moisture. Mall and Arzare (1956) revealed that *Achyranthus aspera* Linn shows heavy reproductive capacity, but its distribution depends to a large extent on biotic operation. The species is found in more or less protected places where grazing is not severe. The seeds show embryonal dormancy. These observations were later on confirmed by Ramakrishnan (1963). Mall and Raina (1957) revealed that seeds of *Tridex procumbens* germinate to the extent of 90 percent under laboratory conditions of temperature and light. However, they need an initial exposure to light; if kept in absolute darkness, they fail to germinate. Kaul and Misra (1957) have shown that how germination and photoperiodic behaviour of the monsoon ecotypes of *Xanthium strumarium* fit the species in its habitat and ecological niches. Mall (1957) studied the autecology of *Cassia tora* and *C. obtusifolia* which shown coextensive and gregarious distribution. No ecological isolation was found between the two species. Joshi and Kambhoj (1959) investigated that *Gisekia pharnaceoides* grows healthiest in sand and most stunted in black cotton soil, further the species shows xerophytic characters in both "drought evading" and "drought escaping". Misra and Ramakrishnan (1960) discovered that *Peristrophe bicalyculata*, an annual tall herb, grows on nitrogen rich soil in shade, and on nitrogen deficient soil in open. Its germination and growth performance respond effectively to such combination of soil and light factors. Ramakrishnan (1960 a) recognised two ecotypes and two ecads for one of the ecotypes in *Euphorbia hirta*. Ramakrishnan (1960 b) studied the autecology of *Echinochloa colonum* Ramakrishnan (1960 c) brought to light some facts of great importance in *Eclipta alba*. Ramakrishnan (1961 a, b, 1965 a, b) established two ecotypes in Euphorbia thymifolia—(1) the red form, and (2) the green form. The former being tolerant of calcarious as well as non-calcarious soils and the latter thriving only in non-calcarious soils. Considering the variations in reproductive capacity, growth behaviour, mineral uptake and inter ecotypic competition of plants growing in soils of differing exchangeable calcium status, he established the existence of green (obligate calcifuge) and red (facultative calcicole) coloured ecotypes; futher he showed three physiologically distinct ecotypes within the red ecotype. Mall and Manilal (1962) showed that soaking in waterlogged mud for atleast 3 weeks and a low temperature of near about 9°C are absolutely essential for the successful germination of the seeds in *Crypsis acquleata*. Ramakrishnan (1963) worked out two ecotypes in *Setaria glauca* - (1) the long panicled form growing in moist localities, (2) the short panicled form growing in drier localities. Ambasht (1963) showed that seeds of *Alhangi camelorum* have a dormancy of 5 - 6 months after which 20-25 percent seeds germinate, but seedlings mortality is 100 percent. Perennation, propagation and reproduction is entirely vegetative through roots and rhizome. Soil conservation value of the species against erosion due to water run off is only between 30-35 percent. Patil and Singh (1963) worked on the ecology and anatomy of *Phyla notiflora* Greene. Saxena (1963) found seed coat impermeability to be the main cause of dormancy in *Hyptis suaveolens*. Manohar and Heydecker (1964 a, b, 1966 a, b) worked on the effect of water potential on germination of Pea seeds, and found that germination was affected not only by the moisture potential but also be the surface area of a seed in contact with liquid water; further the seeds subjected to lower external water potentials germinate at lower internal water potentials than they exhibit in distilled water. Ramakrishnan and Jain (1965) showed that the germination of seeds of the three edaphic ecotypes of *Tridex procumbens* can be correlated with the natural habitat in which they grow. In general, germination improved with increase in pH of the medium. Varshney (1966) observed from variations in the root system of wall and ground populations of *Bidens biternata*. Results of transplant experiments indicate that these variations are ecadic in nature. Choudhuri (1966) reported that the different phases of the life-cycle of *Vallisnaria spiralis* are determined by the environment. In temporary ponds the regeneration depends exclusively on seeds, however, in perennial ponds it is both by seeds and vegetative means. Joshi and Varghese (1967) worked out the reproductive capacity of *Anticharis linearis* to be 2637 and the mortality rate to be 58 percent. Joshi *et al* (1967) found that *Tribulus terrestris* has a average seed output of 3857 per plant and reproductive capacity 1084. Shrivastava (1967) found that *Chenopodium album* become tall in the fields which are highly manured and watered as is practised in potato fields and become dwarf in the fields which are poorly manured and watered as in the wheat, barley and gram fields. Varshney (1967) observed *Arthraxon lacifolius* exclusively on old walls which offer freedom from competition and well-drained substratum. The interspecific competition and water logging on the ground, are determental to the plant. Mullick and Chattergi (1967) reported that germination of *Clitoria ternatea* seeds improved with increase in the period of storage. Physical and chemical pretreatments also proved advantageous. Pandey (1968 a) reported seed polymorphism in *Echinops echinatus* and showed that the larger seeds germinate better and more rapidly than the smaller ones. Pandey (1968 b, 1969) showed that seeds of *Anagallis arvensis* possess embryo dormancy. The optimum conditions for germination are imbibition at 15°C for 3-4 days and one hour photoperiod to white light (1200-1300 lux). Imbibed seeds develop thermodormancy at 30-40°C, prolonged irradiation causes photodormancy. Ramakrishnan and Bisht (1968) observed that the occurrence of ecotypes in *Adathoda vesica* are adapted to the highly calcarious soils of Malla, the moderately calcarious soils of Shiwalik and the comparatively non calcarious soils of Chandigarh. Singh (1968) reported that *Cassia tora* and *C. obtusifolia* exhibit differential growth performance in namture as well as in cultivation. Seeds of both the species possess seed-coat dormancy and scarification favours germination. Shankar (1968) showed that the mechanically resistant seed coat in *Trichoderma amplexicaule* can be broken by scarification after which they germinate fully under all laboratory conditions. Tripathi (1968 a) observed that *Asphodelus tenuifolius* propagates through seed. Germinability of seeds increases with storage, mechanical and acid scarification. The germination in this case is epigeal which is unusual in monocotyledons. Tripathi (1968 b) showed that subfreezing temperature (-13°C ± 2°C) pretreatment considerably influences the suvival and subsequent growth behaviour young plants of *Cyperus sotundus*. The plants subjected to three hours daily exposure for a period of eight days showed 40 percent mortality, while those exposed for one and two hours were all alive.

These studies, as Misra (1967) pointed out, have revealed the biological equipment of plant species tomcope with environmental flux in time and space. The occurrence of habitat adapted populations within a species assist in its wide geographical distribution, as in sexual reproduction the genetic milieu is all the time throwing out micro-populations which are shifted and selected by the habitat. In nature, indeed, the ecotype micro-populations seem to bridge up the gap or discontinuites in forms, by means of ecads, which are produced by the ecotypes on account of phenotypic plasticity. The species - habitat gradients which are very often presented on account of hybridisation of both the population and the habitats correspond to the concept of vegetational 'continuum' of syncological units. These studies of interacting form, function and factor, lead to the belief that the entire flora is patterned on fluctuating micro-populations, which are ever-ready to fill up gradients of micro-habitats, as soon as they arise in time and space. Whatever, be the means of dispersal, the plant propagules reach different areas sooner or later. The viability, dormancy and conditions required for germination of different seed populations set primary limit to the incidence of the species populations in different seasons and habitats.

Aspects of Autecology

In general some of the major aspects of autecology which can be understood are the distribution of the species and the valuation of suitable environmental conditions, adaptation of the species to its environment and the speciation process within them. The basic feature of autecological studies is to understand the regeneration of a plant in its natural environment. This depends on certain aspects of individuals such as - systematics, distribution, dispersal, structural and physiological characters, ecotypic differentiation, seed output, seed size and weight, seed colour, seed viability, seed germination, seed dormancy, reproductive capacity, aggrasive capacity, seedling mortality, seedling growth and seasonal periodicity, vegetative propagation, productivity, food chain, effect of temperature, effect of light, effect of competition etc. All the above mentioned aspects can be broadly divided into following three heads (1) introductory information, (2) developmental history, and (3) ecological relations. The understanding of any individual species on the above lines may reveal a series of interesting features.

Systematics

Systematics is the scientific study of the kinds and diversity of organisms among them. The magnitude of the contribution made by systematics is not appreciated by many biologists, and yet these achievements are extraordinary indeed. They include the description of about half a million species of higher and lower plants, as well as their arrangement in a system. This classification, on the whole, is an immensely useful system of information storage and retrieval.

Taxonomists supply desperately needed identification service for taxa of ecological significance. Taxonomy also supplies the solution of many individual evolutionary problems, for example, isolation, nature of isolating mechanisms, mechanism of speciation, rates of evolution, trends of evolution, and the problem of emergence of evolutionary novelties. From taxonomy, population thinking spread into adjacent fields of genetics, cytology physiology and ecology.

As the interests of the systematists broaden, it is becoming more and more true that systematics has become one of the focal points of biology. It poses the problems in that area of ecology which deals with the phenomena of diversity, the differences in the ricjness of faunas and floras in different climatic zones and habitats and so forth.

Systematics recognised that every classification is a scientific theory with the properties of any scientific theory; it is explanatory, because it explains the existence of natural groups as the products of common descent; it is predictive, because with high probability it can make predictive as to the pattern of variation of unstudied features of organisms and the placement of newly discovered species.

Mayer (1968) feels rather passimistic about the future of taxonomy if it were only identification service for other branches of biology, as is though by some of our less imaginative colleagues. But he who realizes that systematics open one of the most important doors towards understanding life in all of its diversity cannot help but feel optimistic.

Distribution

The distribution of any taxonomic unit (species, genus, or family) of the plant world, is due to the biological peculiarities of plants and to their adaptation to local habitat conditions, which even within the limits of a small territory may vary to a considerable extent.

The area of distribution of a plant is best pictures by maps, on which all its known habitats are indicated by dots. Connecting by a line all the outer points of the distribution of a given plant, we are able to judge as to the shape of its area. The establishment of the areas enable us to arrive at conclusions as to the history and origin of floras.

Of the vital importance for the study and understanding of an area is the determination of that initial territory whence a genus or species began its dispersal whereby it reached the present boundaries of its area. This initial territory is known as the 'center of origin'.

There are no doubts that a new plant species will not extend its area beyond the limits of its origin. It will, without any doubt, begin to spread in all directions open to it, and the region of its origin will constitute the centre of the area being formed.

In order to establish enter of an area and the successive stages of its development, it is necessary to know its past history, which paleobotany alone can give. Unfortunately, its data are very incomplete, and rarely enable one to their basis, to establish the entire past history of an area. Hence, in most cases only the centre of the present day distribution of a given unit may be found within its present area but not the centre of its origin.

In the centre of an area, where, as a rule, the habitat conditions of a given species most nearly approximate the optimum, it can grow under fairly diverse conditions, even on different soils. On the other hand, farther from this region, that is, near to the periphery of the area, not only is an optimum combination of factors of more rare occurrence but often there is lacking even that minimum of conditions required for the normal existence of a species.

Arwidesson (1928) has proposed that those areas entirely included within the limits of one well defined region be called *unicentric*, in contrast to *bicentric* (embrassing two regions), *tricentric*, or *polycentric* areas if there are many such centres.

As an example of how the centres of maximum variation of a genus may be located—we may take the data of Misra (1968) on the area of distribution of the genus *Stipa*. The species and subspecies of this genus are distributed throughout the world as follows :

Table 19 : Distribution of number of species of Stipa.

Country	Number of species
Australia	69
New Zealand	69
Tesmania	2
Asia	100
Japan	2
China	13
Russia	60
India, Burma & Pakistan	27
Afganistan	7
South Africa	3
United States of America	14

From a study of these data it is clear that the centre of origin and development of the area of the genus *Stipa* comprised the Asian countries.

Another example is provided by Shirjaev (1932) on the area of the genus *Onionis* :

Table 20 : Distribution of the genus *Onionis*

Country	Number of Species
Morocco	52
Algeria	44
Spain	44
Italy	24
Portugal etc.	20

On the basis of these data he has concluded that the center of origin and development of the area of the genus *Onionis* comprised the Iberian peninsula (Spain 44, Portugal 20), Moracco (52) and Algeria (44), which at a time formed a united region.

Szymkiewicz (1937) proposed that the centre of the genus should be established not on the basis merely of data as to the total number of species but as to the data on the number of species in each of the following three developmental categories, and points out that by the latter method it is easier to detect a second centre of concentration of species, in case there are two such centres :

(1) Endemic and subendemic.

(2) Species whose area embrace a second natural region.

(3) Widely distributed species.

Distribution of a species may be continuous or discontinuous, Puri (1960) has shown that *Terminalia crenulata*, *Anogeissus latifolia*, *Dicanthium annulatum*, *Botherioclоa pertusa* etc. have continuous distribution. As opposed to this *Adina cordifolia*, *Ougenia oojeinensis*, *Tectona grandia*, *Shorea robusta*, *Dicanthium caricosum*, *Hyptis sucoveolens* etc. have discontinuous distribution.

A few species of vascular plants are widely distributed over the earth, while some of them are cosmpolitan, that is, found every where, others are restricted to certain areas only. Based on their geographical distributior hey can be broadly divided innto four groups :

(1) Arctic

(2) Temperate

(3) Pantropical

(4) Endemic

1. Arctic

Probably no more than 100 species make up this group. They are almost entirely perennial herbs, which occur almost circumboreally around the periphery of the Arctic ocean.

2. Temperate

Species of this group are distributed widely throughout the moist parts of the northern temperate zone, and few species also range to the Southern Hehisphere.

3. Pantropical

Plants of this type are almost universal throughout the tropics. Pantropical species are rare in undisturbed rain forests, but in the grassland, allong the roadsides and sandy shores, in gardens, lawns and pastures they are fairly common. Some of them indeed may have been disturbed by floating or being carried accross the seas by natural means. However, man in his comings and goings of the last thousands of years has propously or accidemtly carried seeds of many of these species to new places.

The most widely distributed tree of the tropics is the cocunutpalm (*Cocos nucifera*). It is found growing where-ever there are sea-shores and enough rain. Much of its wide distribution may be due to its salt tolerance capacity and to the winds and water currents that carry the nut from one place to another.

4. Endemic

Many vascular plants are restricted to a single floristic region. A floristic region is an area, large or small, that has a relatively distinct flora at the species level. Earth has been divided ito 37 floristic regions.

Although, India is connected by land with a number of countries, with distinct floral elements, it has a large portion of endemic flora. Considering only dicots, Chatterjee (1939) showed as many as 61.5 percent endemic to the Indian region. Excluding Maima at least 47 percent of the dicots of India are endemic (Table 21).

Table 21 : Dicotyledonous flora of India.

Total no of dicot spp.	Total no of dicot genera	No of dicot wides	No of dicots endemics Continental India	Himalaya	Maima	General
11124	1831	4273	2045	3169	1071	533
%		38.5	18.2	28.8	9.6	4.9
				61.5		

The highest number of endemic dicots occur in the Himalayas and the least in general area of the country. The number of endemism in Indian flora is quite high as compared to the neighbouring countries in the southern continent. For example, the percentage of endemism is 30 in Ceylon, 72 in New Zealand, 80 in Australia, and 16.5 in Hawaiian island.

The reason for the presence of such a high percentage of endemic plants in India is the presence of lofty mountains of the Himalayas on the north, northeast and northwest of the mainland and sea on other directions.

Consideration of the past and present geographical distribution of various members of a taxon has helped in plotting the dispersal trends and migration rutes of the plant species.

Dispersal

The establishment of the pioneer community on a bare area and the replacement of this community as ecological succession goes on are dependent in the first instance upon the existence of means by which new species can reach the area. Dispersal is a method by which plants keep their descendents separated in space and to provide each with its own site, where it can compete with other plants; it also concerns the meothods employed by exploring new territories, or conversely to maintain a foot-hold on a favourable site. Thus we see that dispersal is a link in the continuity of life on earth.

Dispersal and spread of plants takes place by two main methods, that is, vegetative, and reproductive.

Vegetative Methods

There are a number of methods by which plants reproduce vegetatively :

(*i*) *Offset* - Many of the water plants reproduce by the production of offset or daughter plants from the main stem. Common examples are *Pistia*, *Nelumbo, Eichhornia* etc.

(*ii*) *Coppice* - Many kinds of shrubs and trees copice freely and produce several generations of plants in nature. Plants differ in their capacity and power of coppicing (Table 22).

Table 22. : Coppicing capacity of some plants.

Coppice strongly	Fairly	Badly	No
Tectona grandis	*Pterocarpus*	*Acacia*	*Pinus*
Shorea robusta	*Terminalia*	*Bombax*	*Cedrus*
Dalbergia sissoo	*Quercus*	*Salvadora*	*Picea*
Albezzia sp	*Lannea*	*Madhuca*	*Abies*
Eucalyptis	*Boswellia*	*Tamarix*	*Casuarina*
Salix sp.			

(*iii*) *Branch cuttings* - Many plants are multiplied by branch cuttings. Common wild rose is a good example, other examples are *Boswellia, Lannea, Ficus, Morus alba* etc.

(*iv*) *Root suckers* - Stump is a year old seedling which is pruned both at the root and shoot before planting. Stumps are usually able to withstand much severe conditions of transport than entire plant. Large scale plantations of teak all over India are being raised from stumps. *Dalbergia sissoo* also respond favourably to stump planting.

Reproductive Methods

Most of the plants reproduce by seeds and spores. Dispersal of spores, seeds or entire fruit takes place by the agency of water, animals, or wind, depending upon the various adaptations of the disseminules or propagules, which facilitate the movement. The distribution of a plant is partly influenced by the success of dissemination of its propagules. The mode of dissemination of seeds or fruits is, therefore, of great significance in autecological studies.

(*a*) *Heavy seeds* - Very heavy seeds or fruits drop down straight from the tree and are either washed away by rain or sometimes roll down the slope in mountainous country to long distances. They are also carried away by animals. If the seed is unpalatable, it is rejected and thus transported to some distance. Heavy seeds are usually smooth, without any appendages, for example, *Dillenia sp.*, *Artocarpus* sp., *Mangifera* sp, many legumes etc.

(*b*) *Winged seed* - The development of wings is a device for the heavy seeds to disperse to some distances (for example Dipterocarpaceous and Combretaceous plants). This dispersion mechanism is very efficient in light fruits like *Anogeissus latifolia, Holoptelia integrifolia, Dalbergia sissoo, Pterocarpus indicus, Adina cordifolia* etc.

(*c*) *Hairy seeds* - Another variety of wind dispersal is by hairs or glumes on the seeds or fruits. Examples are found in *Salmalia, Populus, Salix*, most Apocynaceae, Asclepiadaceae and Compositae.

(*d*) *Dispersal by animals*. Seeds or fruits of a number of plants are dispersed by animals. Seeds are either enclosed in edible fruits or have special appendages by which they stick to coats of animals. Examples of the first type are *Acacia* spp, *Ficus* spp, *Prosopis* spp, *Morus alba*, *Dandropthe falcata, Viscum album, Cordia dichotoma* etc. Appendages are rare in tree seeds but are found in many weeds and grasses, for example, *Urena, Heteropogon, Themeda, Aristoda, Xanthium, Trifolium, Acanthospermum*, etc.

(*e*) *Explosive mechanism* - Explosive mechanism of seed dispersal is found in certain herbs and shrubs for example, *Oxalis, Parrottia, Helicteris*; many members of Acanthaceae, *Viscum, Impatiens, Corchorus* etc.

(*f*) *Antitelechoric behaviour* - The desert annual *Gymnarrhena micrantha* Desf, a Composite plant show antitelechoric dispersal, in which seeds have a mechanism preventing their dispersal. The mother plant produces two kinds of flowers : subterranean and above ground. The seeds produced in the subterranean flower are larger and are not dispersed by wind, as are the achenes from the above ground flowers.Thus, they germinate after rains inside the dead mother plant. They have the advantage that the mother plant has already prepared with her tap root a channel in the soil through which rain water percolates more easily, and which can be used by the daughter plant for its roots (Koller and Roth, 1964).

Blepharis persica is a desert species of the Acanthaceae which ejects its seeds with jeculators; after water has penetrated through the apex of the capsule. After the hairs of their coat swell, lifting the seeds to the right position foir root penetration in the soil (Gutterman *et al*, 1969).

Seed Out Put

Seed out put is the potential capacity of a species to reproduce itself. It varies under natural conditions from individual to individual, and in some individuals from locality to locality. Table 23 could reveal the variation in the seed out put of some of the investigated plants. These data suggests that in general trees produce more seeds in comparison to shrubs and herbs. The maximum number of seeds (1907408) per plant are produced by *Anogeissus latifolia*, while a minimum of 172 seeds per plant are produced by *Heylandia latebrosa*.

Table 23 : Average number of seeds per plant.

Species	Average seed output	Average Author
Acanthospermum hispidum	297	Pandeya & Goswami (1962)
Achyranthus aspera	1843	Mall and Arzare (1956)
Anisochilus eriocephalus	21421	Bakshi (1952)
Cassia tora	450	Mall (1957)
C. obtusifolia	374	Mall (1957)
Chrozophora rottleri	552	Mall (1956)
Dicanthium annulatum	2795	Pandeya (1952)
Eclipta alba	909	Ramakrishnan (1960)
Geisekia pharencoides	940	Joshi and Kambhoj (1959)
Lidenbergia polyantha	71731	Misra and Rao (1948)
Mullugo hirta	48121	Ramakrishnan (1963)
Heylandia latebrosa	172	Srivastava (1953)
Anticharis linearis	11774	Joshi and Varghese (1967)
Tribulus terrestris	3857	Joshi *et al* (1967)
Mercurialis perennis	300	Mukerjee (1936)
Butea monosperma	1964	Chourey (1953)
Diospyros melanoxylon	8746	-do-
Lagerstroemia parviflora	12926	-do-
Tectona grandis	31033	-do-
Terminalia tomentosa	8120	-do-
Anogeissus latifolia	1907408	-do-
Boswellia serrata	12367	Sharma (1955)
Shorea robusta	2800	Jain (1962)

It has been reported that seed out put in a plant is dependent upon a number of environmental factors of which habit, habitat, light intensity, moisture conditions, biotic influences, physiological status and age are important. Joshi and Nigam (1970) observed that number of seeds per fruit in *Trianthema portulacastrum* varied with the plant habit (Table 24). Highest number of seeds were reported to be produced by prostrate plant, while the lowest number of seeds were observed from the erect plants.

Table 24 : The number of seeds/fruit in different plant habits growing in the same habitat of *Trianthema portulacastrum* (Joshi and Nigam, 1970)

Habit	Average number of seeds/fruit
Prostrate	8.96 $\pm$ 0.61
Erect	3.80 $\pm$ 0.82
Procumbent	6.22 $\pm$ 0.43

As regards variation in the seed out put in different localities, Joshi and Kanbhoj (1959) observed that the plants growing in sandy areas have got larger number of capsules in comparision to the plants on hard and stabilized areas (Table 25).

Table 25 : Average seed output of Gisekia pharnaceoides growing in and around Pilani.

Locality	Average seeds/plant
Shiv Ganga area	535
Ponda area	195
Birla college area	645
Chandmari area	705
Electronics Institute area	715
Farm area	1160
Police station area	1185
Jheri area	1260
Rajasthan Hostel area	2060

Varshney (1966) reported that *Bidens biternata* grows on wall and ground as well. However, the average seed output is lower in calcarious wall soil (853 seeds/plant), in non-calcarious soils on the other hand it produces more seeds (1572 seeds/plant). Misra and Ramakrishnan (1960) reported that *Peristrophe bicalyculata* grows in open as well as partial shade, the plants in the former habitat produce on an average 2993 seeds/plant while those in the other habitat produce 1013 seeds/plant.

Mostly, high light intensities and dry conditions initiate early seed formation. Sen (1977) observed that *Cucumis callosus* produced 312 seeds/fruit in favourable season, whereas 20 seeds/ fruit in unfavourable season. Bohra (1976) reported that *Crotalaria burhia* produced seeds twice in a year, once in september-December and then in March-May. Most of the fruits produced in September-December were eaten away by an insect caterpillar. However, the seeds produced in this season were healthier and heavier as compared to those produced in the other season. Similarly, the seeds of *Grewia tenax* are regularly eaten by some insects and so the netural regeneration of this species in very poor.

Sant (1964) showed the effect of intensity of grazing on seed output in *Dicanthium aegyptium*, the plant growing in protected, medium grazed and overgrazed fields produce on an average 2184, 2327 and 4128 seeds per plant respectively. Sant (1972, 1974) also observed that grazing intensity increase seed output in *Indigofero linifolia* and *Bothriochlova pertusa* (Table 26).

Table 26. Effect of intensity of grazing on seed output.

Grazing intensity	Average number of spikes or fruits/plant	Average number of seeds per spike or fruit	Author
Indigofera linifolia (summer)			
Protected field	586 ± 102	1 ± 0	Sant
Medium grazed field	375 ± 76	1 ± 0	(1972)
Over grazed field	136 ± 15	1 ± 0	
Botheriochloa pertusa (Winter)		15±3	Sant
Protected field	126 ± 52	17 ± 3	(1974)
Medium grazed field	13 ± 5		
Over grazed field	21 ± 6	14 ± 3	

Ramakrishnan (1963) have shown that competition among plants also influence their seed output, for example, *Mullugo hirta* plant free from competition produce 11076 seeds/plant where as plants under competition have further differences with regard to their location - those growing in the periphery produce more seeds (2816 seeds/plant) over those growing in the centre (793 seeds/plant) of the community.

Plant generally produce more seeds than the habitat can sustain, as many of them are wasted, destructed or consumed in various ways. Annuals generally produce seeds once in their life time, where as perennialsshrubs and trees do so usually once in a year, although there are species that produce seeds once in several years, as well as those that produce seeds several times in a year. Different species differ in respect of their average seed output, which is affected by light intensity, temperature, moisture, biotic interference, etc., certain generalized conclusion in respect of the seed output have been arrived at in some of the above mentioned studies are :

(1) Higher light intensity and dry conditions increase seed output.

(2) Biotic interference such as grazing, insects, etc. reduce seed output.

(3) The age of a given species has a marked influence on the seed output.

Seed Viability

The life-span of seeds has been a subject of conjecture for many years, but the experimental approach for the actual determination of life-span has dated from the beginning of the twentieth century. The maintenance of viability of seeds under various conditions have been observed by Ewart (1908), Crocker (1948), Barton and Crocker (1948), Crocker and Barton (1953), Owen (1956) and Barton (1961).

On the basis of their life-span under so-called 'optimum' conditions, Barton devided seeds into two groups - (1) seeds with a long life-span, and (2) seeds with a short life-span.

Seeds with a long life-span

Under this category the seeds posses a life-span of two or more than two years. Ewart (1908) divided such seeds into three classes; (a) microbiotic—those with a life span of 3 years or less, (b) mesobiotic—those with a life span of 3-15 years, and (c) macrobiotic—those with a life span of 15-100 years or more.

Becquerel (1932, 34) tried to germinate seeds from the Herbaria of National Museum in paris, in order to estimate their life span. He showed that *Mimosa glomerata* seeds remained viable for 221 years, while *Astragalus massilinesia, Dioclea paucifera* and *Cassia bicapsularis* remained viable for 100-150 years. The most extreme case of retention of viability have been reported to be one thousand years old seeds of *Nelumbo nucifera* found in the mud of a lake bed in Manchuria (Mayer and Poljakoff-Mayber, 1963). In similar, but even more extensive experiments Schjelderup-Ebbe (1936) tested viability of 1254 lots and nearly as many species of seeds stored in bottles or paper bags in a cupboard at room temperature for 34 to 112 years. Some of his results are presented in Table 27.

Table 27 : Viability of certain seeds.

Species	Age	% germination
1. *Canna paniculata*	69	7.14
2. *Acacia farnesiana*	—	1.4
3. *Albizzia julibrissin*	70	3.33
4. *Astralagus alpinus*	40	8.67
5. *A. edulis*	43	0.31
6. *A. macrocephalus*	64	6.25
7. *A. utriger*	82	6.00
8. *Baptisia alba*	61	0.25
9. *B. spec*	43	0.75
10. *Coronilla montana*	52	13.75
11. *Cytisus alschingeri*	62	0.5
12. *Glychyrrhiza echinata*	43	1.0
13. *Kennedya apetata*	77	13.0
14. *Lathyrus silvestris*	40	1.05

(Contd.)

Species	Age	% germination
15. *Lupinus pollyphyllus*	40	1.05
16. *Malilotus alba*	40	33.00
17. *Mercurialis annua*	40	0.25
18. *Abuliton reflexum*	40	23.0
19. *Hibiscus californicum*	41	1.25
20. *Sida cordifolia*	41	4.5
21. *Dephe vnezereum*	35	1.0
22. *Covolvulus flavus*	64	15.0
23. *Ipomoea pilosa*	40	0.67
24. *Datura inermis*	40	1.00
25. *Madia spec (M. angustifolia)*	40	0.50

Seeds with a short life-span

Such seeds posses a very short life-span of less than a year, for example, *Ulmus americana*. Arditti (1967) have reported that orchid seeds become viable and are capable of normal development prior to being fully ripe. Longivity of orchid seeds is highly variable. Some of them may loose their viability within 9 months (Humphreys, 1960), 2 months (Brummitt, 1962) or less (Hey, 1963, Lindquist, 1965).

Causes of loss of seed viability

Various theories have been advanced as to the cause of loss of seed viability. These include (1) the intrinsic causes, and (2) the extrinsic causes.

The first theory postulates that the cause of loss of seed viability are some intrinsic factors of the seed's metabolism. Crocker (1948) suggested that intermediate products, accumulating as a result of anaerobic respiration, may be toxic to the seed. Oxley (1949) suggested that the continued life of the seed depends on the use of some labile organic matter present in the embryo, as this substance becomes exhausted, the seed losses viability. D'Amato and Hoffman-Ostenhof (1956) suggested that in some cases the loss of viability might be related to the inactivation of certain anaerobic dehydrogenase enzyme. Spragg and Yemm (1959) observed that in *Pisum sativum* seeds, gradual and irreversible inactivation of sulphydril and other enzymes may be an important factor in the ultimate loss of viability after prolonged storage.

The second theory postulates that loss of seed viability is caused by organisms such as fungi and bacteria, which live in association with stored seed. Semenuik (1954) demonstrated that the decrease in seed viability is related to increase in activity of associated micro-organisms. However, it is difficult to distinguish which is the cause and which is the effect. Went (1957) suggested that loss of viability might be connected with a decrease in resistance against fungi and saprophytic organisms. He showed that there are strong indications, atleast in some species, for example, *Godetia*, *Clarkia* and *Boisduvilla*, that viable seeds produce antibiotics which are not present in non-viable seeds, and he suggested that loss of viability is largely due to the disappearance of antibiotics.

Seed Germination

All viable seeds germinate after a reasonable time and normally develop into seedlings. Evanari (1957) defined as germination, those process starting with the protrusion of the embryonic roots which takes place inside the seed and prepare the embryo for normal growth. Long (1965) listed the following definitions and interpretations which closely follow usages or recommendations of the International Seed Testing Association and which if used constantly, should eliminate any major ambiguity.

(1) *Botanists definition* - Emergence of radical from seed coat in the case of those seeds in which the plumule is first to emerge.

(2) *In seed laboratory practice* (ISTA Rules) - The emergence and development from the seed embryo of those essential structures which, for the kind of seed in question, are indicative of its ability to produce normal plant under favourable conditions.

(3) *Users definition* - Emergence of the aerial part of the seedlings from the soil.

The term seed germination has been frequently treated loosely or interpreted in different ways, it has different meanings for the Botanists, the Seed Laboratory Specialist, and the General user.

Imbibition

Imbibition of water necessarily is the first step in germination. It is always accompanied by an increase in the volume of the seed. However, the total increase in volume is lesser than the volume of water absorbed. The extent to which imbibition occurs is determined by three factors, the composition of the seed, the permeability of the seed coat or fruit to water, and the availability of water in liquid or gaseous form in the environment.

Imbibition is a physical process, related to the properties of colloids. It is in no way related to the viability of the seeds and occurs equally in live seeds and in seeds which have been killed by heat. During imbition molecules of solvents enter the substance which is swelling, causing solvation of the colloidal particles and, in addition, occupying the free capillary spaces and the intermicellar spaces of the colloids. The swelling of the colloid results in the production of considerable pressure called 'imbibition pressure'. The imbibition pressure developed by the seeds may reach hundreds of atmospheres. Imbibition pressure is of great importance in the process of germination as it may lead to the breaking of the seed coat and also to some extent makes room in the soil for seedling development. The magnitude of imbibition pressure is also an indication of the water retaining power of the seed (Table 28) and therefore determines the amount of water available for rehydrating the seed tissue during germination.

Table 28 : Imbibition by seeds at 28°C (% original weight of seeds.)

Time (hr)	lettuce	Wheat	Sunflower	Vicia faba	Zea mays
1	170	114	124	112	111
2	185	120	137	143	116
4	197	127	147	168	—
6	210	133	153	181	124
10	213	140	154	182	—
16	225	—	—	—	136
24	237	151	—	—	137
32	252	155	—	—	—
40	254	—	—	—	—
48	270	161	—	—	—

Imbibition is effected by different environmental factors for example, temperature, light, dry storage, exogenous chemicals, etc.

Effect of Dry Storage on Imbibition

Freshly harvested seeds of legumes show very poor imbibition rate. Agarwal and Vyas (1970) have shown that the imbibition increases with dry storage period in *Indigofera astragalina*. Seeds after 53 weeks of dry storage gave 10 percent imbibition in continuous light (Table 29). However *Indigofera cordifolia* seeds gave 48 percent imbibition after the same storage period (Vyas and Agarwal, 1972).

Table 29 : Percent imbibition after different dry storage periods.

Storage period (Weeks)	*Indigofera* Light	*astragalina* Dark	*Indigofera* Light	*cordifolia* Dark
Nil	4	0	22	14
4	4	0	28	16
8	4	0	28	20
12	4	2	32	24
18	8	6	34	24
24	6	6	38	26
28	8	6	40	28
32	8	6	42	30
36	8	4	46	30
40	10	4	46	36
53	10	8	48	38

Effect of temperature on imbibition

The effect of temperature on imbibition is probably complex. The viscocity of water decreases with increased temperature and its kinetic energy increases. The kinetic energy is directly proportiobal to the absolute etemperature, while the molecular velocity varies as the square root of the absolute temperature. *Indigofera astragalina* seeds subjected to 20, 25, 30, 35, and 40°C show a maximum of 42 percent imbibition at 40°C (Agarwal and Vyas, 1970). The observations presented in Table 30 show that *I. cordifolia* seeds shoed 44 percent imbibition at 30°C only (Vyas and Agarwal, 1972).

Table 30. Efffect of temperature on percent imbibition.

Temperature °C	Indogofera astragalina	Indigofera Cordifolia
20	0	24
25	4	32
30	8	44
35	12	42
40	42	36

Effect of light on imbibition.

Besides temperature light is another important factor influencing the imbibition of seeds. Agarwal and Vyas (1970) showed that *Indigofera astragalina* seeds when subjected to different photoperiods (0, 8, 16 and 24 hours daily) at different temperatures (20, 25, 30, 35 and 40°C) show maximum imbibition (4 $\pm$ 0%) in continuous dark at 20°C, as the temperature is raised the imbibition value gradually increases with increase in photoperiods (Table 31), attaining a maximum value 42 $\pm$ 6 percent in continuous light at 40°C.

Table 31 : Effect of photoperiod on imbibition percentage in *Indigofera astragalina* seeds

Temperature °C	Photoperiod (hrs) 0	8	16	24
20	4 $\pm$ 0	2 $\pm$ 2	0 $\pm$ 0	0 $\pm$ 0
25	2 $\pm$ 0	2 $\pm$ 2	2 $\pm$ 2	4 $\pm$ 0
30	4 $\pm$ 0	6 $\pm$ 2	10 $\pm$ 2	8 $\pm$ 0
35	8 $\pm$ 0	8 $\pm$ 0	10 $\pm$ 2	12 $\pm$ 0
40	16 $\pm$ 0	18 $\pm$ 2	30 $\pm$ 6	42 $\pm$ 6

Effect of exogenous chemicals

The application of certain exogenous chemicals enhance the rate of imbibition in some seeds, for example, in *Indigofera astragalina* sulphuric acid treatment increases the rate of imbibition (Agarwal and Vyas, 1970). These seeds attain maximum imbibition

(100 percent) after 30 minutes pretreatment. Similarly, Agarwal (1971) showed that *I. tinctoria* require 45 minutes pretreatment for a maximum (100 percent) imbibition. Any further increase in the duration of pretreatment causes embryo injury and consequent low germination (Table 32).

Table 32. : Effect of sulphuric acid pretreatment on Indigofera tinctoria seeds.

Pretreatment (hrs)	Imbibition	germination
0	0	0
5	4	4
10	28	28
15	58	58
20	72	72
25	80	80
30	80	80
35	85	82
40	90	90
45	100	100
50	100	86
55	100	80
60	100	60

In addition to the means already mentioned, which effect imbibition, we can cite that of the *Atriplex dimorphostegia* in which the high salt content of the bracts, which enclosed the single seeded fruit produce such a high osmotic pressure that imbibition by the seed is impossible. A likely further means of control of imbibition is evident from the characteristics of the three phases of imbibition of isolated embryos. In *Lupine* the first phase is stricity physical and is complete in 45 minutes. In the second phase, which lasts 7-8 hours, there is no uptake of water. Respiration, however, which beings to increase rapidly within 15 minutes of the start of imbibition continues to increase through this second phase. The third phase of water uptake begins at about 10 hours and continues to about 40 hours. During this phase, there is no cell division — only cell enlargement, an increase in respiration, and a resumption of water uptake. Anaerobiosis, as well as a wide variety of metabolic inhibitors prevent water uptake in the third phase.

Germination

The observation that the seeds of succulent fruits such as tomato fail to germinate whilst they are in the intact fruit, but are capable of immediate germination if removed, washed and planted in a moist medium, have aroused much curiocity to many horticulturists and botanists in the past. Germination while the seed is still in the fruit is prevented by the presence in the tissues of the fruit substances which inhibit germination. The inability to germinate because embryo is removed by further development of the embryo during an 'after ripening' period. The barrier to germination provided by the mechanical strength of the seed coat, and the presence of chemical agents are removed over a period of time. Each year a small percentage of seeds germinate as these barriers are gradually breached by natural agents, such as freezing and thawing, forest fires, passage through an animal digestion tract, the leaching by a heavy rainfall, a specific light regime, or some combination of these.

Factors Affecting Germina..on

In order that a seed can germinate, it must be placed in environmental conditions favourable to this process. Among the conditions required are an adequate supply of water, a suitable temperature and light. The requirements for these conditions varies according to the species and variety, and is determined both by the conditions which prevailed during seed-formation and even more by hereditary factors. Frequently it appears that there is some correlation between the environmental requirements for germination and the ecological conditions occurring in the habitat of the plant and the seed. In the following, these various factors will be considered in detail. *Effect of water potential on germination.*

The first process which occurs during germination is the uptake of water by the seed. Kaul and Manohar (1966) observed that with decrease in water potential from 0.700 to 1400 Jules/kg there is a corresponding decrease in the cumulative germination of *Acacia senegal* seeds at successive days after sowing under P_0 (94.5), followed by P_7 (7.0) and P_{14} (10.5). These results confirm the earlier findings of McGinnis (1960), and Manohar and Heydecker (1964 a, b) that germination of seeds was considerably reduced with the reduction in the availability of mositure from field capacity (P_0) to near the wilting point (P_{12}). The minimal moisture requirement studies during germination on three arid zone grasses (Lahiri and Kharabanda, 1963) indicate that the requirement for germination, in the seeds of *Cenchrus setigerus* is 50 percent, the highest requirement of 120 percent is for *Lasiurus sindicus, Cenchrus ciliaris* which have a requirement of only 80 percent, occupies an intermediate position in this respect. Ramakrishnan (1963) kept *Mullugo hirta* seeds at 10, 20, 30, 40, 50 percent mositure levels and obtained best germination (48%) at 30 percent moisture level in the soil. The germination of different crops seeds at 25, 50, 75 and 100 foot candle soil moisture indicate that best germination is obtained at 100 FC in wheat, barley, potato and pea. With higher value of soil moisture the seed germination decreases (Table 33).

Table 33. Effect of varying soil mositure on germination of different crops (Ramakrishnan, 1963).

Soil moisture (foot candle)	Percent germination			
	Wheat	Barley	Potato	Pea
100	91.8	88.8	100	84.0
75	91.6	90.6	100	82.8
50	93.2	87.8	100	91.6
25	10.2	32.0	85.0	56.0
C.D. at 5%	11.10	12.21	9.81	18.62

The power to germinate in low mositure level does not necessarily mean that optimum will be achieved.

Effect of substratum on germination

It would seem advisable if the sensitivity of the seeds to the substratum is considered in the design of germination experiments. When filter papers are used as substratum, they should be from a single lot number in any experiment. Control of the substratum may be of major importance in germination experiments extending over long periods of time, for example, those designed to show after ripening.

Rehwaldt (1968) observed a striking difference in filter paper by the average germination of Whatman No. 1, lot 0 was 11.6, while that of lot N was 91.7 percent. Lot O and N were manufacturer's lot numbers 13302333 and 402713 respectively.

Verbena bipinnatifida seeds show sensitivity to soil types in their germination. Sandy clay loam was observed to be most favourable (56 ± 4%) while the sandy soil the least favourable (14 ± 2%). The germination energy index in sandy clay loam and brick-klin was higher than that in other soil types (Table 34).

Table 34. Effect of soil types on seed germination of Verbena bipinnatifida (Vyas, Agarwal and Garg, 1971).

Parameter	Sand	Sandy-clay loam	Sandy loam	Brick-Klin
% Germination	12±2	56±4	42±8	28±0
Germination energy index	0.3437	0.75	0.5359	0.6857

Indigofera cordifolia seeds germinate to a maximum level (44%) in gravel soil, while to a minimum level (10%) in brick-klin (Table 35).

Table 35. Effect of soil types on seed germination of Indigofera cordifolia (Vyas and Agarwal, 1972).

Soil types	% Germination
Sand	14
Gravel	44
Loam	22
Brick-klin	10

The nature of soil affects the germination process. Some seeds respond to calcium content of the soil, while others respond to the sodium content in their germination behaviour. Seeds of *Hypericum perforatum* germinate in soils containing calcium. *Avena fatua* germinate well in sandy soils. *Atriplex halimus* germinate well in saline soils.

Effect of Temperature on Germination

Temperature is one of the most important factor governing the germination of all seeds. There is usually an optimum temperature range where germination of a species is most rapid, is most complete or results in the greatest number of healthy seedlings. The so-called "optimim" temperature depend on the type of temperature treatment used and on the critarion selected for measurement of growth response. Edward (1932) used "optimum temperature" with the following meanings :

1. The upper limit of a range of temperatures at which approximately the same final percentage of germination is attained when tests are continued long enough to make sure maximum germination has occurred.
2. The temperature yielding the highest germination percentage at the end of a given incubation period.
3. The temperature at which the least time is required for the appearance of the first or last of the seedlings to germinate.
4. The temperature corresponding to the least average incubation time.
5. The temperature giving the largest average height of the seedlings of a set at the end of a specialized period for germination and subsequent growth.
6. The temperature resulting in both the highest rate of seedling productivity and highest final germination percentage.

On of the regulating mechanisms which determine the timing and locality of germination acts as a built-in temperature gauge. In its simplest form it restricts germination of a species to a specific temperature range that is often narrow and precise. A more complex response mechanism enables the seed to distinguish between constant, unchanging temperatures and those which alternate between warm and cold in a daily rhythmic cycle.

Temperature relations of germination were given special attention in early work with seeds. Maximal, minimal and optimal temperatures were defined first by Sachs (1860). Toy and Willingham (1966) tested ten species and varieties of *Limnanthes*, for germination at temperatures of 40, 50, 60, 70°F and at room temperature (72-78°F). The germination percentage of each species at different temperatures, is shown in Table 36.

Table 36. Seed germination (%) in 29 days of Limnanthes spp at various temperatures.

Species	40°F	50°F	60°F	70°F	Room temp
L. alba var *alba*	83.0	76.0	50.0	11.3	0.0
L. alba var *versicolor*	95.7	91.3	87.3	35.7	0.3
L. bakeri	96.3	97.0	94.0	16.0	0.0
L. douglasis very *douglassii*	90.7	93.7	96.7	68.7	2.3
L. douglassii var *nivea*	95.3	95.3	96.7	63.7	2.0
L. douglassii var *rosea*	48.3	68.0	95.0	38.7	1.0
L. floccosa var *floccosa*	97.3	99.3	85.3	11.0	0.0
L. gracilis var *pariskii*	83.7	76.3	49.7	1.3	0.0
L. montana	75.3	78.7	52.7	10.0	0.3
L. striata	91.0	83.0	76.7	19.7	1.0

The data show that 4 out of 10 varieties germinated best at 40°F, 3 at 50°F, and 3 at 60°F. Germination was relatively poor at 70°F and almost zero at room temperature. Sen and Chatterji (1966) reported that optimum temperature for *Ruellia tuberosa* ranges between 32-38°C.

Freshly harvested seeds to some cereals were found to have a low temperature requirement of 10 to 15°C (Attenberg, 1907). As the needs aged over several weeks, this maximal value was raised by about 10°C. The seeds of many plants show a sharp inhibition of germination with increase of temperature. Seeds of *Brassica juncea*, when tested immediately after harvest, showed 97 percent germination at 10°C. 97 percent at 15°C, 63 percent at 20°C, 8 percent at 25°C. Three weeks later, germination was 95 percent at 25°C. Thompson (1968) did a comparative study in *Primula* species, between freshly harvested seeds and seeds which had been stored. His results incorporated in Table 37 indicate that fresh seeds of *P. florindae, P. secundiflora, p. sikkimensis, P. polyneura* did not do too badly when sown fresh, where as *P. aurantiaca, P. cockburniana, P. chungensis* and *P. sinopurpurea* produced very poor rates from fresh seed, failures were much less frequent with the seed sown in December and only *P. sinopurpurea* gave very good results when sown after 8 months.

Table 37. Germination (%) of Primula seeds at different storage period, contant and durnal alternating temperatures.

Storage period	0		4		8	
Temperature °F	68/50	59/59	68/50	59/59	68/50	59/59
P. smithiana	94	96	98	96	94	92
P. prolifera	93	92	90	93	—	—
P. helodoxa	90	88	89	56	—	—
P. florindae	74	49	84	86	—	—
P. secundiflora	57	52	70	60	—	—
P. sikkimensis	29	46	94	84	—	—
P. polyneura	35	24	71	53	—	—
P. aurantiaca	13	39	42	42	—	—
P. cockburniana	—	2	92	78	63	89
P. japonica	53	1	3	4	6	0
P. chungensis	7	4	38	15	35	10
P. sinopurpurea	8	5	1	0	75	84

Alternating rather than constant temperatures appear to be favourable to the germination of many seeds and are apparently necessary for germination of the seeds of some plants. The most unusual cases of such alternations are diurnal ones, between a low and a high temperature. Thompson (1968) also indicated that in *P. halodoxa, P. florindae* and *P. polyneura* a diurnal alternating temperature favours maximum germination as compared to constant temperatures. Other species of *Primula* investigated by him to not show any significant difference between constant and alternating temperature.

The thermoresponses of *Portulaca oleracea* indicate that the seeds can germinate well at short duration high temperature (50°C) exposure, which is otherwise inhibitory to germination process. However, a longer high temperature exposure results in the inhibitory effect Table 38.

Temperature can influence the germination process in at least three ways : (1) by removing a block to the light response, (2) by changing the sensitivity of the photoreaction, and (3) by changing the pathway of the germination process (Toole *et al.*, 1955). A single shift of temperature from 15°C to 25°C at the time of exposure to light can remove some block that prevents full response

Table 38. High temperature thermoresponse of germinating Portulaca oleracea at 50°C (Singh, K.P. 1968).

Duration of exposure	% Germination
0 hour	94.3 ± 2.0
2 hour	94.6 ± 1.7
4 hour	94.3 ± 2.3
1 day	86.0 ± 3.2
3 day	76.6 ± 2.9
5 day	20.3 ± 4.1

of seeds of *Lepidium virginicum* to the stimulation of red light. Similarly, after preliminary imbibition at 20°C, a two hour period at 30°C followed by return at 20°C removes some block and allows many more of the seeds to respond to light, than would respond if they were held continually at 20°C. Furthermore, if the two hour period at 35°C preceds exposure to red light, the sensitivity of the photoreaction is increased by the high temperature so that less light energy is required to promote to a given percentage germination.

While an alternating temperature stimulates conditions experienced by seeds on soil, no hypothesis adequate to explain its beneficial has appeared. It has been suggested that these temperature changes in some way stimulates respiration and catalase activity (Davis, 1939).

Effect of Light on Germination

Light is another important factor which regulates germination of seeds. Some seeds require light for germination (positively phtotoblastic). Lettuce seed if kept at a relative humidity of 60-70 percent in light and darkness, the subsequent germination of those kept in light is considerably higher. Similarly, light sensitive seeds of *Launea sarmentosa, Solanum surantanse, Ficus aurea* and *Cryptostegia grandiflora* do not germinate in the absence of light. However, some seeds germinate in dark (negatively photoblastic). *Nigella sativa, Allium* species and members of Amaranthaceae do not germinate on exposure to light (Sen, 1988).

Not all light responsive seeds complete germination after a single exposure to light, a fact that is well recognized in seed testing methods. Isikawa (1954), Black and Wareing (1955), and Evanari

(1965) showed that germination of some kinds of seeds depends upon duration of light and dark periods, suggestive of photoperiodic control of seed germination.

Seeds of *Nerium oleander*, germinate to a very low extent, in continuous light. By increasing the period of darkness and decreasing the photoperiod, the germination was improved. Under absolute darkness a high rate of germination was observed. Maximum germination (100%) was observed at a photoperiod of 6 hours, while a 24 hour photoperiod gave minimum (20%) germination Table 39.

Table 39. Photoperiodic response of germinating Nerium oleander seeds (Datta, 1961).

Photoperiod (Hrs)	% Germination
0	94
2	96
4	98
6	100
18	88
20	82
23	38
24	20

Indigofera linnaei seeds showed a favourable germination response to 8 and 16 hour photoperiod. The most suitable response was observed at 8 hour photoperiod. However, *I. cordifolia* showed maximum germination at 16 hour period. On the other hand, *I. astragalina* was observed to be indifferent to photoperiod Table 40.

Table 40. Effect of duration of light on seed germination.

Photoperiod (hrs.)	Indigofera linnaei	Indigofera astragalina	Indigofera cordifolia
0	49	97.3	63
8	93	96	87
12	71	92	77
16	85	98.6	91
24	77	98.6	88
	Vyas and Agarwal (1970)	Agarwal and Vyas (1970)	Vyas and Agarwal (1972)

Besides photoperiodic response, spectral sensitivity during germination of seeds have also been well observed. Continuous irradiation with white, blue, green, red light promoted the germination of the seeds of *Artemisia monosperma*. Red light was observed to be the most promotive and blue light the least promotive. Similar spectral response of the seeds of Dalbergia sissoo was also observed (Table 41).

Table 41. Spectral sensitivity of germinating seeds.

Light quality	*Artremisia monospernna*	*Dalbergia sissoo*
Blue	74 ± 5	80
Green	87 ± 2	80
Yellow	—	100
Red	89 ± 6	100
White	73 ± 4	100
Dark	3 ± 2	100
	Koller et al (1964)	Vyas and Agarwal (1970)

Observation on the action of light in the germination of seeds of Asclepiadaceae show that the absolute requirements for light in the germination of seeds was satisfied as well by radiant energy in the continuous blue, red and far-red regions of the spectrum, as by unfiltered light. *Cryptostegia* grandiflora showed maximum germination in total darkness (Table 42).

Table 42. Spectral responses of germinating seeds of Ascepiadaceae (Sen, 1969).

Species	Blue	Red	Farred	White	Dark
Calotropis procera	55.0 ± 19.8	91.6 ± 10.6	91.6 ± 4.0	88.3 ± 5.0	86.6 ± 12.0
Cryptostegia grandiflora	66.0 ± 29.4	76.6 ± 14.9	66.6 ± 26.2	56.6 ± 28.1	95.0 ± 5.0
Leptadaenia pyrotechnica	100 ± 0.0	100 ± 0.0	93.0 ± 4.0	96.0 ± 5.0	94.0 ± 8.0

Spectial sensitivity of different species of the same genus show variations during their germination. *Merremia aegyptia* completed its germination in all light spectra, within 24 hours. However, *M. dissecta* showed highest germination (25%) under red light exposure within first 24 hours and lowest germination (5%) in white light. Subsequently, the germination value progresses gradually and attains maximum level on the fourth day. The maximum germination values of both the species indicate that they are indifferent to spectral exposure in their final germination of sees. (Table 43).

Table 43. Spectral sensitivity of germinating sees of Merremia species (Sharma and Sen, 1975).

Spectrum/ Days	M. aegyptia		M. dissecta			
	1	2	1	2	3	4
White	100	—	5	40	55	75
Blue	100	—	15	65	80	95
Green	100	—	10	70	80	85
Red	100	—	25	60	85	95
Far-red	100	—	10	35	50	75
Dark	100	—	0	70	90	100

Phytochrome is a widely distributed biliprotein that controls many aspect of plant growth and development. The mutual photointerconvertibility of its two forms Pr and Pfr, absorbing maximally at about 665 and 725 m respectively, provides a logical explanation for its role in photomorphogenic processes. Despite our growing understanding of the physical and chemical characteristics of the molecule, we still have no clear understanding of how it acts. Two major theories have emerged :

(1) One considers that phytochrome acts through a gene depression mechanism, on the basis that certain biochemical transformations occurring soon after phytochrome conversion are blocked by the usual inhibitors of RNA and protein synthesis.

(2) The other attributes to phytochrome a role in the regulation of membrane permeability, on the basis that certain relatively rapid plant movements, known to be due to turgor changes in the motor cells, show changes in response within minutes after phytochrome conversion and are insensitive to actinomycin D.

Choosing between these two rival theories is rendered more difficult by a relative lack of information on the intracellular localization of phytochrome. Phytochrome as reported earlier, is present in the plastids and mitochondria, but the evidence has been objected. Recently Galston (1968) has suggested the probable existence of it in the nucleus especially near the nuclear membrane, thereby strengthening the view that the pigment inter acts in some way with genetic material.

Studies on the physiology of seeds leading to a fuller appreciation of the effects of light on seed germination has been discussed for nearly a hundred years, but it was the work of Flint and McAlister (1937) that first demonstrated a clear cut promotion of germination by red-light and an inhibition by far-red-light. Reviews of much of the voluminous literature on the role of light in seed germination have been given by Crocker (1936), Evenari (1956) and Toole *et al* (1956).

The photoresponsive mechnisms are intriguing but complex responses, for light may be inhibitory, promotive or non-promotive to germination, the response varying with the species and/or proportional light quality. Toole *et al* (1955 b) pointed out that fully inbibed, viable seeds of many plants may fail to germinate where one limitation on germination may be a specific light requirement which can be satisfied by the correct poising of a reversible photoreaction.

Borthwick *et.al* (1952) proposed the following reaction to account for the red-light promotion and far-red inhibition of lettuce seed-germination :

$$\text{Pigment} \underset{7350\ A^\circ}{\overset{6500\ A^\circ}{\rightleftharpoons}} \text{Isomeric pigment} \quad \text{(i)}$$

$$\text{Pigment + reactant} \underset{7350}{\overset{6500}{\rightleftharpoons}} \text{changed pigment + changed reactant} \quad \text{(ii)}$$

$$\underset{\text{(dormancy)}}{\text{inactive Pigment } 6500} \underset{\text{far-red}}{\overset{\text{red}}{\rightleftharpoons}} \underset{\text{(Germination)}}{\text{Active pigment } 7350}$$

Based on these observations Toole *et.al.* (1955 a) later concluded that the same reversible photoreaction exists in lettuce seed germination as that which accounts for floral initiation in cocklebur (Xanthium strumarium). In another paper Toole *et.al.* (1955 b) states that this reversible photoreaction mechanism is present in all seeds but it is not always obligatory for germination.

Since the discovery of a reversible photoreaction in the light sensitive responses of plants, a search has been underway to determine the nature of the pigment involved in such processes and to uncover the mechanism of the photochemical reactions basic to the responses. Recognising a balance between promotive and inhibitory substances in physiological processes, Thimann (1956) suggested that a Ciz-trans-, configuration may be the key to the biological activity of such compounds. This theory has interesting implications, for it is well known that the cis-trans isomerisation of carotene is responsible for the photoexcitation stimulus in vertibrate vision. More recent evidence, however, indicate that phytochrome is a proteinaceous substance (chromoprotein).

The pigment is present in the green parts of plants as shown by physiological responses, but it has not been detected in such tissues because of interference by chlorophyll. The pigment named phytochrome has been extracted and concerntrated by partial purification because it occurs at very low concentration in the plant cytoplasm, it can not be detected in living tissues by eye, but the partly purified extract is blue-green (Butler *et.al.* 1963, 1964 a, b ; Bulter, 1961 ; Hendricks *et.al.* 1962 ; Lane *et.al.* 1963 ; Siegelman and Firer, 1964). The photoreversible changes in the absorbance of phytochrome are the properties of a specific chromophore which is present in very small quantity (6 μ g of the pigment in 5 ml of purified phytochrome solution) as reported by SIEGELMAN *et.al.* (1966).

Of the two forms of Phytochrome one has a maximum absorption at 660 mμ (pr 660) and the other at 730 mμ (P fr 730).

$$\text{Pr} \underset{\substack{730\ \text{m}\mu \\ \text{Darkness}}}{\overset{660\ \text{m}\mu}{\rightleftharpoons}} \text{Pfr} \xrightarrow[\substack{\text{darkness or} \\ \text{Prolonged irradiation}}]{\text{destruction}} \text{x}$$

The red absorbing form (Pr) is stable in dark. It is converted by exposure to red-light into the far-red absorbing from (Pfr). Pfr can be recovered to Pr by exposure to far-red light (Borthwick *et.al.*

1952 b). High temperature and/or high osmotic concentration also reverse Pfr to Pr.

Pfr is the physiologically active form. It is a highly active enzyme or hormone (Hendricks, 1964). It is effective only in a limited range of concentrations below which it is inadequate to promote response and above which it is inhibitory (Borthwick *et.al.*, 1956). Pr itself does not act as an inhibitor (Scheibe and Lang, 1965).

It has been demonstrated that red light does not covert all the Pr to Pfr, but set up an 80 : 20 equilibrium of Pfr : Pr and far-red light sets up a photo stationary equilibrium with approximately 99% of the phytochrome in the red-absorbing form (Butler *et al*, 1963, 1964 a, b ; De Lint and Spruit, 1963 ; Kasperbauer *et.al.*, 1963 : Siegelman and Firer, 1964 : Pratt and Briggs, 1966). Even 1% Pfr is sufficient to promote germination in some seeds (Downs and Piringer, 1958).

The results complied in Table 44 show that seed germination is promoted by short exposure to red light is reversed by subsequent exposure to far-red light. And explanation of why germination is not inhibited by one brief exposure to far-red or even by moderatly long continuous far-red radiation might be that new Pfr is formed from a precursor, as fast as the original is converted to Pr.

There are some photoresponses which are controlled by the phytochrome and the high energy reaction (Mohr 1957, 1959, 1960 : Siegelman and Hendricks, 1956). While the two photoreactive systems are independent, they may act antagonistically (Mohr and Haub, 1962) or synergistically (Mohr and Nes, 1963).

The Ecological implications of low energy reaction (Phytochrome system) lies in the fact that seeds with low requirement of light, may germinate under shade of crop plants or even below a certain depth of soil where the length of the light period may not be so limiting as seen under the intensity of light.

The special composition of light reaching the seed after filtration through the above ground vegetation or soil layers also remains an important factor in germination of these seeds. Toole *et al* (1968) has shown that red-light penetrates the soil to a greater depth than the blue light and far-red has still higher percentage transmission than red, and seals the stock of deeply burried seeds, to dormancy.

Table 44. Repeated reversal of germination (Percentage) response by alternating Red (600 mμ) and Far red (730 mμ) radiations.

Plant species	Amara-thus arini-cola Johnson	Pinus teada L.	Pinus strobus L.	Pinus virgani-ann Mill.		Erag-rostis (Schrad) Nees.	Lectuca curvula sativa L.	Lec-tuca sativa L.	Lec-		L. sativa L.		Ocimum americ-annum L.
Time (minutes)	1 each	R-4, FR-16	R-16, FR-64	R-8, FR-16		2	8	R-1, FR-4	R-1,FR-4		R-1,FR-4		5 each
Irradiation/Temperature oC treatment/		25	25	25	L.P.62	29 L.P.63	29	20	27	7	26	6.8	
Dark (control)	56	21	3	4	14	38	6	8.5	14	11	—	—	6.0
R	41	75	81	92	89	96	30	96	70	72	70	72	62
FR	13	—	—	—	—	59	6	—	—	—	—	—	—
R + FR	17	19	16	4	2	62	10	54	6	13	6	13	4.0
R + PR + R	—	75	73	94	93	92	30	100	74	74	74	74	61.0
R + FR + R + FR	—	20	13	3	1	—	—	43	6	8	6	8	4.0
R + FR + R + FR + R	—	78	83	93	92	—	—	99	76	75	76	75	61.4
R + FR + R + FR + R + FR	—	—	—	—	—	—	—	54	7	11	7	11	4.0
R + FR + R + FR + R + FR + R	—	—	—	—	—	—	—	98	81	77	81	77	65.0
R + FR + R + FR + R + FR + R + FR	—	—	—	—	—	—	—	—	7	12	7	12	—
Author	Hendricks *et.al.* (1968)	Toole *et.al.* (1962)	Toole *et.al.* (1962)	Toole *et.al.* (1961)		Toole *et.al.* (1968)	Borthwick *et.al.* (1952)		Toole *et.al.* (1953)		Borthwick *et.al.* (1954)		Varshney (1968)

Seed Dormancy

Dormancy in seeds may be commonly defined as their temporary suspension of growth activity. They do not germinate/sprout under conditions of moisture, temperature and oxygen favourable for vegetative growth. Sussman and Halvorson (1966) defined dormancy as.... "any rest period or reversible interruption of the phenotypic development of an organism". Determination of viability of seeds has long been a difficult problem because of the lack of knowledge concerning the various causes of seed dormancy, and the methods to overcome the dormant condition.

Bunning (1947) distinguished two basic kinds of dormancy-ectogenous, influenced by external factors like light, temperature, water, chemicals etc. and edogenous, influenced by the internal constitution of the seed, such as coat impermeability, a metabolic block or the production of a self-inhibitor. Sussman (1965) also recognised a similar distinction, but he used the term constitutive instead of endgenous. The above classification does not always hold good. To be more specific, Portar (1949) recognised seed dormancy to be of two types, that is, seed coat dormancy, and embryo dormancy. Misra (1968) dealt seed dormancy under the following three heads :

(*a*) Seed coat dormancy

(*b*) Hormonal dormancy, and

(*c*) Embryonal dormancy or due to a combination of these three.

The causes of dormancy may be several such as mechanical resistance offered by the seed coat to the expansion of the embryo and the impermeability of the seed coat to water and gases, causes seed coat dormancy. The outstanding studies of Crocker (1907) showed that a number of aquactic plants produce viable seeds, but that seeds of most of these plants fail to respond to normal germination conditions because their embryo remain covered by an impermeable seed coat. If the seed coat is removed or ruptured the seeds germinate promptly.

The presence of chemical inhibitors in the seed coat, pericarp, endosperm or the embryo, and the absence of growth promoters cause hormonal dormancy. Evanari (1949, 1957) and Wareing (1965) reported a great many inhibitors of germination : hydrogen cyanide, ammonia, ethylene, mustad oil, organic acods, unsaturated lactones, aldehydes, coumarin, essential oils, alkaloids etc. Not only are there naturally occurring and functional inhibitors, but also a large number of inducers or activators and promoters of germination. They include such agents as kinetin, gibberllic acid, ammonium nitrate etc (Radley, 1958; Murakami, 1959; Blumenthal-Goldschmidt and Lang, 1960).

A rudimentary or immature or underdeveloped embryo and unactivated enzyme system may cause embryonal dormancy.

Under natural conditions, seed coat dormancy is removed by the seed coat decomposition or by mechanical scarification. A treatment much like the chemical scarification, results from the action of soil micro-organisms such as bacteria, fungi and possibly soil acids. The chemical inhibitors which cause dormancy are generally removed by washing of the seed by rain water. All unripe embryo may become fully mature in time, when favourable conditions for seed germination reapproach.

For many decades emperical research has sought for various types of physical and chemical treatments to break dormancy, by only in the last few years has there begun to evolve the moderate physiological understanding of dormancy such that it can be controlled (Amen 1968).

The seed coat dormancy may be removed by mechanical scarification—breaking or removal of the seed coat, inter-mittant flashing of light (Crocker and Barton, 1953 ; Crocker, 1936), X-rays (Maxwell and Pidgeon, 1933), dry heat (Evart, 1908 ; Muller, 1912 ; Harrington, 1916 ; Lute, 1928 etc), Boiling water (Midgley, 1926), High pressure (Davis, 1928). The most widely used laboratory method is treatment with concentrated sulphuric acid (Barton, 1965) which renders the seed coat permeable to water.

Some chemical inhibitors which cause dormancy may be removed by washing the seeds by running water for several days. The growth promoters such as Colchicine, gibberellic acid, Kinetin, Hydroxyl amine, Thiourea, Potassium nitrate, Ascorbic acid,

Table 45. Showing various types of seed-dormancy, and most suitable treatment for their removal in some of the investigated Indian Plants.

Species 1	Removal treatment 2	Author 3
Seed Coat Dormancy		
Abrus precatorius	Sulphuric acid 90-120 minutes	Vyas and Agarwal (1973)
Acacia senegal	Excision of seed coat	Kaul and Manohar (1966)
Achlypha malabarica	Mud treatment 1½ month	Bhat (1968)
Alhangi cameloram	Sulphuric acid 3-5 minutes	Ambhasht (1963)
Alysicarpus bupliwrifolius	,, 30-120 minutes	Ketkar (1965)
A. pubescens	,, ,,	,,
A. rugosus	,, ,,	,,
Bidens biternata	,, 60-120 minutes	Varshney (1966)
Cleome chellidoni	,, 30 minutes	Hiremath (1955)
C. viscosa	,, ,,	,,
Clitoria ternata	,, 30-40 minutes	Mullick and Chatterji (1967)
Croton sparsiflorus	Mechanical scarification	Kaul V.N. (1959)
Cyperus rotundus	Mud treatment 1 week or sulphuric acid 2 minutes	Shukla (1964)
Datura innoxia	Nitric acid 5 minutes	Saxena M.D. (1964)
Indigofera astragalina	Sulphric acid 30 minutes	Agarwal and Vyas (1970)
I. cordifolia	Sulphric acid 10 minutes	Vyas and Agarwal (1972)

(*Contd.*)

1	2	3
Hormonal Dormancy		
Anagallis arvensis	Leaching with water	Singh K.P. (1966)
Alternaria sessalia	"	Kaul A. (1968)
Cuminum cyminum	Indol acetic acid or alphalenaecetic acid 10 ppm	Gandhi and Bhargawa (1961)
Dictylcetenium aegypticum	GAA 50 ppm 5 weeks	Gupta K.C. (1966)
Embryonal Dormancy		
Achyranthus aspera	Stratification at 15 – 20°C for 8 weeks	Mall and Arzare (1967)
Arthraxon lansifolius	Alternating temp of 20 – 40°C every 12 hr	Varshney (1967)
Chrozophora rottleri	Stratification at 10°C	Mall (1956)
Erigeron linifolius	Dry storage at rom temperature for 1 month	Ramakrishnan (1963)
Verbena bipinnatifida	Continuous red light at 25°C	Vyas and Agarwal (1969)
Vitis vinifera	Stratification at 40°C for 3 months	Singh S.N. (1961)
Oscimum americanum	Red light for 5 minutes	Varshney (1968)
Seed Coat + Embronal Dormancy		
Crypis aculeata	Mechanical scarification + 50°C for 1 week	Mall (1958)
Hormnal + Embryonal Dormancy		
Anagallis arvensis	Washing + 15°C for 3 days + 1200 – 1300 lus for 2 minutes	Pandey (1968)

Cytosine, Guanine, Adenine etc. either individually or in some suitable combination, may be applied exogenously to break dormancy and enhance seed germination (Black and Naylor, 1959 ; Evanari, 1965 ; Timson, 1966 ; Bachlard, 1967).

Dormancy due to immature embryo may be removed by after-ripening and by keeping the seeds at low temperature for various periods (Stockes, 1965 ; Capon *et al*, 1967). Enzyme systems may be activated by irradiation of the dark imbibed seeds with red or far-red light, and then removed to darkness again (Flint and McAlister, 1937 ; Borthwick *et al.* 1952 ; Toole *et al*, 1955 ; Hendricks and Borthwick, 1959 ; Downs *et al* 1961.).

Recognition of seed dormancy in Indian plants has received appreciable attemption of Ecologists in India. A review of some of the available literature has been presented in Table 45. A glance at Table 45 reveals that dormant seeds are more frequent in some families than in others. However, there does not appear a clear relationship between taxonomic position, occurrence, and type of seed dormancy. Different genera of the same family and even different species of the same genus may exhibit either different types of seed dormancy or no dormancy at all.

The Ecology of plants and the occurrence of seed dormancy also do not suggest too obvious relationships. However, more often seeds of cultivated plants are largely non dormant, but weed seeds usually exhibit dormancy in an active manner, thus assuring the continuence of the species.

Dormancy, therefore, has a survival value because it restricts germination to a favourable period of growth. If the seeds were to germinate inside the fruit that contained them, one of their main biological functions, dispersal of the species over a wide area, would not be fulfilled. Such premature germination is prevented by the inhibitors of the seed themselves. The beneficial effects of dormancy in increasing the survival value of a species can be illustrated for certain species seeds in the Indian climate. Most of the forest species set their seeds in winter, but the seeds remain dormant till the next monsoon sets in. In India, the monsoon season is wet, but brief and followed by a longer dry period. Even if few seedlings are able to emerge at the time of seed dispersal due to favourable conditions, further growth is greately hampered because of low temperature, want of adequate mositure etc., prevelent in the winter season. It is therefore advantageous for the species to have a period of dormancy till the next wet season sets in when the moisture conditions for the establishment of the new plants is favourable.

Reproductive Capacity

The potentiality of a species to colonise, perpetuate and establish itself depends upon a number of factors. Salisbury (1942) stated that the potential reporductive capacity of a species is a measure of its susceptibility to natural mortality and that the process of natural selection has brought about a nicety of adjustment between the seed output and mortality. Of the average number of seeds produced per plant the mean number that can germinate normally, is referred to as the reproductive capacity. the values of average seed output and reproductive capacity of some Indian plants have been given in Table 46.

Reproductive capacity is of much significance in the physiognomy and phytosociological relations of a species. According to salisbury the average size of a reproductive propagule is determined by the nutritional needs for the establishment of a new young individual of the species in its typcial habitat. A high reproductive capacity usually ensures a better chance for the survival and dispersal of a plant species.

Seedling Growth

This is the most critical phase in the life cycle of a plant. Shortly after germination, seedling have to face the adverse environmental factors such as pathogens, humidity, temperature etc. If any of these conditions becomes adverse seedlings die.

Seelings represent the juvenile stage of plant life and since they are delicate, they have to face extremes of environmental factors. Germination of seeds, the growth and development of plants and their ultimate capacity to produce flowers, fruits and seeds are regulated by genes at the sub-cellular level. Any modification in this internal environment alters the ultimate function of the plant. The internal cellular environment is constantly influenced and modified by its external environment. External physical factors such as temperature, precipitation, humidity, wind, light and substratum regulate the rate of cellular reactions and determine if they will proceed normally and optimally.

Table 46. Reproductive capacity of some Indian plants.

Plant species	Average seed output	Reproductive capacity	Author
Anisochilus eriocephalus	21421	1328	Bakshi (1952)
Hypericum actum	23000	18454	Salisbury (1942)
Lindenbergia polyantha	71731	70266	Misra & Rao (1948)
Mollugo cerviana	1689	777	Bakshi & Kapil (1954)
M. nudicaulis	940	526	Bakshi & Kapil (1952)
Mercurialis perennis	300	15	Mukherjee (1936)
Gisekia pharnaceoides	940	226	Joshi & Kambhoj (959)
Desmodium pulchellum	3169 (Forest patches) 3482 (Dense forest)	752	Tripathi & Srivastava (1970)
Borreria articularis	1938-90	321	Baby & Joshi (1970)
Asphodelus tenuifolius	2302	2302	Tripathi (1968a)
Bothriochloa pertusa	1157 (Protected area in (P) 1704 (Rainy)	46 (Winter) (Winter) 767 (Rainy)	Sant (1974)
Indigofera linifolia	497 (Summer) 427 (Rainy)	144 (Summar) 196 (Rainy)	Sant (1972)

In the seed stage, the embryo is protected, biological activity is at a minimum level and it can withstand the extremes of environmental factors easily. However, once germination is over, in the young seedling biological activity like cell division, tissue differentiation, biological reactions etc. are at a rapid rate. High temperature, low humidity and high wind velocity may dessicate seedlings if they do not have any protective device against these. The young root is soft and deligate and has to bear the hard and granular soil through which it makes its way. A large variety of soil fauna and micro-organisms also find it easy to damage the soft roots. Even a temporary dryness around the roots may cause to them permanent damage. On germination, the reserve food in the seeds is quickly exhausted, and the shoot has to meet the high food requirements for a rapid growth.

Bohra and Sen (1977) observed that in *Crotalaria burhia* and *Convolvulus microphyllus* leaves on juvenile plants were longer and broader as compared to adult plants. After producing first few leaves the shot growth stops for certain duration ; while the root system penetrates deep into the ground, thus enabling the seedlings to receive soil moisture from deeper soil layers. Gupta and Prakash (1975) observed that with 10 mm or less of rain, not a single seedling appeared, with 15 mm rain a limited number of seedlings germinated, whereas at 25 mm rain extensive germination took place. Many of these seedlings die down if subsequently there is a dry period. Excessive moisture or humidity increases chances for fungal infection. Beside the influence of environmental physical factors on the seedling establishment, biotic factors such as grazing, trampling browsing, nibling, pathogenic micro-organisms, inter-and intra-specific competition are other important factors which play their role in the establishment of seedlings.

Silvertown (1987) observed prolonged arrest of tree seedlings, resulting in a stunted growth, awaiting the death of older trees, especially where there is closed canopy stand as in temperature belt with *Tsuga canadensis*, *Picea abies*, and *Quercus alba*. He called this phenomenon as *'Oskar syndrome'*. In this phenomenon the seedlings remain more or less dormant and wherever the canopy becomes open the dormant seedling growing there, rapidly elongates, produce branches, flowers and seeds.

The vegetative growth of any plant species is influenced by the intensicity, duration and quality of light, temperature, water and soil conditions. A well developed root system fulfils mineral and water requirement of plants whereas a well developed shoot system meets the food requirements for growth of new tissues, increase in size and biomass. Every species has its own *ecological amplitude*—which is a range to tolerance of that particular species. In autecology root-shoot studies are important to gain an insight of the growth performance of the plants. In grasses and some weeds, the vegetative growth, such as the root depth, length of shot, number of nodes, internaode length, number and size of leaves, stomatal frequency, thickness of leaf cuticle etc. are affected by environmental conditions.

The reproductive growth includes flowering, pollination and fruiting of a species. Angiosperms are best adapted to the multiplicity of habitats. It is only the sexual reproduction which gives chance for intermingling of genes and producing variations (mutations). For a sussessful growth in time and space species must flower and fruit.

Biological Clocks

Each species in the life-cycle of a plant is greatly influenced by a number of environmental factors. More over in turn the species also modifies the environment of a place continuously. The study of life-cycle of a species in nature in association with other species, completely bathed in environment both above ground and underground is the core of autecology as illustrated in Figure 42.

Each stage in the life cycle in studied in detail, both in the field and in cultures grown under controlled environmental conditions and then the whole picture of autecology is built by putting observed and interpreted facts in proper sequence and perspective.

The requirements of germination, growth, flowering, fruiting, etc. of different species are met at the same place, but at different times of the year. There is so much synchronisation of the phenological behaviour of the species and the different factors of the environment that plants are spoken of as

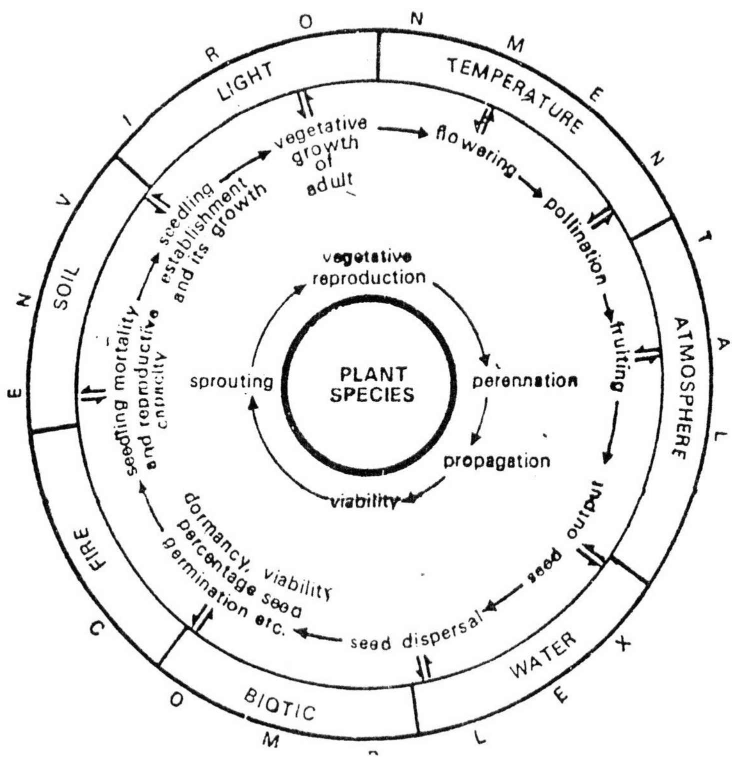

Fig. 42 : Biological clock diagrammatic.

'Biological clocks' or 'phenological clocks'. Biological clocks are regulated by external singnals from the environment, and the ecologically much advantageous since these couple environmental and physiological-rhythms and enables the organism to anticipate daily, seasonal and other periodicities in environmental complex of a pleace.

12. *Genecology*

The theory of evolution developed independently by Charles Darwin and Alfred Wallace states that natural selection favours those individuals whose traits render them best able to survive and reproduce. This laid down the foundation of the concept of constant and gradual change as characteristic of all life. To this geneticists contributed the finding that characteristics favoured by selection are passed on from generation to generation according to Mendelian laws of inheritance, thus preserving genetic variation in a population. Further, individuals do not evolve, populations do. That is the genetic make-up (genetype) of an individual remains constant throughout life, while the types and numbers of gene flow in a population of organisms can vary. The field of genecology deals with gene flow among individuals in inbreeding group.

Populations comprise groups of individuals in one locality or within defind geographic boundries capable to freely inbreeding. All of the genes of these individuals form a single *gene pool* and the way in which each allele is represented in the gene pool determines a population's variation or *gene frequencies.*

Variations in both phenotypes and genotypes can be seen. The range of both ultimately is determined by the individual genotype. Phenotypic variation represent adaptation by the individual, where as genotypic variation represents adaptation by the population, enabling it to change over time. Genotypic variation is both continuous or regular, and discontinuous or sharp.

Sexual reproduction permits recombination of genes on crossing over at meiosis is gamete formation and in the pairing of parental chromosomes at zygote formation. Mutations take the form of changes in the gene's DNA and changes in chromsome number and structure. Recombination and mutation are regarded basic source of evolutionary change.

Migration operates as a force of evolution when it occurs between relatively isolated populations with distinct gene pools, as immigrants from one population to another can contribute totally new alleles. This exchange of genes is known as gene flow, and functions to increase variation in a population.

Operating upon genetic variation are the forces of selection and genetic drift. Natural selection occurs in nature, where as artificial selection can be engineered in the laboratory; both however, result in differential reproduction of the genotypes of a population. Selection acts upon the individual phenotype to increase or decrease the representation of the corresponding genotype in a population. The following three types of selection have been categorized (1) Directional selection, which describes the change that occurs when a population shows a particular trend through time, (2) Stabilishing selection which describes the change when extreme individuals are eliminated from the population. Most selection that occurs in populations is stabilising and homeostatic because it tends to maintain status quo, and (3) Disruptive selection which describes the process where certain type of character has a high survival value, where as intermediate types are not advantageous.

Genetic drift is most likely to become an evolutionary factor in a small population, especially one characterised by a high degree of inbreeding. Due to more frequent pairing, homozygour recessives may appear in relatively high frequencies. On the other hand some genes may be carried by so few members that thy become lost by chance. When such changes occur as a result of a non-representative original population, they are ascribed to *founder effect.*

The experimental work of Turesson (1922) on population genetics, embodying the genetical basis of ecological differentiation occurring within populations of species, is a land mark in the field of ecology. This work was aptly described as 'genecology'. Genecology is an independent study of adaptive properties of the race of any level in relation to its environment.

Genecology deals with assortment of phenotypes. This exerts an indirect influence on the frequency of biotypes, and thus on gene-pool of the population. In this way the influence of the environment will prove to be inherited by the population. Genecology is different from experimental taxonomy, morphology and biosystematics, because the purpose in these cases is not primarily the study of the relations between populations and their environmental factors, but the determination of taxonomic affinities and evolution (Langlet, 1971). Similarly, genecology is not the study of the genetics and ecology of plant groups as suggested by Nobs (1961), because the term genecology is not constructed of the 'gene' in genetics combined with 'ecology'. Bennett (1965) rightly stated that it is the central concern of genecology to understand and control the interaction between adaptive genetic variability and the forces of natural selection.

Considerably older than the term genecology, is the observation that there exists within the species racial hereditary adjustment to diverse environments. Linnaeus (1739) stated that habitat also plays a great part in plants, so that a plant grown in the southern part of the world ripens later for the first few years after it has been brought over here, but in time, and from habit, it hurries up, and in the succeeding summer it ripens much more rapidly.

Kerner (1875) observed the occurrence of a genotypical response of the populations to their warmer or colder original habitats with regard to date of flowering. Osborn (1897) stated that ontogenetic adaptation is often of very profound character, it enables organisms to survive in very critical changes in their environments. Thus, all the individuals of a race are similarly modified over such long period of time, that very gradually, congenital variations which happen to coincide with the ontogentic modifications are controlled and become phylogenetic.

In a population studies a species is treated as genetically pure population of similar individuals. However, many species consists of micropopulatios base on genetic differences. Since, many of the morphological characters are linked with their physiological and development origin, a close examination of the population in different habitats generally reveals ecotypes.

Species showing variation governed by environment are known as *ecads*, controlled by heredity are known as *ecotypes*. Ecads are plants of the same species which differ in appearance such as size, nature, reproductive vigour etc in different environmental conditions. These variations are not genetically fixed and when transplanted to natural conditions the variations vanishes. an ectoype is a population, or a series of like populations which are in selective adjustment with their selective environment. An *ecospecies* is a sum of ecotypes which are able to exchange genes. The ecospecues may be several in a unit geographic area and all of them together constitute *coenospecies*. Coenospecies are therefore plants of common evolutionary origin and it may correspond to a group of taxonomic species or in some cases a entire genus.

The concept of ecotype was put forward any Turesson (1922). He was quite familiar with the phenomenon of the different environmental tolerances of various species, and that different populations of the same species occupying distinctive habitats frequently showed morphological and physiological traits, often of an adaptive nature. The various species that he studied were grown in identical conditiôns in his experimental garden at Akarp in Sweden, in order to know whether the variations were modifiable. He did not get it. This clearly ment that a new individuals adaptability to meet another condition resides in a preaaapation, which may not have been affected by the environment. In other words, the role of the environment is to select individuals cytological and gene combinations favourable to that environment, and resulting individuals not suited to it. Gene exchange and recombination in the population operates according to the law of probability,. In this analysis of the genotypical response of the plant species to its environment Turesson developed the idea of ecotype to explain his results. In his words, 'an ecotype is a unit geographic race which is in selective adjustment with the total environment'.

Guided by the Turesson's concept Clausen, Keck and Hiesey (1940, 1948) conducted transplantation experiments on different plants like *Potentilla glandulosa* and *Achillea lanulosa-borealis* comp..., from the sea level to above timber line in California into three different field stations—sea level (at 100 ft in Standford), intermediate (at 4600 ft at Mather), and timberline (at 10000 ft), having an average growing season of 282, 145 and 67 days respectively. In *Potentilla glandulosa* they recognised four climatic ecotypes : (1) *P. glandulosa nevadensis* was of shortest height flowered early in season and was frost tolerant, (2) *P. reflexa* was taller and occurred at mid elevations in dry habitats, (3) *P. hanseni* was tall and occurred at mid elevations but on moist or wet meadow lands, and (4) *P. typica* was shorter than (2) and (3) in height but grew best in low elevations at Stanford and poorly at mid elevations and absent at higher elevations. The differences between the four ecotypes were in both morphological features such as stature, habit, leaf area, form of the inflorescence etc. and physiological features such as seasonal growth rhythm, time of flowering, frost resistance etc., features which were related to their capacity to survive in their particular habitats. In *Achillea lanulosa-borealis* showed eleven climatic ecotypes and they suggested that the total number of ecological races within the three constituent species of the *A. millefolium* complex may run into hundreds. They have shown that each climatic representative of a species reproduces its kind through breeding and differences existing before and after transplantation appear to result from both genetic and chromosomal differentiation.

Gregor *et al* (1936) observed ecotypic variations in a wide range of maritime hibitats from water logged coastal mud, through salt marsh to more or less draine coastal mud. In *Plantago maritima* they distinguished populations from one another by quantitative characters. They showed that the pattern of edaphic variation was continuous corresponding to the gradients shown by habitats, and termed that continuous ecotypic variations should be termed as an *ecocline* and redefined the ecotype as a range of an ecocline.

Besides the climatic and geographic ecotypic differentiation the edaphic and biotic ecotypes have also been found to exist. Stapledon (1928) recognised a number of biotic ecotypes in *Dactylis glomerata* in relation to intensity of grazing. Kruckberg (1951) differentiated two edaphich ecotypes in *Achllea borealis californica* in relation to serpentine and non-serpentine soils.

A number of ecads in different species have been reported in India. Ramakrishnan (1960a) observed two ecads in *Euphorbia hirta*, one growing in dry hard soils (prostrate type), and the other growing along the foot paths under trampling (prostrate compact type). Pandeya (1962) observed that in *Bothrioclova pertusa* and *Dicanthium caricosum* variation occurs in habit, number of culms, number of spikes per culm, number of spikelet per spike and seed output. Such variation have been found to be largely governed by the intensity of grazing and soil mositure. In each species there were found two ecads, (1) those under protection showed basket form habit, and (2) those in overgrazed areas acquire a saucer shape. Grazing brings about reduction in size of erect stem, number of spikes per raceme, number of spikelets per spike, and in the length and breadth of lower glume of spikelet. Besides some physiological variations also took place, which included the development of red pigment, and initiation of early flowering.

Ecotype differentiation in plants has received very meagre attention of ecologists in India, inspite of the fact that the vegetational types of India range from the desert vegetation of Rajasthan to the wet evergreen forests of eastern Himalayas and western Ghats. Soil types also show wide range of variations. Biotic interference probably most unchecked in our country. Such a wide range in climate, soil and vegetation offers a vast scope for investigations on climatic, edaphic and biotic races. Table 47 gives a summary of the available literature on this aspect.

Origin of Ecotypes

Origin of ectoypes still remain a difficult question but it is largely accepted that it is due to a consequence of hybridization between taxa giving the initial variability necessary for selection to act upon. Hybrids between related taxa may be produced on contact and these would have a great variety of habitats in which to become established. Continued back crossing from such hybrid populations might introduce adaptive genes from one species into the other and further migrations might then lead to ecotypic differentiation (Curtis, 1959). Harberd (1961) stated that, "if however, the parental taxa are completely submerged by hybridization, and so not describable, a situation might arise having many of the characteristics of differentiation by selection". He further argues that if all species show genecological differentiation, then not all would be the result of hybridization, but if only a few species do so, then we should be suspicious of selection as an adequate interpretation. Turesson (1922) considered that ecotypes do not originate through sporadic variation preserved by cance isolation, they are on the contrary to be considered as products arising through the sorting and controlling effect of habitat factors upon the heterogenous population. His work may be summerised as follows : (1) wide ranging plant species show spatial variations in morphological and physiological characters; (2) much of this intraspecific variation can be correlated with habitat differences;

Table 47. Ecotype differentiation in some plants.

S.No.	Species	Ecotypes	Author
		EDAPHIC ECOTYPES	
1.	*Adathoda vasica* Ness	3-in relation to soil calcium	Ramakrishnan and Bisht (1968)
2.	*Boerhaavia diffusa* Linn	2-robust and normal (Calcium)	Srivastava and Misra (1968)
3.	*Corchorus acutangulus* Linn	2-erecto and prostrate(do-)	Dixit (1966)
4.	*Eleusine indica*	2-erect and prostrate	Gupta and Srivastava (1969)
5.	*Euphorbia thymifolia* Burm.	2-red and green (do-)	Ramakrishnan (1961a)
6.	*Gomphrena Celosioides*	2-ascending and decumbent (do)	Srivastava (1966)
7.	*Heteropogon contortus* (Linn) p.Beauv.	2-(substratum wall or rock)	Varshney (1964)
8.	*Lindenbergia polyantha* Royle	2-(Calcium)	Misra and Rao (1948)
9.	*Mecardonia dianthera*	2-(do)	Koul (1965)
10.	*Tridex procumbens* L.	2-(do)	Ramakrishnan Jain (1966)
		SOIL MOSITURE ECOTYPES	
11.	Echinochloa colonum (Linn) Lank	2-erect and Prostrate	Ramakrishnan (1960b)
12.	Euphorbia hirta Linn.	2-(do)	Ramakrishnan (1960a)
13.	Lecas aspera Spreng.	2-	Dixit (1966)
14.	Setaria glauca Beauv.	2-long and short panicled	Ramakrishnan (1963)
		LATITUDINAL ECOTYPES	
15.	Anagalis arvensis Linn.		Singh K.P.1967
16.	Cassia tora Linn		Singh J.S. 1967
		THERMO AND PHOTOPERIODIC ECOTYPES	
17.	Ageratum conyzoides	2-Winter and monsoon form	Koul (1965)
18.	Xanthium strumarium	4-monsoon, winter, winter & summer and summer forms	Koul (1965)

(3) to the extent that ecologically correlated variation is not simply due to plastic response to environment, it is attributable to the action of natural selection in moulding locally adapted populations from the pool of genetical variation available to the species as a whole.

Muntzing (1954) is of the opinion that an ecotype is a result of interaction between recombination and natural selection. Genetical variability which is the basis for formation of an ecotype will constantly undergo natural selection before its establishment. The so-called mutation and recombination of genes cannot be spontaneous but only the residual effect of action, co-action and raction between environmental complex and the organism. Misra (1962) stated that introgressive hybridization of ecotypes is a free phenomenon resulting in intermediate forms and physiologically distinct ecotypes. According to Misra (1967) a population of a species enters into a community as an interacting unit. Volume-spacing of individual plants of different species provides a community with a structure (form) which has arisen on account of their interactions (function) in relation to their environmental factors. Pandeya (1962) metaphorically described a newly formed ecotype as a biological crystal ready to get dissolved in a suitable environmental solvent to form a 'whole' solution called 'ecosystem'.

Precondition for Ecotypic Differentiation

Davis and Heywood (1963) have outlined certain characteristics, which are important in studies pertaining to ecotypes. They are :

1. In several cases the adaptive features of the ecotypes do not lend themselves to taxonomic treatment or are not correlated with features which do. Ecotypes are often distinguished by flowering time, seasonal growth rhythm, first resistance or other physiological differences.
2. Most morphological differences between ecotypes are quantitative, depending on polygenic inheritance ; they may require statistical treatment for their detection and it may be found that the variation within the ecotypes obscures their racial differences.

3. Geneecological variation in populations may frequently occur at a lower level than formal nomenclatural classification can usefully do.

4. Much ecotypic variation is continuous, following a clinical pattern, often of a complex nature. The principal factor involved in determining whether the genecological variation pattern will be broken up into discrete units or be continuous is the breeding system operative in the population.

5. A final series of problems arise from the different modes of formation and distributional patterns of ecotypic variants. Some ecotypes appear to have arisen polytopically in satially separate but ecologically similar habitats where populations of a species have been subjected to similar selective forces.

13. *Organisation of Communities*

The community concept is one of the most important principles in ecological thought and practice. A biotic community is an assemblage of a number of organisms usually of different species, living in a prescribed area of a physical habitat. When only assemblage of plants is a habitat is considered, it is called *plant community*. Similarly, when assemblage of animals in a habitat is considered, it is called *animal community*? However, in any biological organisation, plants and animals are very closely related and interdependent and share the same set of conditions. The community is thus essentially a *biotic community*. It is a loosely organised unit because it has characteristics of its own, in addition to the living part of the ecosystem. The idea of community concept emphasizes the fact that diverse organisms live together in an orderly manner and the impact of the community to be recognised is that the organisms grow as the community grows.

The organisation of the community is intricate enough. The concept regarding its organisation that have developed firstly by observation and secondly by experimentation, reveal that a biotic community is an organised or a super-organised unit. This particular idea is difficult to prove and disprove. There are variation in biotic communities as we observe in nature.

The concept of community organisation is by no means a recent one, and may be traced back to the time of Theophrastus (370-250 B.C.), who recognised the existance of plant communities or associations of species in different climatic areas. It was Grisebach (1938) who recognised the *plant formation* as an important unit of vegetation. Although the concept of community was first visualised by Forbes (1844), while studying the molluscs of Aegean Sea, when it was observed that in different areas (*i.e.*, at different depths) there might be distinguished different types of assemblages of species. Mobius (1877) recognised the biotic nature of the community. He considered the oyster bed as a community of living beings "where the sum of species and individuals mutually limited and selected have continued in possession of certain definite territory".

The occurrence of a large number of species in the same area is explained by difference in their ecological requirements and functions. According to Gause (1934) no two species with similar requirements can exist together. In this context, we may refer here to the *niche* concept. Grinnell (1917) who used the word niche for the first time, considered it as a specific portion of the habitat occupied by each species. Elton (1927) defined niche as an animals place in the biotic environment and its relation to food and enemies. Niche is also the functional position of an organism in the community. Thus, several organisms with different functions occupy the same habitat. Lack (1944) showed that closely related birds could occur together due to differences in their nesting or feeding habits. Later, Hutchinson (1951, 1958) introduced the concept of ecological niche referring it to the totality of biotic and abiotic factors to which a given species is uniquely adapted. The various environmental factors were conceived to obtain a multidimentional hyper volume within which each species is located at a given position occupying a small hyperspace which was considered to be the niche of the species. Several theories have been put forward to explain distribution of the species in the community according to the niche hypothesis. In terms of niche concept, the community is defined as a "functional system of interacting niche differentiated species populations that tend to complement one another rather than directly competing, in their utilization of one community's space, time, resources and possible kind of interactions (Whittaker, 1970).

Community Structure

A community as certain structural attributes which can be made use of in distinguishing the diverse communities. The community characters may be conveniently classified into two main categories—analytical and synthetical. The analytical characters includes quantitative and qualtitative ones. Some of the community characters which deserve critical evaluation are the following :

Analytical Qualitative

(a) Floristic composition
(b) Stratification
(c) Physiognomy
(d) Dispersion and Sociability
(e) Vitality
(f) Association of species

Analytical Quantitative

(a) Frequency
(b) Density
(c) Cover
(d) Height of plants
(e) Weight of plants
(f) Volume of plants

Synthetic

(a) Presence
(b) Fidility
(c) Dominance
(d) Indices of species structure
(e) Index of dominance
(f) Index of diversity
(g) Importance Value Index
(h) Ecotone or the concept of edge effect
(i) Homologous series of twin association
(j) Naming of the biotic community

Vegetation Analysis

(a) Successive approximation
(b) Individualistic concept
(c) Continuum concept
(d) Gradient analysis

Floristic Composition

A complete list of species growing in a community is the first step towards its study. In preparing a list care should be taken to include the rare species also which may be of value as indicators to certain environmental conditions. Besides higher plants, the lower plants like the ferns, mosses, liverworts, lichens, epiphytes etc., should be listed. In order to have a complete list, inspection and collection throughout the growing season are required so that all species appearing in different season are included. The plant identification should be confirmed by systematic botanists and herbarium specimens prepared for future reference.

Floristic lists are valuable for characterization because each species has its own range of ecological amplitude. For example, 36 species were present in a certain number of non-grazed sample areas in virgin tallgrass prairie in Central Oklahoma, while 64 species were found in the same number of sample areas in grazed part, indicating that a decrease in abundance of dominant species had permitted many invadors to be established in the later.

A decrease in the number of species from one area to another may indicate increasingly adverse conditions. For example, the number of species in xeric grasslands in Colorado changes greatly with increasing elevation ; 160 species were found in the mountain front, at 5300 ft, 139 in the foot hills at 7500 ft ; 130 in the upper foothills at 8400 ft and only 50 in the sub-alpine zone at 11500 ft (Hanson and Churchill, 1965).

Agarwal (1971) while studying the floristic composition of the forests of Gogunda and Prasad forest range, Udaupur observed 600 species belonging to 95 families and 365 genera. The number and percentage of families, genera and species in Dicotyledons and Monocotyledons are given in Table 48.

Table 48. Distribution of dicots and monocots in the flora of Gogunda and Prasad.

Particulars	Dicotyleadons		Monocotyledons		Total
	No	%	No	%	
Families	83	87.37	12	12.63	95
Genera	297	86.08	48	13.91	345
Species	514	85.67	86	14.33	600

From the above table, it is clear that the proportion of Monocots to Dicots is 1 : 7 of families, 1 : 6 of genra and 1 : 6 of species. This shows a great poverty of monocots whether we consider families, genra or species. On the whole the ratio of families to genra and species is 1 : 3. 6 : 6. 3. The ratio of genera to species is 1 : 1 : 7 which clearly indicates the very poor proportion of species to genera.

Stratification

It refers to profile structure or vertical disposition of organisms, or their parts at different levels in a community. All plants in a community are not of the same size and do not occupy the space there. The stratification of a community is determined largely by the life-form of plants - their size, branching and leaves- which in turn influences and is influenced by the vertical gradient of light. The stratification allows the plants to exploit the incoming radiation and the space to the maximum according to their differential requirements. The crowding effect is mitigated through adjustment in height.

A well developed coniferous and broad leaves forest has three principal layers of vegetation. From top to bottom, they are overstory stratum or mature forest canopy ; understory stratum or intermediate forest canopy or shrub layer ; and groundstory stratum or forest floor or herb layer (Figure 43).

The mature forest canopy is the major site of energy fixation and has a major influence on the rest of the forest. In the dry climate, where it is fairly open, considerable light reach the lower layers and the understory strata develop fairly. However, in the tropical rain forest, where the canopy is closed, the understory strata show poor development. Further by reaching the canopy, the forest trees may gain advantage, where aboundant sunlight supports photosynthesis, but the tree must spend much of the energy of photosynthesis in the growth of woody tissues of stem and branches to support the foliage in the canopy. There may be apparent disadvantage in the low light intensities in which the forest herbs must live, but the herbs need not spend its more modest photosynthetic profit on woody supporting tissue. Forest structure thus involves a gradient of growth forms - upper and lower trees, upper and lower shrubs, upper and lower herbs and soil surface mosses - in adaptation to the gradient of light intensity.

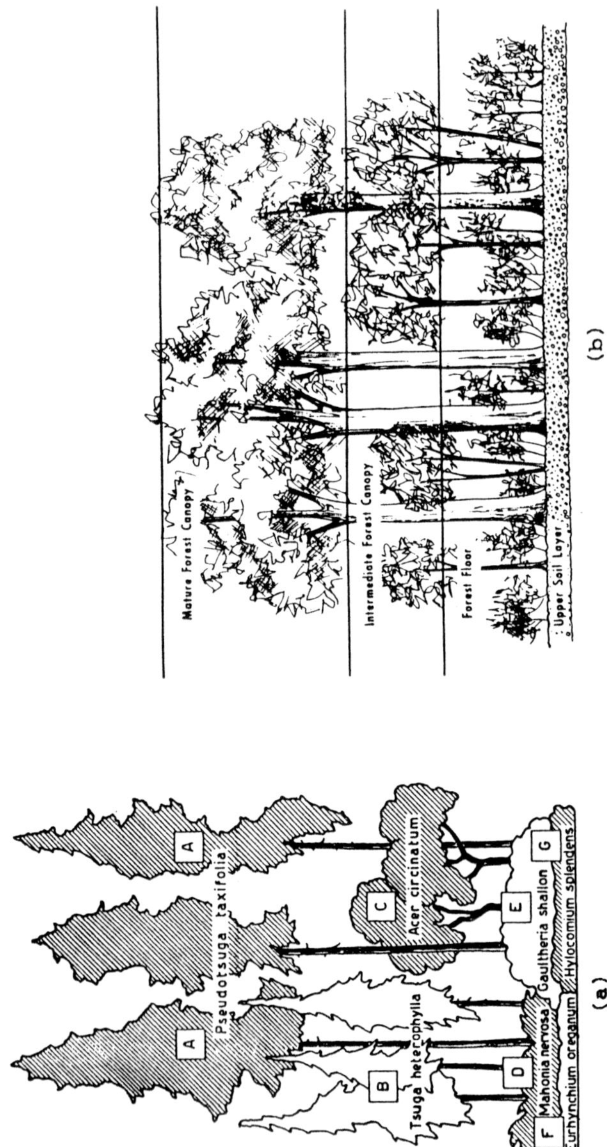

Fig. 43 : Stratification in forest communities (a) Coniferous forest, and (b) Broad leaved forest.

The understory stratum consists of tall shrubs, understory trees and younger trees, some are the same as those of the crown, others are of different species. Species that are unable to tolerate shade and competition will die, others eventually reach the canopy after some of the older trees die or are harvested.

The herb layer depends on the soil moisture conditions, slope position, density of overstory etc., which vary from place to place through the forest. Even within a herbaceous community the stratification can be easily noticed as there are plants of different sizes and there are also some prostrate plants. Beneath the herbs, liverworts and mosses on the ground may form still another vegetation layer. The final layer, the forest floor, is the site where the important processes of decomposition of the forest litter takes place and where nutrients are released into the nutrient pool.

Stratification is readily seen above ground, but is present also in the underground parts, for example, roots and rhizome. The roots may be spread nearer the ground or may be deep penetrating. This spacing among the roots permits the plants to draw their water and nutrient requirements from different layers of the soil without affecting each other.

The number of strata above ground vary according to the kind of community. In the early stages of succession, usually one stratum is present and is comprised by lichens, mosses or annual herbs, but as succession proceeds new strata are gradually added.

In the south-eastern Rajasthan, the top layer at 6-10 meter consists of trees with wide spreading crowns. The middle layer at 3-4 meter consists of many shrubs or small trees and rounded crowns. The herb layer, not over 2 meter in height, is poorly developed and in places missing. The composition of vegetation in each layer may vary place to place. These different groups of plants of similar life-form are called *synusiae*. For example, synusiae are stands of *Tectona grandis* in the tree layer of Baldia and a stand of *Butea monosperma* in Prasad forest of Rajasthan.

It is interesting to note that associated with the plant community stratification is also observed in animals community. Various insects, birds and other animals occupy different positions along the vertical space in the plant community. Different groups of birds may be found feeding and nesting near the ground, in the shrub and small tree foliage beneath the canopy and in the canopy itself. Different arthropod species occur at different levels from the canopy downwards to the herb layer and on the ground or below the ground surface.

In aquatic communities also there is elaborate stratification. The upper layer which is known as epilimnion is dominated by phytoplankton and is the site of photosynthesis, while the lower layer known as hypolimnion is the site of decomposition.

Physiognomy

Physiognomy or the life-form of a plant is the vegetative form of the plant body that is generally believed to be a heriditary adjustment to the environment. The plants that shows the same general vegetative features belong to the same life-form irrespective of their systematic position in the plant families. Certain major life-forms have been recognised by man even in the past ages of history, such as : (1) herbs, (2) shrubs, (3) trees, (4) lianas, (5) grasses, (6) mosses etc.

Humboldt (1805) originally put forth the concept of life-form. He attempted to group vegetation types on the basis of external form and appearance, considering that there is relation between environment and the major external forms of plant species. He named 15 groups of life-forms among plants in relation to landscapes :

1. The banana form : *Musa. Heliconia.*
2. The palm form : *Cocos, Mauritia.*
3. The tree fern form : *Alsophila, Cyathea.*
4. The aroid form : *Arum, Pothos.*
5. The conifer form : *Taxus, Pinus, Picea.*
6. All the sharp leaf form : *Araucaria, Juniperus.*
7. The tamarisk form : *Mimosa, Porlieria.*
8. The mallow form : *Sterculia, Hibiscus.*
9. The liana form : *Vitis, Bauhinia.*
10. The orchid form : *Epidendrum, Serapias.*
11. The cactus form : *Opuntia, Cereus.*
12. The casuarina form : *Casurine, Equisetum.*

13. The grass form : *Andropogon, Panicum.*
14. The moss form : *Bryum, Sphagnum.*
15. The lichen form : *Cladonia, Parmelia.*

Following Humboldt's work many systems of life-form classification have been introduced with or without the presumption of a direct relation between vegetative form and environment and systematic understanding. Grisebach (1884) described 60 vegetative forms and his classification like that of Humboldt concerns the remarkable plant types which decide the physignomy and help in differentiating the important floristic regions of the world. After these appeared the systems of Drude (1897, 1913), Kraus (1891), Pound and Clements (1898), Warming (1909), DuReitz (1931), Raunkiaer (1909, 1911, 1934). The only system that has presently received worldwide use is that of Raunkiaer. Raunkiaer's system is based on the principle of the protection of bud during the unfavourable season or seasons is the most compact and consistent probably because he has given definite names to practically every type. The broad principles laid down by Raunkiaer in his work are as follows :

1. Plants have different ecological amplitudes or tolerances.
2. In a plants successful existance it makes an automatic physiological integration of the total environment.
3. There is often a correlation between morphology and adaptation.

In addition to these assumptions Raunkiaer employed three guiding principles in his selection of life-forms characteristics for the recognition and classification of relationship between plant life-form and climate :

(*a*) The characters used must be structural and essential, that is, they must represent important morphological adaptations;

(*b*) The characters must be sufficiently obvious so that one can see nature to which life-form of a plant belongs; and

(*c*) The life-form collectively must constitute a homogeneous system. It is this last requirement, especially that has made Raunkiaer's system usable and popular.

According to Raunkiaer it is the duration and number of unfavourable periods which effect the physiognomy of vegatation, and thus the apparent morphological features of the plant species growing in different areas can be correlated with the environment in which they grow and therefore can be used as indices of the environment. On the basis of this idea Raunkiaer recognised five life-form classes according to the degree of protection afforded to the perennating buds of plant species : (1) Phenerophytes, (2) Chamaephytes, (3) Hemicryptophytes, (4) Cryptophytes, and (5) Therophytes. Further divisions have been proposed by Fassett (1930), DuRietz (1931), Braun-Blanquet (1932), Dansereau (1945), Oosting (1956), and Aubreville (1963).

Phanerophytes

This group of life-form corresponds to woody trees and shrubs in which the perennating buds are located on the aerial shoots. Depending upon the height they are sub-divided into :

Megaphenerophytes (Mg)—Trees which are over 30 meters.

Mesophenerophytes (Ms)—Trees which are between 8-30 meters.

Microphenerophytes (Mc)—Trees and shrubs between 2-8 meters.

Nanophenerophytes (N)—Woody plants between 25 cm - 2 meters.

Each of the above sub-classes of phenerophytes may be further sub-divided into various groups on the basis of buds, that is, whether they are provided with bud-scale or not, and whether the plants are evergreen or deciduous.

In the tropical flora the phenerophytes comprise the major life-form of plant species, and some species that grow here cannot be grouped under any of the above categories and therefore they have to be given special treatment, for example :

Lianas (L) - Luxurient woody climbers with stems of anomalous structures - *Bouganvillia, hiptage* etc.

Parasites (P) - Plants which live at the cost of another plant (host) - *Striga, Orobanche, Cuscuta* etc.

Succulents (S) - Juicy plants - members of the families Cactaceae, Euphorbiaceae, Sterculiaceae.

Epiphytes (E) - Plants which grow on another plant, but not parasitically - members of Bromiliaceae, Orchidaceae etc.

Chamaephytes

This group includes woody or semi-woody perennials which are of the nature of a low shrub, and the height above the ground surface is less than 25 cm. The perennating buds are located above the ground surface, but they often get protection from the fallen leaves or by snow cover in certain types of climates. Four principle sub-divisions have been recognised :

Semishrubs (Ss) - A woody perennial less than 25 cm in height e.g. *Alhangi camilorum.*

Decumbent (Dc) - Plants in which stem after tailing on the ground for some distance tends to rise at the apex e.g. *Tridex procumbens.*

Stolon (St) - Plants in which a slender lateral branch originating from the base of the stem and bends down on or into the ground and grows horizontally outwards for a shorter or longer distance. It is often provided with nodes and internodes. Their ends emerge out of the ground and develop into a new plant e.g. *Passiflora, Tecoma* etc.

Cushion (Cu) - Plants in which shoot system is much branched and densely packed to form hemispherical cushions e.g. *Raoulia, Silene acaulis* etc.

Hemicryptophytes

The perennating buds are located at ground level, all above ground parts dying back at the onset of unfavourable conditions e.g., grasses and herbs. The three principle subdivisions are :

Rosette (R) - The leaves are restricted to a rosette at the base of the aerial shoot, for example *Bellis perennis, Taraxcum officinale* etc.

Subrosette (Sr) - The best developed leaves form a rosette at the base of the aerial shoot, but some leaves are also present on the aerial stems e.g. *Ajuga reptens.*

Non-rosette (Nr) - Lowermost leaves on the stem are perfactly developed than the upper ones e.g. R *Rubus idaeus.*

Cryptophytes

Types included under this life-form have their perennating buds beneath the soil or in water or in soil under water. Subdivisions are :

Geophytes (Ge) - With underground rhizome, bulb, corn, stem tuber etc., which perecennate during unfavourable conditions.

Hydrophytes (Hy) - Plants with their perennating buds under water, and with their leaves submerged or floating. The buds may occur on rhizome as in *Nuphar, Nymphaea*, or winter buds may become detached from the plant and sink to the bottom of water for example *Potamogeton obtosifolium, P. pusilus* etc.

Helophytes (He) - Those plants which have their perennating buds in soil or in mud below water level with aerial shoots above water level e.g. *Typha, Alisma.*

Therophytes

Annual species which complete a life-history from seed to seed during the favourable season of the year. Their life-span can be as short as a few weeks, and they are characteristic of desert regions and cultivated soils where the interference of man protects them from their natural competitors, and this is to be considered as the ultimate type of life-form developed in the plant kingdom (Figure 44).

Raunkiaer attempted to define the main plant climates of the earth according to the percentages of the various life-forms. Thus, he divided the vegetation of the earth into the following main plant climates :

1. Phanerophytic climate in the tropics.
2. Therophytic climate in the deserts.
3. Hemicryptophytic climate in the cool temperate zone.
4. Chamaephytic climate with a fair proportion of geophytes in cold areas.

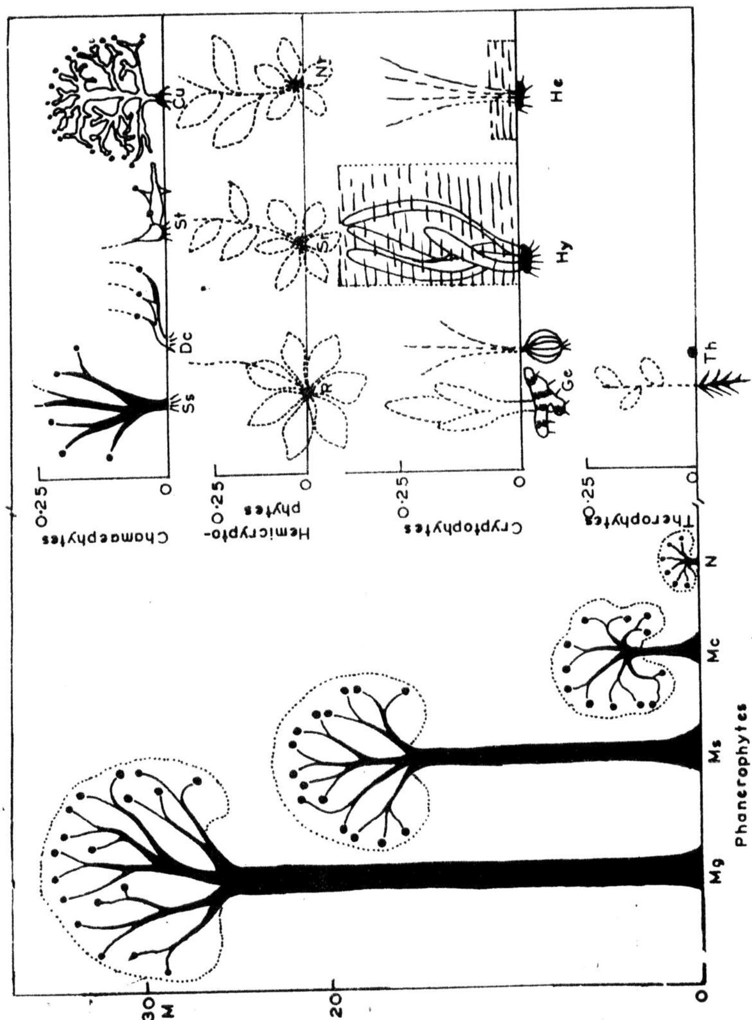

Fig. 44 : Life-forms in plant Kingdom

Application of Raunkiaer's system to India

A review of the relationship between different climatic zones of India and the phytoclimate on the basis of life-form have been presented in Table 49.

In the arid regions of India like the North-western parts of Rajasthan, characterised by rainfall under 250 mm and 11-12 months of dryness, the percentage of therophytes is 40 to 50 as against 13 percent in the normal spectrum; the plant climate is therophytic (Das and Sarup, 1951; Agarwal, 1975).

In the northern semiarid region of India which is contiguous with the Indian desert, and extends in the eastern and southern Rajasthan, Punjab and northern Gujarat. Climatically, this semi arid tract receives an annual rainfall of 250-700 mm spread over 2-3 months, leaving 9-10 months totally dry. The semi-arid zone of southern India comprise the Deccan plataeu and parts of the districts of Coimbatore, Ramnathpuram and Tirunelveli of the Karnataka state. In this zone dry season lasts for 8 months and annual precipitation is of the order of 500 to 700 mm. The spectra of both these zones results as therophytic-chamaephytic because in these regions the percentages of the two life-forms exceeds the corresponding figures in the normal spectrum (Agarwal, 1974; Siddiki, 1972, Rao, 1968; Meher-Homji, 1964). The higher percentage of phenerophytes in the northern semiarid regions as compared to the western arid regions are in consonance with the drier nature of the western arid region than its eastern counterpart. Similarly, higher percentage of phenerophytes in the northern semiarid regions as compared to the southern one are in confirmity that the former is of drier nature than the later (Meher-Homji, 1960, 1963; Meher-Homji and Misra, 1973).

The region of Poona with 7 dry months indicates phenerophytic - chamaephytic phytoclimate and the Andhra region with a rainfall of 750-1200 mm and 5-6 dry months presents a phenerophytic plant climate (Ferriera, 1940). North Kanara with 7 dry months and 4000 mm rainfall presents a phenerophytic plant climate (Arora, 1966).

Table 49. Life-form spectra of some of the investigated localities in India.

Region	Rain fall mm	Dry months	Ph	Ch	H	Cr	Th	Plant Climate	Author
World			46	9	26	6	13	Normal	Raunkiaer(1934)
Thar desert	73	11-12	33.2	18.9	2.2	6.2	39.5	Therophytic	Agarwal (1975)
Mount Abu	1860	7-8	50.1	2.4	2.4	-	46.7	Therophytic	Sarup (1952)
Alwar	667	7-8	33.0	4.3	7.6	3.4	51.7	Therophytic	Vyas (1962)
Gogunda & Prasad	624	7-8	40.8	2.8	3.5	10.7	42.5	Thero-Cryptophytic	Agarwal (1974)
Gorakhpur	834	6-8	43.0	4.4	7.4	12.4	32.6	Thero-Cryptophytic	Siddiki (1972)
Karamnasa	1040	8	40.0	6	1	10	43	Thero-Cryptophytic	Rao (1968)
Allahabad	1060	7	38.0	9.2	3.4	7.8	41.6	Thero-Cryptophytic	Shrivastava (1944)
Lucknow		7-8	25.8	3.3	10.2	8.5	52.2	Thero-Cryptophytic	Trivedi & Sharma (1965)
Poona	790	7	45	20	4	3	26	Therophytic	Lakshmanan (1962)
Andhra	1200	5-6	61	12	4	4	18	Phanerophytic	Ferreira (1940)
North Kanara	4000	7	46.3	14.8	4	11.1	22	phanerophytic	Arora (1966)
Karnataka	670	5	52	31	2	-	12	Chamaephytic	Ferreira (1941)
Tamilnadu	1260	6	43	23	9	5	18	Chamaephytic	Bharucha and Ferreira (1941)
Lolabvalley	660	2	32.5	17.2	17.6	24.2	7.9	Geo-Chamaephytic	Wali (1966)
Bhaderwah	1117	2	32.1	23.9	17.5	6.6	11	Geo-Chamaephytic	Kaul & Sarin (1976)

The phytoclimate of the Lolab velley in Kashmir (Wali, 1966), Bhaderwah in Kashmir (Kaul and Sarin, 1976), is geo-chamaephytic and this is in concordance with the classical concept of Raunkiaer's system : prevalence of chamaephytes with a high proportion of geophytes in cool climates.

The biological spectra drawn for different altitudinal zones of Yusmarg, Kashmir (Gupta and Kachroo, 1983) show an interesting feature (Table 50). Between 2400 m and 2650 m the biological spectrum indicates a geotherophytic type of climate as compared to that of Yusmarg forest as a whole. Therophytes and geophytes are more than double of those in the normal spectrum while there is an increase in the number of chamaephytes (16.7% as against 9% in the normal). The altitudinal zone between 2650-2900 m exhibits only a minor change in the percentage of life-forms. Geophytes and chamaephytes start increasing while therophytes show a decrease. Between 2900-3150 m geophytes constitutes the highest class and nanophenerophytes and phenerophytes also increase in numbers. There is a considerable change in the life-form composition in the altitude zone extending from 3150-3400 m. Hemicryptophytes are dominant, accounting for 38.6% of the flora as against 26.0% in the normal, while the percentage of nanophenerophyte and phenerophytes is reduced. Thus in this zone the phytoclimate is geo-chamaephytic-hemicryptophytic. Above 3400 m the therophytes decline (10.7%) as compared to the normal spectrum (13%). The chamaephytes show a five-fold increase as compared with the normal spectrum. Nanophenerophytes show the highest percentage as compared to the other altitudinal zones. Thus the plant climate of this zone is chamaephytic.

Table 50. Biological spectrum (% of all life-form) of various altitudinal zones in Yusmarg, Kashmir (Gupta and Kachroo, 1983)

Altitudinal zone (m)	No of spp	Th	HH	G	H	Ch	N	Ph	L	E	P
2400-2650	145	32.8	0.9	25.4	12.4	16.7	4.6	4.4	0.9	0.5	0.8
2650-2900	168	30.4	0.8	26.2	13.4	18.8	4.2	6.0	0.8	0.5	0.7
2900-3150	135	25.2	0.9	28.1	18.6	12.2	6.4	6.8	0.9	0.7	0.8
3150-3400	72	10.7	0.9	14.9	38.6	28.4	3.8	2.6	-	-	-
Above 3400	42	6.2	-	10.2	26.4	46.4	10.4	-	-	-	-

Meher-Homji (1960) pointed out some abnormalities. For example, the spectra of Poona, Karnataka and Tamilnadu show the same phytoclimate - phanerophytic - Chamaephytic (Ferreira, 1940; Bharucha and Ferreira, 1941), from climatic point of view Tamilnadu receives 1260 mm rainfall in 6 months whereas Karnataka and Poona 760 and 790 mm respectively in 7 and 5 months. Similarly, Mount Abu (Sarup, 1952), Alwar (Vyas, 1962), Allahabad (Srivastava, 1944), Vihar lake forest, Bombay (Lakshmanan, 1962), and Karamnasa in Varanasi district (Rao, 1968) show high percentage of therophytes. These studies have led to the recognition that where the biotic influence is very active as around the urban centres, the spectrum do not reflect the true phytoclimate.

When the spectrra are applied to particular plant communities which are only the fragments of total vegetation, the results are expected to bring out only the operative environmental factor and not the climate (Bharucha and Dave, 1944; Pandeya, 1954, 1964; Shah, 1956; Jindal, 1956; ansari, 1956; Chaphekar, 1967), Pandeya (1954, 1964) employed the biological spectra to estimate the intensity of grazing in the grasslands of Sagar, he observed :

(*a*) the life-forms of each association are maintained by the extent of grazing,

(*b*) the percentage of geophytes and therophytes are about four times higher than in the normal spectrum,

(*c*) the highly therophytic character of the grassland is not only due to the periodic climate but also due to grazing which maintains the vegetation open for further invasion of animals.

Raunkiaer asserted that the changes in the flora of a country never affected the proportion of the life-forms in his biological spectrum as : (1) spectrum of the naturalised species always had the centre of gravity in the same part of the spectrum as the indegenous flora, and (2) that even eradication of all the individual trees and shrubs in Denmark could not change the dominance of the hemicryptophytes in the biological spectrum of Denmark. As regards history, Raunkiaer said that the immigration of new species would alter the spectrum. Lastly, concerning the importance of soils, his biological spectrum were founded on the total flora of certain region and not upon the flora of certain regional communities. The above inference does not offer satisfactory explanation, since the cover provided by the upper strata influences the vegetation on the lower layers greatly and determines the composition of its life-forms to some extent (Kaul and Sarin, 1976). Wali (1966), and Kaul and Sarin (1976) pointed out that the physiognomic dominance of coniferous forest was not brought out in the spectrum as the trees are fewer in number than the herbaceous and shrubby elements. Similarly, Agarwal (1975) pointed out that dominance of shrubs was not brought out in the spectrum as they were fewer in number than the herbs. Therefore, it will be better if due importance is given to the percentage cover of each life-form in a community or an area and not to base the spectra only on the number of species of a life-form. Raunkiaer's life-form spectrum is most useful as an ecological descriptive device when the number of individuals as well as the number of species are considered (Cain, 1945; Carles, 1948; Stern and Buell, 1951; Misra and Puri, 1954).

Dispersion and Sociability

The dispersion refers to the distribution of the individuals in the horizontal space. This distribution may be uniform, random, or clumped (grouped) as shown in Figure 45. The sociability expresses the relation of individuals to each other and indicates the closeness between individuals. Braun-Blanquet (1932) recognised five arbitrary cetegories of sociability :

S 1 - plants growing singly.

S 2 - plants growing in small groups.

S 3 - plants occurring in small patches.

S 4 - plants forming large patches.

S 5 - plants occurring in essentially continuous populations.

In a natural community, the individuals of a species tend to aggregate because the offsprings are more numerous near the parents than elsewhere. In other words, they are clumped or overdispersed. Organisms under some conditions are uniformly spaced (under dispersed).

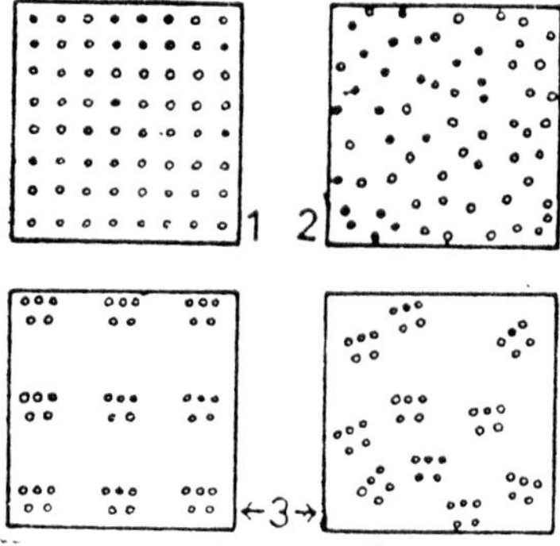

Fig. 45 : These possible ways of dispersion in a community : 1. Uniform, 2. Random, and 3. Clumped.

Vitality

It is concerned with the normal growth and reproduction ability of the species which helps it in maintaining its position in the community. The vigours indicates the state of health. The vigour of a species in field may indicate the end or the beginning of a new stage of succession, a change in soil moisture, attack by parasites, effect of grazing etc. Various authors have recognised three to five categories of vitality. These are :

V 1 - plants which germinate but die soon without reproducing,

V 2 - plants which inger after germination but cannot reproduce,

V 3 - plants which reproduce but only vegetatively,

V 4 - plants which reproduce sexually but rather feebly,

V 5 - plants which reproduce very well sexually.

Association of Species

It is the growing together of two or more species in close proximity to one another as a rather regular occurrence; for example *Boswellia serrata* and *Dandropthe falcata* in the Aravalli mountain of south-eastern Rajasthan. Association of species may be brought about by the similarity in ecological amplitudes of two or more species; similarity in geographical ranges; differences in life-forms so that extensive competition can be avoided; dependance of one species upon another for shade, or for food as parasites; dependance for protection from grazing. Association may be so pronounced that a certain species may indicate the presence of other species in the stand.

When environmental conditions change, the species that are associated will vary. A species growing as a dominant in one stand usually has different associates when growing as a sub-dominant in another stand; for example *Butea monosperma* is associated with *Tectona grandis* is Prasad forest range, while with *Eugenia hayneana* in Gogunda forest range of Rajasthan. The presence or absence of certain species indicates severe competition, presence of disease, or change in one or more environmental factors.

A number of methods have been used to measure the degree of association. The association index, is obtained by dividing the number of random samples of a given stand in which a species (A) occurs in the number (h) of the samples in the same stand in which species (A) and (B) occur together, for example, if species A occurs in 40 sample areas and species B occurs together with A in 30 of these sample areas, the association index of species A will be :

$$AiA = \frac{A + B}{A} = \frac{30}{40} = 0.75$$

Frequency

It is concerned with the homogeneity of the occurrence of the individuals of a species within an area. It is expressed as the percentage occurrence of individuals of a species in a number of observations.

The distribution of species is rarely uniform or regular in a area. Variation is caused by many influences, such as micro-habitat conditions of topography or soil, vegetative propogation, quantity and dispersal of seeds, time of invasion, grazing, activity of rodents, loss by insects or diseases. As a result, some areas may have higher frequency while others may have lower frequency. Such differences are more pronounced in irregular topography or where soil varies within short distances.

Many species having low cover or population density also rate low in frequency, but some may have high frequency because of their uniform distribution.

Raunkiaer (1934) was the first to classify the species in a community into five frequency classes as follows :

Class A - Species with frequency from 1-20 per cent.

Class B - Species with frequency from 21-40 per cent.

Class C - Species with frequency from 41-60 per cent.

Class D - Species with frequency from 61-80 per cent.

Class E - Species with frequency from 81-100 per cent.

The normal distribution of the frequency percentages, derived from such classification is expressed as A > B > C > D > E and has been named Raunkiaer's "Law of frequency". He found that in a homogeneously distributed vegetation, the proportion of the species will be 53, 14, 9, 8 and 16 per cent respectively in the five classes. The normal frequency diagram (Figure 46) is therefore a J-shaped curve. The frequency ratio is the result of the effects of the dominant species which, by their superior competitive capacity, prevent others from equalling them in frequency; but they cannot prevent many species from invading some of the species.

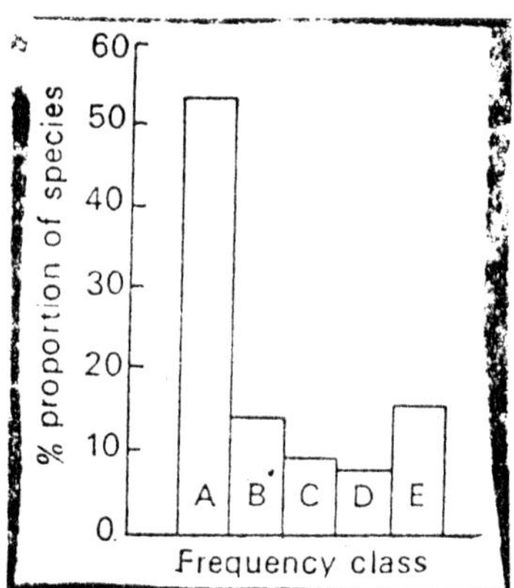

Fig. 46 : Normal frequency diagram.

Density

It indicates numerical strength of plants is a community and implies number of plants per unit area. The unit area may be square meter or hectare. Density values are significant because they show the relative importance of each species in the community, when they are of similar life-form and size. However, where plants are different such as grasses, forbs and dwarf shrubs, density alone in insignificant for comparison. With increasing density the competition stress increases which is reflected in poor growth and lower reproduction capacity of the plants.

Misra and Kothari (1971) studied the effect of population density on growth behaviour of *Crotalaria medicaginea* and observed that the dry matter content per plant increases up to a density of 200-250 per square meter, while at higher density the dry matter content decreases. The seed output per plant and reproductive capacity was observed in the plants growing in a density of 16 plants/meter^2, and compared with field data with a density of about 200 plants/m^2. It was observed that crowding decreases the individuals reproductive capacity and vegetative growth as well. Thus, intraspecific competition caused as a result of increasing density imposes an autoregulation of the population.

Agarwal (1980) observed that species density was maximum on the slopes irrespective of the aspect in Baldia, Parai and Prasad forest blocks. However, in Kharbar block it was maximum on the top of the hill (Table 51). The maximum value (8.16) was observed on the eastern slope in Prasad, and the minimum value (2.90) at the eastern foot hill in Kharbar block.

Table 51 : Species density 1/ha in different stands of Prasad forest range, Udaipur.

Locality	Aspect	Species density (1/ha) Foot	Slope	Top
Baldia	East	1610	800	520
	West	1810	1250	600
Kharbar	East	640	1130	1370
	West	790	960	1850
Parai	East	1880	1950	1800
	West	1340	1510	2250
Prasad	East	1350	1800	1350
	West	1300	1310	1300

The species density indicate a dissimilar trend in different stands. The value of density were maximum at the foot hills in Baldia, on the slopes in Prasad block, and on the top of hills in Kharbar block. In Parai block the interference of aspect has been observed, consequently the density on the eastern aspect was more on the foot hills and slopes while on the western aspect it was maximum on the hill top.

The low species density on the foot hill in spite of the favourable environmental factors like more moisture content, better soil formation, and richer nutritional status of the soil might be attributed to excessive biotic interference chiefly feeling and overgrazing due to poor protection, the lower values on the hill top might be explained on the basis of comparatively more exposure and lesser available moisture as compared to that of the middle zone of the hill.

Cover

Cover is the area occupied by a plant and can be expressed as either basal cover or canopy cover. The basal cover is the land area occupied by the cross-section of the stem, and canopy cover is the total land area under the canopy of a plant (Figure 47). The basal area can be only a small fraction of the total land area under a community, but the canopy cover of a single species may be several times the total land area because of overlapping canopies. Stratification in a forest causes increase in canopy cover by more than 150-200 per cent of the ground area. On the basis of foliage cover in a community the species are grouped into five classes :

Class A - Species with 5% foliage cover or less.

Class B - Species with 6-25% cover

Class C - Species with 26-50% cover

Class D - Species with 51-75% cover

Class E - Species with 76-100% cover

Plants with greater canopy cover are capable of intercepting more solar energy and cause deeper shade which influences the plant distribution of the ground flora.

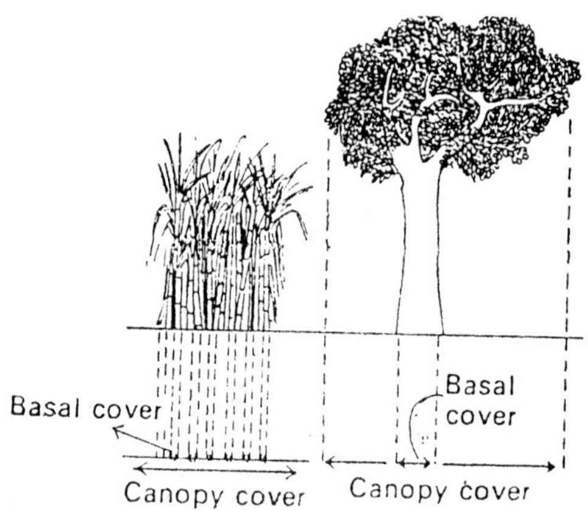

Fig. 47 : Relation between canopy cover and basal cover.

Basal area increases with age of the plant up to a certain extent, and indicates the dry matter accumulation in the shoot and their relative importance in the community. To foresters, basal area is an important criterion for evaluating the timber production in trees.

Agarwal (1971) established correlation between basal area and biomass production of *Butea monosperma, Boswella serrata, Lannea coromendelica* and *Tectona gradis* in the deciduous forests of Gogunda and Prasad in Udaipur area. Such correlations are very useful in studies as we can know the biomass of the tree without felling them and damaging the vegetation.

Height of Plants

The height of plants usually indicates their vigour, and a measure of the environmental conditions. It is sometimes difficult to secure accurate measurements because the height attained by the stem and leaves varies with individual plants of the same species growing under similar conditions.

Height measurements have given us many interesting and important results, for example, interspecific competition between *Artimissia tridentata* and range grasses, resulted in 4 inch tall *Agropyron cristatum*, but 7 inch tall where competition was lacking (Robertson, 1947).

The height of the plants and the depth of the root system show a relationship. For example, *Buchloe dactyloides* in Nebraska, with an average height of 5.5 inch had a working depth and maximum depth of the root system at 12 and 20 inch respectively (Weaver and Darland, 1947).

Weight of Plants

Weight is one the most important quantitative characteristics of plants. It is the total weight of food substances, protoplasm and other substances that constituties the forage value of the herbage. Most of the researches on weight has been done on the forage parts, but since the seventies some work has been done on the weight of stem and branches. Very meagre amount of work has been done on root system probably because of the difficulty of sampling and exploiting the entire root system.

Volume of Plants

While weight is a more important characteristic when plant growth or productivity is being considered, but the volume occupied by plant parts of great importance in understanding the structure of the vegetation. Except in forestry where the volume of trunks of trees is measured for yield of lumber, little work has been done on volume.

Fidility

It is a characteristic of certain species which show a degree of exclusiveness towards a particular association. A species with low fidility occurs in a number of communities, in contrast to those with high fidility which occurs in a few or in only one kind of community. This is because species differ in ecological amplitude, or in a capacity to grow in a wide range of ecological conditions, or because some species are able to associate with others to compete. Five classes of fidility have been distinguished by Braun-Blanquet (1932) and Pavillard (1928) :

Fidility 5 - *Exclusive* or *true*. Completely or almost completely confined to one community. They are present in 81 to 100 per cent of the quadrats.

Fidility 4 - *Selectives*. Species found frequently in a certain community but also rarely in other communities. They are present in 61-80 per cent of the quadrats.

Fidility 3 - *Preferential.* Species which occur in several kinds of communities but more abundantly in some. They are present in 41-60 per cent of the quadrats.

Fidility 2 - *Indifferent.* Species which occur in any community without showing preference to any particular kind of community. They are present in 21-40 per cent of the quadrats.

Fidility 1 - *Stranger* or *accidental.* They are more or less rare and accidental intruders from other communities or are relicts from earlier stages of succession. They are present in 1-20 per cent of the quadrats.

Poore (1955) is correct is saying that the only situation in which degrees of fidility can be properly assessed is that which exists when all the vegetation of a region has been described. Species with high fidility may have considerable value in indicating ecological conditions, for example, the restriction of certain species to particular soil conditions.

Vyas (1965) studied the vegetation of hills around Alwar, North-east Rajasthan and classified it into the fidility classes.

Presence

It refers as to how uniformly a species occurs in number of stands of the same community type, or in other words presence of a species is often expressed as the proportion of the communities that contain the species. Species may be classified into five classes of presence according to the percentages of stands in which they occur as follows :

Class 1 - found in less than 20 per cent of the stands

Class 2 - found in 21-40 per cent of the stands

Class 3 - found in 41-60 per cent of the stands

Class 4 - found in 61-80 per cent of the stands

Class 5 - found in 81-100 per cent of the stands

A fairly large number of species in Class 4 and 5 indicates floristic homogeneity in the community. A high degree of presence indicates that a species has wide ecological amplitude and is therefore capable of growing in various microhabitats. Similarity increases between community types as the number of constant species present in common becomes greater. Thus, it forms an important criterion in community classification.

Dominance

Species which exert controlling influence on the community by virtue of their size, numbers, production or other activities are called dominant. A relatively few species or species group exert the major controlling influence on the entire community, such a species or species groups which control the physical habitat and the community are known as the ecological dominants. In a grassland, dominant species are few, in tropical rain forest, many, for example, the grassland community in the college campus at Kota is dominated by *Bothrioclova pertusa*, which is a perennial grass. In a teak forest community the teak plants dominate overall other three species. Cover and population density are the chief qualities determining the dominance. The removal of dominants would result important changes in a community. Clements and Shelford (1939) characterised the ecological dominants as :

1. They receive the full impact of the climate.
2. They are best adjusted to the climate and habitat, therefore, they are more abundant in terms of density as well as coverage.
3. They react directly upon the climate and modify the water and light relations on land, and gas and salt contents in the sea.

In order to evaluate the concentration of dominance in a species within a community Simpson (1949) has devised a formula for calculating *index of dominance*, which shows importance of each species in relation to the community as a whole :

$$C = \sum \left(\frac{ni}{N}\right)$$

Where ni = Important value of each species or number of individuals or biomass of each species.

N = Total importance value of all the species or total number of plants or total biomass.

C = Index of dominance.

Whittaker (1965) calculated dominance index for forests on the basis of net primary production of each tree species, and for redwood forest dominated by single species, the index was 0.99, while at Smoky mountains where several species share dominance the value was 0.12 only. Thus, the dominance is more concentrated in one species the values are high and when several species contribute equally values of the index are low.

Species Diversity

This is a fundamental characteristic of plant communities, yet it is very little studied and its meanings understood. There is often an attempt to associate species diversity with the environment as is obvious in the familiar decrease in the number of species from the tropics—northwards. The relationship of species diversity to the environmental favourableness do not seem to be however simply correlated.

The older and more stable the community, is the more will be species diversity. Nature favours high species diversity while man prefers monoculture and brings about uniformity. Natural communities with high species diversity are less vulnerable to environmental vagaries, while man-made communities, for example, crop fields, orchards, nurseries etc. are open to greater damages by environmental hazards or epidemics and may be completely destroyed.

Species diversity is avery useful parameter for comparison of two communities especially to study the influence of biotic disturbance or to know the state of succession and stability in the community.

Species diversity is quantified by calculating *Species diversity index*, which is the ratio between the number of species and importance value or number or biomass or productivity of the individuals. Margalef (1968) used a formula for calculating Shanon Index of General Diversity as :

$$H = -\sum \left[\left(\frac{ni}{N}\right) \log \left(\frac{\frac{ni}{ni}}{N}\right) \right]$$

Where $\bar{H}$ = Shanon index of general diversity

ni = Importance value of relative dominance or biomass of each species

N = Total importance value or biomass of all the species.

Odum (1971) considered diversity aspect in relation to energy flow for the entire community through metabolic activities. In thermodynamic system, for instance, living organisms or community, there is a constant change untill it reaches stability through self-regulating mechanisms. The energy leaves the system through respiratory activities of the organisms and the accumulated energy remains in the form of biomass, which brings forth 'order' and stability in the structure of the community. In the community, the ratio of the total community respiration (R) to the total community biomass (B) can be considered as maintenance to structure ratio (R/B) or *ecological turnover* or as a thermodynamic order function. Considering the above ratio, which is also known as Schrodinger ratio, we find that the diversity increases with a decreases in the R/B ratio.

Higher diversity is found in communities which have longer food chains, that is, natural mature communities, where more cases of symbiosis and parasitism exist.

Man is rapidly changing the face of the Earth and is reducing natural diversity from the terrestrial and aquatic communities. Through pollution and resource depletion the diversity is diminishing and the ecosystems are being influenced. Man is influencing the biotic and abiotic environment by his activities, as a result of which many of the plant and animal species have become extinct and many are on the verge of extinction.

Due to differences in interspecific associations the biotic composition of two plant communities are never exactly alike. They may resemble in physiognomy, may have the same family, and show differences in species composition. In order to compare two communities which resemble each other in appearance we calculate the *index of similarity* as under :

$$S = \frac{2C}{A + B}$$

Where A = Number of species in sample or community A

B = Number of species in sample or community B

C = Number of species in sample common to A and B

S = Index of similarity

Index of dissimilarity = 1 – S.

Importance Value Index

The characters like frequency, density, cover or biomass cannot be used singly to show the relative position of a species. A single tree with a large cover and biomass is of little importance in a grassland community and likewise, grasses although numerous have little importance in a forest community. Curtis and McIntosh (1950) proposed an Importance value index (IVI), which is the sum of the relative values of the three quantitative characters :

IVI = Rn + Rd + RD

Where Rn = Relative frequency

Rd = Relative density

RD = Relative Dominance

Agarwal (1971) observed that IVI has permitted a clear-cut assessment of each of the 36, 36 and 26 tree species that were encountered in some 250, 231 and 231 quadrats at three altitudes between 250 and 725 meter at Prasad forest range. Within 250-400 meter range of altitude the two species that contributed most to the vegetation in the order of their importance value are *Tectona grandis* and *Butea monosperma*, which together contribute 1391.08 units of possible 3000 IVI units attributed to this region. Between 400-600 meter *Wrightia tomentosa* and *Lannea corromendelica* contributed 998.59 units of a possible 3000 units. While above 600 meters the most obvious change was the increased importance of *Boswellia serrata, Lannea corromendelica* and *Anogeissus latifolia* which approached their upper limit of growth and contributed most to the IVI. These three species contributed together 1797.96 units of a possible 3000 units.

The importance value index is purely a measure of the contribution of a species to that vegetation which is present, regardless of whether the ground is completely or sparsely covered.

Ecotone of the Concept of Edge Effect

An ecotome is a transition between two or more communities (Figure 48). It may be considered as a junction zone or tension belt. In extent the ecotone is usually narrow, that is, it occupies a smaller area as compared to the areas occupied by the adjoining communities. The communities of the ecotone area commonly contain many of the organisms which are characteristic and often restricted to the ecotone area itself. The frequency and density of some of the species is often greater in ecotone than in the adjoining communities. There is therefore, a tendency for increased variety and density at the community junction, and this is generally known as the *edge effect*.

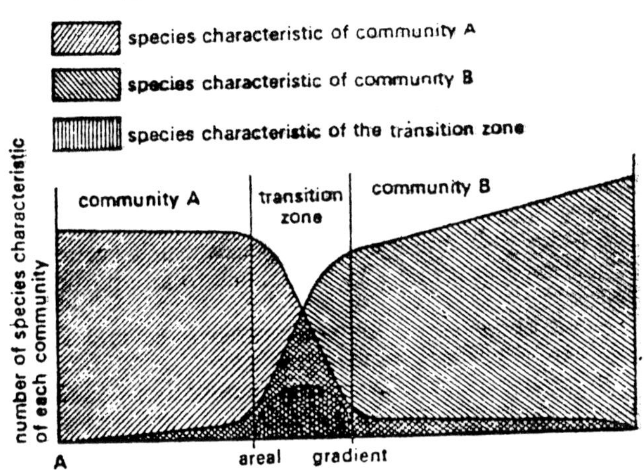

Fig. 48 : Ecotone between two communities (after Clapham, Jr. 1973)

A very illustrative example of the edge effect can be seen in the forest edges. A forest edge may be defined as an ecotone between forests and grass or shrub communities. The effect of man's settlement is genrally obvious in the edge effect, for example, if man sattles in a forest he reduces the forest to small scattered areas, grasslands, croplands and other such open habitats. If man sattles on the plains, he plants trees and creates a similar pattern. Some of the original organisms of the forests or the plains are able to survive in the man-made forests, where as others are specially adopted to the forest edge, such as weeds, birds, insects and mammals often increase in number and expand their ranges. Increase in density at ecotone is however, not a universal phenomenon, for example, the density of trees will be obviously less in a forest edge ecotone as compared to the forest itself.

Homologous Series of Twin Association

It is another interesting concept that has come into practice. It reflects on the structure and ecology of the community. The orginial idea about this concept was formulated by Hult as early as the year 1881. The characteristic basis of this idea is that we often find different communities may have one or more layers in common, while other layers in the vegetation are different, for example, in the desert rock formations of India with rigorous climate two to three species that could be stated in common to these formations are *Euphorbia nivulia*, *Grewia flavescens* and *Rhus mysorensis*. However, in the communities in different rocky areas of the desert, the other species will be different, for example, in Udaipur basin the rock formation are characterised by the presence of *Anogeissus latifolia*, *Tectona grandis* and *Boswellia serrata*. While in the Haroti region *Anogeissus pendula* is very common. In the western Rajasthan particularly Jodhpur, *Jizyphus nummularia* associations are frequent, while in the central basin of the arid Rajasthan more moisture loving species such as *Anogeissus* are frequent. The above formations or vegetation which possess certain species in common within them and others which are not common to each one of them are known as 'twin associations'.

Braun-Blanquet (1932) sites examples of occurrence of homologous series of twin associations in the desert formations of the world in general, for example, Californean desert and the African desert. The Thar desert of India, though widely separated geographically, yet develop plant associations of identical nature comprised of a shrub substratum, dominated by *Calligonum poligonoides* and *Tamarix diocia*. The ground cover is comprised of grasses and compositeae which are different in different areas. The alternating layer is generally comprised of succulent cacti or Chenopodaceous members such as *Atriplex*.

The utility of the concept of homologous series of twin association has been critically examined by Cain in a series of papers on the tropical and temperate flora of the world. He felt that changes in dominance in any layer of vegetation is of extreme significance. It should not however, be taken as a means of vegetational classification and naming of associations. On the other hand such changes in dominance in anyone layer should be able to provide us knowledge of local factors which are responsible for such changes. Sometimes, it may indicate the occurrence of particular pests or pathogen, at other times, the changes in the site or substratum may be indicated and these could be of significance in planning of agriculture and forestry.

Naming of The Biotic Community

The naming of the biotic community is generally done after the major dominant species or species groups, the major life-form of the species, the habitat occupied by the community, for example, aquatic, halophytes, lithophytes etc. The choice in naming of the community is generally done on the basis of the cleareast concept that the name can provide.

Vegetation Analysis

The most important decision made by an ecologist is that made, when he stops his car. In other words, the choice of a place to study is made likely to affect the results than anything the ecologist does subsequently.

The ecologist who wishes to select appropriate methods for the investigation of particular problems is faced with the choice of a variety of systems, the theoretical basis of which he may not be in a position fully to understand, and some of which are insufficiently tested.

The following working principles has been used in many instances dictated by different goals to be achieved :

(1) Successive approximation
(2) Individualistic concept
(3) Continuum concept
(4) Gradient analysis

Successive Approximation

The concept of successive approximation in descriptive ecology was proposed by Poore (1956). By this process, a series of inferences are drawn, each closer to the truth than the preceding.

By the continued examination of similarities and differences gradually build up classification of vegetation types and environmental situations. As his experience grows he will be able to distinguish more finely and draw more detailed inferences. No analysis can expect to be the ultimate truth, and all hypothesis are flexible and may be modified if new and contradictory evidence comes to light. The method of successive approximation would appear to be the most economical way of obtaining comprehensive understanding of vegetational variation (Poore, 1962).

Individualistic Concept

The seperate species within the community act very independently of each other, each species reacts to its own complex of environmental needs. Furthermore, each individual of a species is still very much of an individual, and to a degree act independently of others. Gleason (1926) firmly held that it was relatively un-integrated, its parts (the individual plants) still remained very much individual.

It is to be supposed that Gleason went to the other extreme, and said that "communities" do not exist. Individualistic can be used is dual sense : both for individual of a species, and for separate species of a community. Modern evidence lends little support to the individualistic concept or discrete units in vegetation.

Continuum Concept

The continuum concept is widly held to be a legitimate off-spring and development of Gleason's individualistic concept of plant associations. The continuum concept as laid down by Curtis (1955) accepts the continuity and discontinuity between major communities of Wisconsin.

The term 'continuum' implies an uniterrupted series of elements passing into one another, asserting that so sharp transitions are obvious between communities and that species composition changes gradually from place-to-place or time-to-time. The basic advantage of the continuum concept lies in the possibility of studying vegetation through a wide and comprehensive approach. Here the existance of descrete community type is not postulated *a priori* it does not deny the interactions among species within the plant communities either. It is objective and unbiased approach to the vegetation. Vegetation is not so integrated as a single biological organism, nor it is totally unintegrated phenomenon. A forest is an entity worth of study, as distinct from grassland; it is a whole that is somewhat more than its parts, even as a molecule is more than the sum of atoms comprising it.

Gradient Analysis

It is an alternative approach of classification of vegetation of a landscape. In general the elaboration of community types has been used in the descriptive phase of an ecological study, whereas gradient analysis has been used for causal analytical studies. Gradient analysis means that the distribution pattern and association of plant species in the form of communities is ruled by the environment, species populations and characteristics of communities. As a result we find repeatedly similar compositions which usually allow us to identify community type. The various communities appear like links in a chain. They show, however, a variable degree of intermixture which allows us to treat that the natural occurrence of the various communities in the form of a gradually changing continuum (Whittaker, 1967).

Both continuum concept and the gradient analysis approaches are valuable as working tools; neither one alone, however, can serve as the one and only principal basic philosophy to elaborate the nature of communities on a world-wide basis. The community type system should describe the general pattern of distribution and the continuum concept might explain the fine structure on a provincial level and would enable us to homologize community pattern from distinct and distant regions with no species in common.

14. *Community Dynamics*

Although a typical community maintain itself more or less in equilibrium with the prevailing environment, in nature this is hardly true. Communities are never stable, but dynamic changing more or less regularly over time and space. The plants and animals that grow in one area show a series of changes among themselves, which can be compared with the developmental changes an individual organism, that is, a living organism starts as a juvenile and reaches an adult form by passing through a series of stages. A population also undergoes the same changes, that is, it makes a juvenile appearance and develops over a period of time and gradually reaches the final form, characteristic of itself in that locality. In other words, the population shows a series of progressive reactions. It not one population that starts in one area, the different populations that make their beginning progressively reacts with one another and therefore some populations may be selected out in such a way resulting in the survival of only one community in that area. This process continues and successive communities develop one after another over the same area, until the terminal final community again becomes more or less stable for a period of time.

This orderly development of relatively definite sequence of communities over a period of time is known as *ecological succession*. Hult (1885) used the term succession for the first time, for the orderly changes in communities. Clements (1907, 1916) thereafter putforth various principles that governed the process of succession. He proposed the monclimax hypothesis of succession. During the later years certain other hypothesis were proposed by ecologists to explain the nature of climax com-munities : for example, polyclimax hypothesis by Braun-Blanquet (1932) and Tansely (1939); Climax pattern hypothesis by Whittaker (1953), McIntosh (1958) and Sellack (1960); and stored energy theory of information theory by Fosberg (1967) and Odum (1969).

Succession is the 'birth' of an ecosystem and subsequent 'ageing' process of its biotic and abiotic features. Odum (1971) preferred to call this orderly process as ecosystem development. He defined it in terms of the following three parameters : (1) It is an orderly process of community development that involves changes in species structure and community processes with time, it is reasonably directional and therefore, predictable. (2) It results from modification of the physical environment by the community, that is, succession is community controlled even though the physical environment determines the pattern, the rate of change, and often sets limits as to how far development can go.

It culminates in a stabilised ecosystem in which maximum biomass (or high information content) and symbiotic function between organisms are maintained per unit of 'available energy flow'.

Kinds of Succession

Succession takes place under a variety of habitats such as terrestrial, freshwater and marine; and the different sub-habitats and micro-habitats of the two major environments. Succession may be of the following types :

Primary succession

It is the process of species colonization and replacement in which the habitat is initially virtually free of life. That is, the process starts with bare rock, sand dune, river delta, glacial debris and end when climax is reached.

Secondary Succession

It is the process of change that occurs after an area under colonization has been cleared by whatsoever agency (like burning, grazing, clearing, felling of trees etc). It involves ecosystem diruption but not obliteration. In this situation, organic matter and some organisms from the original community will remain, thus, the successional process does not start from scratch. Consequently, secondary succession is more rapid than primary.

Autotrophic Succession

It involves predominance of green plants in the initial seral stages. Green plants are greater in quantity than animals and this takes place in amdium rich in inorganic substances.

Hetrotrophic Succession

It involves the predominance of heterotrophic organisms like fungi, bacteria and animals at the initial stages. Such a succession begins in a medium rich in organic matter.

Autogenic Succession

It involves replacement of one community by another due to modification of the environment by the communities themselves. There seems to be a very reasonable point in considering succession as a characteristic of the community itself, because the community acts on the habitat and tends to make the area or the habitat less favourable for itself but more favourable for other communities.

Allogenic Succession

It is the replacement of one community by another due to forces other than the effects of communities on the environment.

Induced Succession

Activities such as overgrazing, frequent scrapping, shifting cultivation or industrial pollution may cause succession. In agriculture, man makes an effort to stop the normal developmental series of a community and to maintain it at a certain point. He prevents the invasion of a field sown with some crop by other plants, that is, the so-called 'weeds'. Agricultural practices are retrogression of a stable state to a young state by man's deliberate action. In a natural stable state the community respiration balances community production and there is little left, which can be harvested. Therefore, man tries to control the succession in such a way that a managed steady state is maintained which is different from a natural steady state and wherefrom a good amount is harvested.

Retrogressive Succession

It means a return to simpler and less dense or even depauperate form of community from an advanced or climax community. Most of our natural forests are degrading into shrublands, savanna or even more depauperate desert like stands by the severity of grazing animals brought from the surrounding villages. Extensive removal of wood, leaves and twigs also leads to retrogressive succession.

Cyclic Succession

It refers to repeated occurrence of certain stages of succession, wherever there is an open condition created within a large community. In some forests, the juvenile plants of the dominant plants forming overstory are found in the understory at an arrested stage of growth. As soon as an old tree dies, the juvenile plants of the same species, on getting adequate light and space, suddenly show active growth and cover the open canopy space.

Yeaton (1978) observed that on bare desert *Larrea tridentata* invades and forms a shrubby cover, under its protection *Opuntia leptocaulis* succeeds to grow. When *Larrea* is eliminated, soil erosion is accelerated, surface soil is lost, the roots get exposed and *Opuntia* dies. Again the bare stage is reached, the cycle of same successional stages are repeated. (Figure 49).

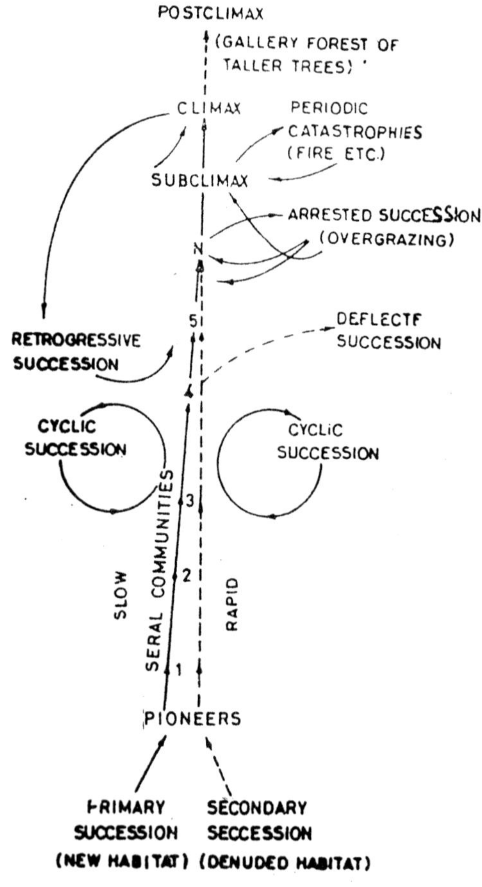

Fig. 49. Diagramatic representation of successions (after Ambasht, 1990).

Succession process

Succession is a universal process by which all communities develop from beginning or bare areas to maturity (Figure 50). It is directional, that is, it has a beginning and an end. The beginning is characterised by certain pioneers which gradually change as time passes and the end product is the final stable community, so to say the climax community of that area which is capable of self-perpetuation and which is in equilibrium with the physical environment. The complete process of succession involves the following sequential steps, which follow one another :

Nudation

Production or creation of bare areas are the intital causes of succession. Bare areas may be produced in a variety of ways, for example, *erosion* by the action of water, wind, gravity and ice. *Water* may produce a bare area that is entirely without plants along a stream bank. *Wind* brings about land slides especially in sandy soil resulting in bare areas. *Gravity* produces bare areas by pulling down material freed by weathering on the steep slopes of hills. *Glaciers*, *ice* and *snow* cause extensive deposits, presenting an extreme condition for rock succession. *Volcanoes* produce bare areas by deposits of lava, such bare areas are characteristic features of Yellowstone National Park. *Fire* caused by lightening in vegetation creates a bare area for colonization. Bare areas may occur as a result of increased water content of the substratum. *Man* is the most destructive of the biotic factors and destroys vegetation over the largest areas.

Invasion

It is the movement of one or more organism from one area into another, which is already occupied by other organisms. It is going on at all time and in all directions. The invasion includes following three steps : (1) *Plant migration* - the movement of plants from one area to another constitutes plant migration. Plants must migrate to that area where nudation has taken place. Of the various migration agents, for example, wind, water, gravity, glaciers, man and animals, man by migrating into practically his entire potential geographic range, conciously or accidently has made possible the rapid migration of many kinds of plants. For example, black mulbery was brought to India from Iran, **and peanut from** Portugal. Besides these a large number of **flowering trees, shrubs** and herbs were brought by English **travellars, investigators** and

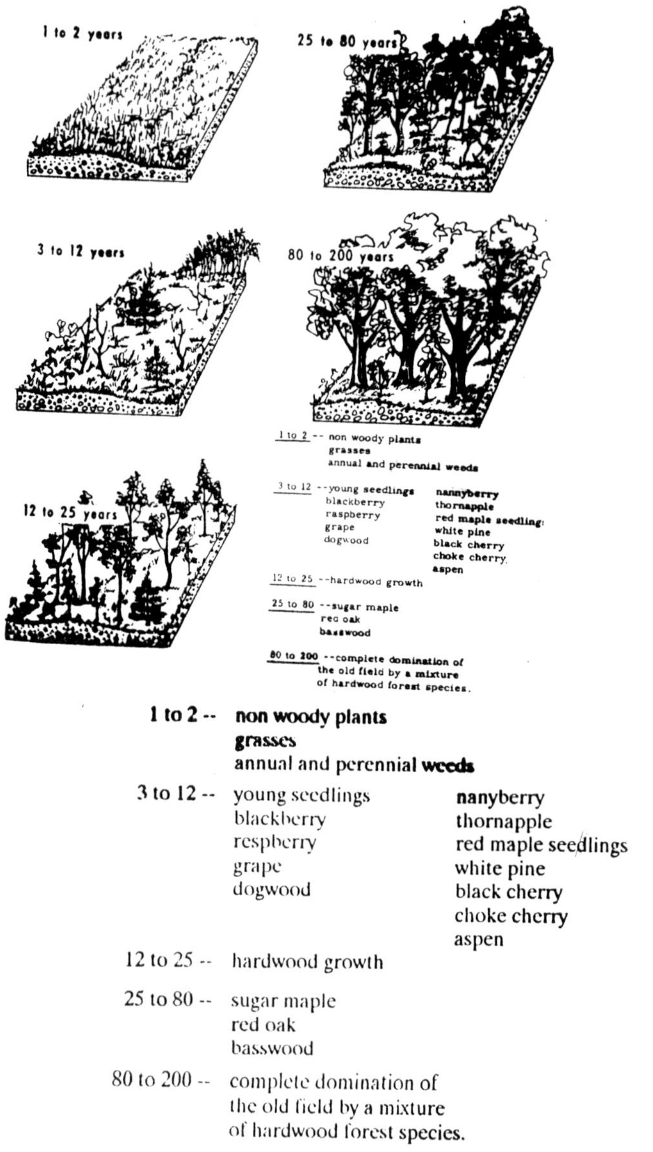

1 to 2 -- **non woody plants**
grasses
annual and perennial **weeds**

3 to 12 -- young seedlings; blackberry; respberry; grape; dogwood; **nanyberry**; thornapple; red maple seedlings; white pine; black cherry; choke cherry; aspen

12 to 25 -- hardwood growth

25 to 80 -- sugar maple
red oak
basswood

80 to 200 -- complete domination of
the old field by a mixture
of hardwood forest species.

Fig. 50. A generalised field succession **process.**

administrators. (2) *Ecasis*—Migration must be followed by ecasis before invasion can be said to have completed. Ecasis is the adjustment of the plant to a new home. It consists of three processes—germination, growth and reproduction. All plants cannot successfully establish themselves in a new area, few would be destroyed and the remaining would be able to establish themselves there. (3) *Aggregation*—After the ecasis of the migrators, the development of families and colonies is due to aggregation. The simplest form of aggregation is the grouping of young seedlings about the parent plant. Aggregation usually modifies the composition and structure of existing communities. Its influence is especially important in communities destroyed by fire, cultivation etc. Migration brings the species of each stage and aggregation followed by ecasis causes them to become characteristic of fominant.

Competition

It occurs between individuals of different species or between individuals of about the same size and having the same general environmental requirements such as water, light, gases etc. Competition tends to produce plant diversity in both structure and function in most vegetational types. Such diversity leads to higher yield of plant material per unit area of land surface, because of greater utilization of the available resources. There is no competition between a parasite and the host.

Reaction

Compitition results in reaction. When plants grow together and compete for the necessary factors, they profoundly react upon the place in which they grow. The area once fully lighted becomes more or less densely shaded. It it was wet, the large amount of water absorbed from the soil and lost through transpiration makes it drier. If it was dry, the accumulation of humus by the decay of dead roots, stems and leaves adds to the water retaining power of the soil. The area gradually becomes moister. The vegetation checks the wind movement near the soil where seedlings grow. Owing to shade, temperature becomes lower and the moisture content of the air in the vicinity of the plants increase. The soil becomes richer due to the accumulation of humus, bacteria, fungi, etc., and thus more favourable for plant growth. As a result, the vegetation is kept in a dynamic state and succession proceeds towards the climax.

Coaction

Coaction refers to the influence of two or more species or individuals upon each other. Coaction upon woody plants are less frequent, but they are not without importance. Plant parasite or animals may retard succession by destroying certain kinds of plants, grazing animals often produce this effect. Fire which may be either biotic or a climatic factor is largely destructive though it may be only retarding in its effect.

Stabilization

The progressive invasion typical of succession produces stabilization. The dominant species of the climax prevent the entrance into the community, of many new species of plants. The adjustment between the dominant species of a climax and environment approaches so near to perfection that the community is relatively stable.

Attainment of Climax

When the habitat becomes mesic, the dominant species of the climax succession appears and in time gain control over the environment. The climatic climax is the same throughout a climatic region. The course of development and its ultimate climax can be usually predicted.

Concept of Climax

The concept of climax has since long been a subject of much controversy and discussion. The following are the three theoretical approaches to the climax :

Monoclimax Theory

This theory was largely developed by Clements (1936). This theory recognises only one climax, determined solely by climate, no matter how great the variety of environmental conditions is at the start. All seral communities in a given region, if allowed sufficient time, would ultimately converge to a single climax. The whole landscape would be clothed with a uniform biotic community. All other communities than the climax are related to the climax by successional developmental and are recognised as subclimax, disclimax, preclimax, postclimax and so on. A *subclimax* is a stage in succession of forests just preceding the climatic climax community. The *disclimax* is the particular type of vegetation maintained in an area as a result of recurrent disturbance, chiefly biotic, thus preventing a successful establishment of climatic climax community. The *preclimax* is a vegetation of lower life-forms than the one adjacent to it and results from different edaphic conditions. The *postclimax* is a strip of vegetation of higher life-forms occurring within a climatic climax, for example, a forest along a stream in a grassland community.

The most objectionable point and subject of controversy about the climax has been the intimate relationship with the climate. In many conditions, in an area of uniform climate, it is common to observe different types of climax communities according to the soil, topography and other factors. Under such situations it would not be correct to consider only the climate as determining factor in climax. For example, Hult (1885, 1887) in Finland, long before the Clement's concept of climax, described as many as seven different habitat types under similar climatic conditions. Tansley (1935) too disagreeing with Clements put forth his concept that climax communites are infact controlled by more than one factor and not solely by climate.

Polyclimax Theory

This theory was developed by Tansley (1935). It considers that the climax vegetation of a region consists of not just one type but a mosaic of vegetational climaxes controlled by soil moisture, nutrients, topography, slope exposure, fire and animal activity. Accordingly, Tansley's concept became popularly known as polyclimax theory. The advocates of polyclimax theory ppreferred to call each stable community as a climax and described them with a prefix as edaphic climax, topographic climax, biotic climax and fire climax.

Climax Pattern Hypothesis

This theory was developed by Whittaker (1953). According to this theory, the composition, species structure and balance of a climax community are determined by the total environment of the ecosystem and not by one aspect, such as climate alone. Involved are the characteristics of each populations, their biotic inter-relationships, availability of biota to colonise the area, the chance dispersal of seeds and animals, the soil and the climate. The pattern of climax vegetation will change as the environment changes. Thus, the climax community presents a pattern of populations that corresponds with and changes with the pattern of environmental gradients, integrading to form ecoclines. The central and most widespread community in the pattern is the prevailing or climatic climax. It is the community that most clearly expresses the climate of the areas.

Information Theory

This theory was proposed by Fosberg (1965, 1967) and Odum (1969). It considered succession and climax in terms of ecosystem development. In autotrophic succession, diversity of species tends to increase with an increase in organic matter content and biomass supported by the available energy. Thus in a climax community, the available energy and biomass, which is called *information content,* increase. In contrast to it, in a heterotrophic succession occurs a gradual depletion of energy, because the rates of respiration always exceeds that of production. However, in an ecosystem, both the autotrophic and heterotrophic successions operates in a co-ordinate manner. The autotrophic individuals drives minerals elements from the soil and the atmosphere, while the heterotrophic individuals carry on the return of the nutrients to the soil and the atmosphere, through decomposition of complex dead organic matter. Thus, succession reaches a stage, the climax stage, when the amount of energy and nutrients received from the environment by the plants is again returned in more or less similar amount to the environment by decomposition through heterotrophs.

Examples of Succession

Ecological succession can be best illustrated by studying the following examples in different habitats :

Hydrosere (Aquatic Succession)

The various stages in a hydrosere are well studied in ponds, pools or lakes of a limited area (Fig. 51). In a man-made pond, where the soil has been dug out, there are little or no nutrients in the sub-

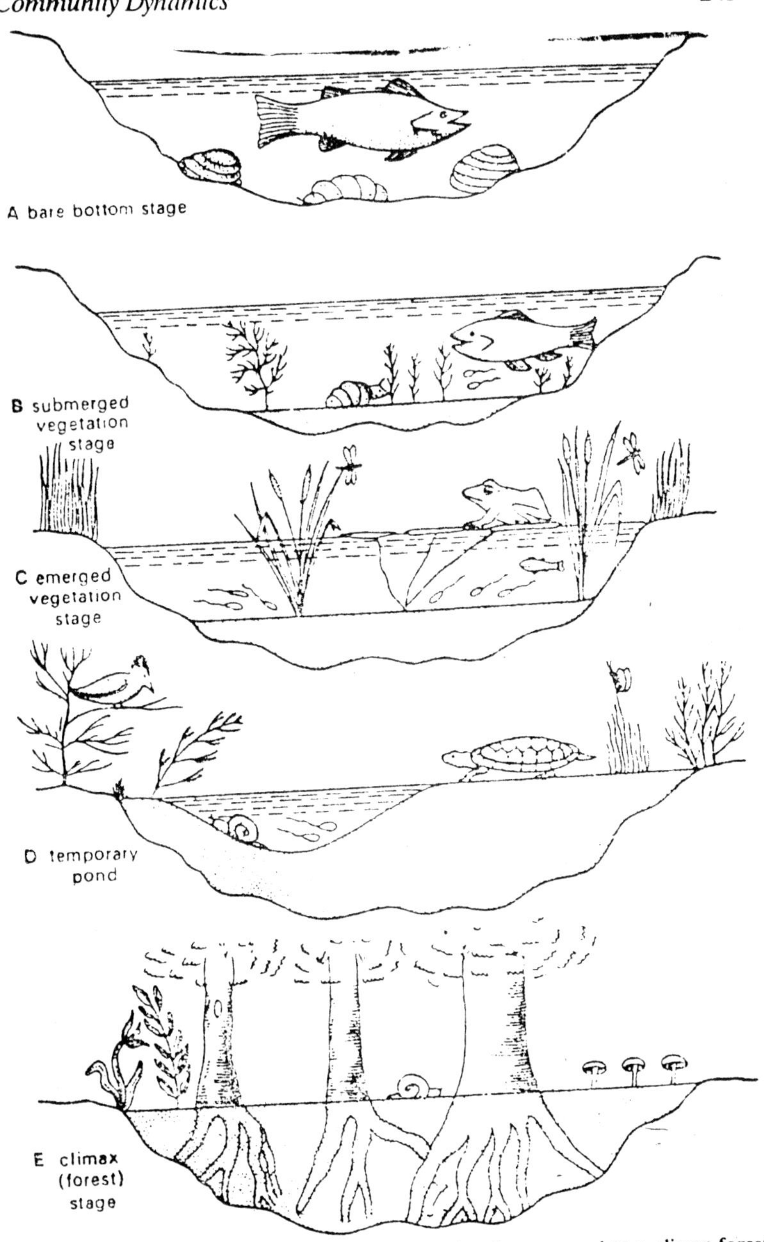

Fig. 51. Diagram showing ecological succession from a pond to a climax forest (after Cockrum and Mc Cauley, 1965).

stratum below the water and the water itself does not containany nutrients. Due to this fact, this stage is characterised by a bottom barren of plant and animal life. In such a new and virgin pond, hydrosere starts with the colonisation of some phytoplanktons which forms the pioneer plant community and finally terminates into a forest (the climax community). The complete process of hydrosere includes the following stages :

Phytoplankton stage

During the pioneer stage of succession algal spores may be brought by wind along with soil particles and deposited on the water. The unicellular and colonial phytoplanktonic forms begin to multiply and quickly become the pioneer colonisers. In the presence of traces of phosphorus in the medium, large blooms of blue-green algae appear as they could utilise atmospheric nitrogen. Later on filamentous algae like *Spirogyra* and *Oedogonium* appear.

Simultaneously, certain pioneer zooplanktons including protozoon like *Amoeba, Paramecium, Euglena* etc. also make their appearance.

With the death of phytoplanktons and zooplanktons, the population of decomposing organisms like bacteria and fungi increases in the pond mud. Decomposition results in release of minerals and enrichment of aquatic habitats.

Submerged stage

As a result of death and decomposition of planktons, and their mixing with the silt, brought from the surrounding land by rain water and by the wave action of pond water, there develops a soft mud at the bottom of the pond. This new habitat which tends to be a bit shallower and where light penetration may now occur easily becomes now suitable for the growth of rooted submerged hydrophytes like *Vallisneria, Ceratophyllum, Potamogeton Hydrilla, Utricularia* etc., on the shallow regions where light reaches the bottom in sufficient quantity. The death and decay of these plants further contribute to the enrichment of the substratum. The water level, also decreases making the pond more shallower. This new habitat now replaces these plants giving way to another type of plants which are of floating leaf type.

The deeper zones are then occupied by such species which are rooted in mud but whose leaves reach the water surface and float. *Trapa bispinosa, Nelumbo nucifera, Nymphaea stellata,* etc. are common examples of this type.

Some free floating species such as *Azolla, Lemna, Wolffia, Pistia, Salvinia* also become associated with the rooted plants, due to the availability of salts and other minerals in abundance. If *Eichhornia* happens to invade such ponds, it most rapidly spreads all over the sheet of water due to its high rate of productivity and rapid vegetative reproduction and aggressiveness. The water level now becomes very much decreased, making the pond much shallower. The decomposing organic matter formed due to the death of these plants brings about further buildup of the substratum. Thus, floating species sooner or later disappear from the area.

Reed-swamp stage

This is also known as *amphibious stage* as the plants are rooted but most part of their shoot remain exposed to air. Species of *Scirpus, Typha, Sagittaria, Phragmites, Cyperus, Fimbristylis* and *Polygonum* etc. are the chief plants of this stage. They have well developed rhizomes and form a very dense vegetation. The water level is by now very much reduced and finally becomes unsuitable for the growth of these amphibious species.

Sedge-meadow stage

Due to successive decrease in water level and further changes in the substratum species of some Cyperaceae and Gramineae such as *Carex, Juncus, Cyperus* and *Eleocharis* colonise the area. They form a mat-like vegetation towards the centre of the pond with the help of their much branched rhizomatous systems. As a result of high rate of transpiration, there is much rapid loss of water, and sooner or later the mud is exposed to air. Thus, mesic conditions approach the area and marshy vegetation disappears gradually.

Woodland stage

By the time of disappearance of marshy vegetation, soil becomes drier for most time of the year. This area is now invaded by terrestrial plants, which are some shrubs. By this time there is much accumulation of humus with rich flora of micro-organisms. Thus mineralisation of the soil favours the arrival of new tree species in the area.

Forest stage

This is the climax community. In tropical climates with heavy rainfall, there develops tropical rain forest, where as in the temperate regions, there develops mixed forest. In tropical regions and moderate rainfall, there develops tropical deciduous or monsoon forests. Since the particular type of trees can tolerate the environmental conditions they create, the forest cover which becomes stabilized.

Xerosere (Psammosere)

Under arid conditions, succession is a very slow process, mature soils are slow to develop and easy to destroy. Accumulation or organic matter is very slow and leaching processes are limited by the scanty precipitation. Because of the serious large scale destruction of the original vegetation, it is very difficult to form a clear notion of what the climax ought to be and its theoretical reconstruction is possible only after comprehensive investigations on the phytosociology of the arid regions. The existing vegetation on sandy plains and dunes are due to the extreme degradation of natural vegetation through cultivation, overgrazing, lopping and burning. Under such conditions it is extremely difficult to infer the hypothetical order of succession.

In arid zones, successional process is usually faster in the initial stage, but becomes very slow after the first few phases. Extensive ecological studies in western Rajasthan for the last two decades have led to build up the probable successional stages (Saxena, 1977) on various habitats, which are described as follows :

Hills

The denuded rocks, in initial stages are colonised by tough grasses and forbs e.g., *Oropetium thomaeum, Tragus biflorus, Enneopogon brachystachys, Melanocencharis jacquemontii, Lepidogathis trinervis* and *Blepharis sindica* (Fig. 52). With the accumulation of substratum shrub species such as *Euphorbia caducifolia, Grewia tenax, Zizyphus nummularia* get foothold.

Simultaneously, higher grass species such as ***Eleusine compressa*** start dominating. The shrub species pave the way for the tree species *e.g., Maytenus emerginatus, Morringa cancansis* and *Cordia gharaf.* If these species remain undisturbed, they appear to end up ultimately in the climax community to reach its climax with *Schima nervosum.* The slow buildup of the soil arrests the grassland community to the *Eleusine* stage only. In the higher rainfall zone, *Hackelochloa, Bothriochloa* and *Brachiaria* maintain the disclimax for ***Sehima nervosum*** type of grassland.

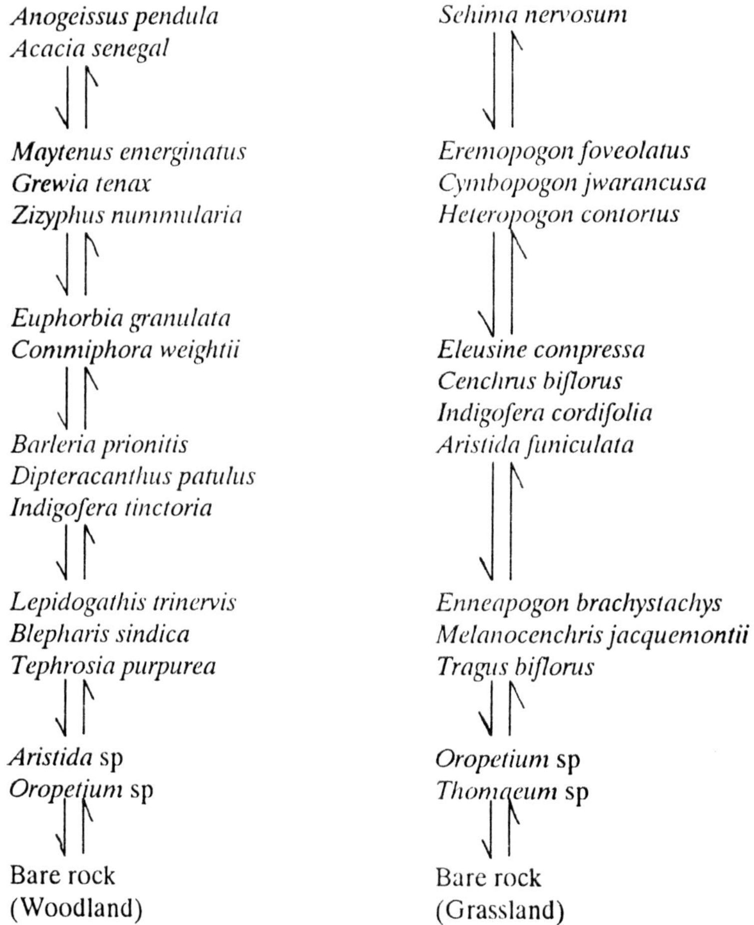

Fig. 52. Succession on rock in Indian arid zone.

Older Alluvial Plains

Flat alluvial plains with loamy sand to sandy loam soils get the invasion of *Aristida, Melanocenchris, Tephrosia, Indigofera* and *Crotalaria* in the early stage of colonisation. Under protection perennial grasses and undershrubs gain the upper hand in 3-4 years. If the area continues to be undisturbed spiny species e.g.*Zizyphus* and *Capparis* gather ascendency, which simultaneously culminates into the climax stage of Prosopis cineraria. (Fig. 53).

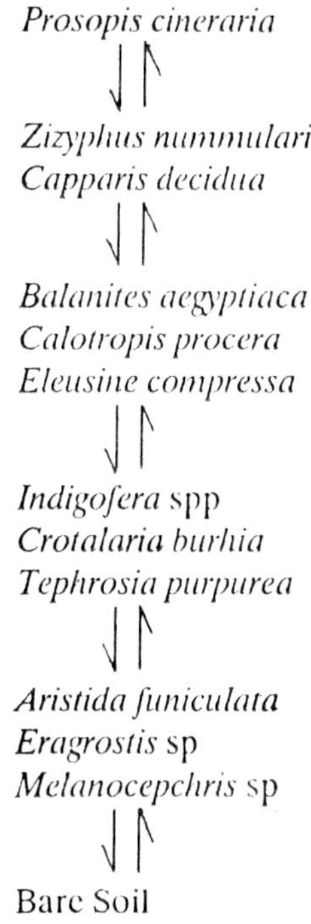

Fig. 53. Succession on sandy loam in Indian desert.

Lithosere

It is a typical xerosere and it occurs on bare rocks. Lithosere includes the following stages :

Crustose lichen stage

The rocks are dry and hard, fully exposed to sun, violent fluctuations in temperature and rapid changes in moisture content.

The soil is absent for penetration of roots and supply of nutrients. Such a habitat is unsuitable for most plants but for blue-green algae and lichens, the pioneer species. The blue-green algae such as *Scytonema* are found to adhere to the rock by their mucilaginous cell walls. These algae tolerate extremes of temperature and moisture contents can utilise atmospheric nitrogen. On the other hand in cooler climates crustose lichens such as *Rhizocarpon, Rinodina* and *Lecanora* are common pioneers. They produce some acids which bring about weathering of rocks. The dead organic matter of algae and lichens become mixed with the small particles of rocks to form the very thin layer of moist soil on the rocks. The crustose lichens are then replaced by foliose type of lichens.

Foliose lichen stage

They appear on the substratum partially build up by the crustose lichens. This community includes species like *Dermatocarpon, Umbilicaria* and *Parmelia,* which have large leaf like thalli. They can absorb and retain more moisture, are able to accumulate dust particles and produce more organic matter which help in the further buildup of the soil on the rock.

Moss stage

The development of this humus rich soil layer on rock surface favours the growth of certain xerophytic mosses such as *Grimmia, Tortula, Polytrichum, Bryum, Barbula, Funaria, Weissia, Hypnum* and *Fissidens.* These mosses grow and compete with the lichens and add more organic matter to the soil after their death and decay. The accumulated thick layer of soil retains greater amount of water and provides suitable habitat for the invasion of herbs, grasses and ferns.

Herb stage

This stage is characterised by plant species like *Aristida, Festuca, Poa, Sporobolus, Lindenbergia, Solidago, Adiantum, Apslenium, Cheilanthes, Actinopteris, Justicia, Tridex* etc. The weathering of the rock by physical, chemical and biological means provides more soil and nutrients. At this stage this herb layer is followed by shrub layer.

Shrub stage

It is characterised by small and large shrubs such as *Zizyphus, Capparis, Zygophyllum, Rhus* etc. in moderate climates and *Rubus, Fragaria, Physocarpus, Symphoricarpus* etc. in the cooler climates. The shrubs shadow the herbaceous vegetation and produce more organic matter and thus increase the amount of organic matter and consequent humidity in the soil. The species diversity is also increased. The shrubs are finally replaced by trees which form the climax community.

Forest stage

Forest climax community commense by invasion of certain xerophytic plants and this is determined by climate of the region. In dry climates, due to slow weathering of rocks, comparatively this layer of soil is formed which supports small trees like *Acacia, Prosopis, Balanites, Anogeissus, Boswellia* etc. In moist and wet climates dense climax forest develops.

Plagiosere

It refers to deflected succession, that is, the normal development of the vegetation is prevented by anthropogenic influences. However, the normal development is not completely checked. The anthropogenic influences may be in the form of grazing or fire or some other factors, for example, the succession towards a forest may be directed towards a grassland community by grazing. In the deflected succession in Marshland, Fenland and Bogland in England, there operates a master factor continuously and results in a deflected climax vegetation, that is, the climax communities that develop in the above localities are a deviation from the normal course of development and such climax communities as we find in Marshland, Fenland and Bogland are therefore plagioseres.

In Marshland the master factor that is in continuous operation is the soil which is mainly comprised of silt. In Fenland also the master factor involved is soil, but here the soil is formed of peat with considerable quantities of lime or other bases. In Bogland the master factors as again is soil which is formed of peat, but very poor in bases.

The plagiosere influences which bring about plagioseres result in vegetation which is very little truely natural to that area. Other outstanding examples of anthropogenic influences are cultivation, forest cutting etc. It is therefore reasonable to treat plagiosere as sub-climaxes which are brought about by the prevailing physical or biotic factors of the environment there.

15. *Ecosystem*

Living organisms and their non-living environment are insaparably inter-related and interact upon each other. Those organisms occupying the same area make up a biotic community, a naturally occurring assemblage of plants and animals that live in the same environment, are mutually sustaining and interdependent, and are constently utilizing and dissipating energy. The biotic (living) and abiotic (non-living) components in a given area interact so that a flow of energy leads to clearly defined trophic structure, biotic diversity and material cycles within the system is an ecological system or ecosystem.

The concept of ecosystem was put forth by A.G. Tansley (1935). It is a major ecological unit. It has both structure and function. The structure is related to species diversity. The function is related to flow of energy and cycling of materials through structural components. According to Woodbury (1954), ecosystem is a complex in which habitat, plants and animals are considered as one interacting unit, the materials and energy of one passing in and out of the others. According to E.P. Odum (1971), ecosystem is the basic functional unit of organisms and their environment, interacting with each other and with their components.

An ecosystem is a discrete structural-functional life sustaining system. The biotic and abiotic components in a habitat constitute an interacting environmental system in which inorganic constituents are synthesized into organic structure and through energy exchange processes life of various forms exists in the system. Such a dynamic environmental system is known as ecosystem (Misra, 1980).

Ecosystem Structure

Ecosystem has structure, and the structural components are four : The first is the *abiotic* (materials, elements and compounds of the environment, such as soil, water, air and mineral salts). The second is *autotrophic primary producers,* chiefly green plants, which synthesise inorganic constituents into organic structures through photosynthesis and solar energy is utilised in the process. The third is the *heterotrophic secondary producers,* consisting of herbivores, who feed on herbivorous animals. Green plants and herbivorous animals become so'urce of energy for the heterotrophic organisms. Animals of all types grow and add organic matter to their body weight in the form of complex organic compounds. The fourth are the *Decomposers* and *Transformers.* They are chiefly bacteria, fungi and a lot of microbes that break down the complex organic compounds. The transformers decompose simple organic compounds into the inorganic components and liberate them into the environment (Fig. 54). During the process of decomposition and transformation of organic molecules, the energy which kept the inorganic components bound together in the form of organic molecules gets liberated and dissipated in the environment as heat energy.

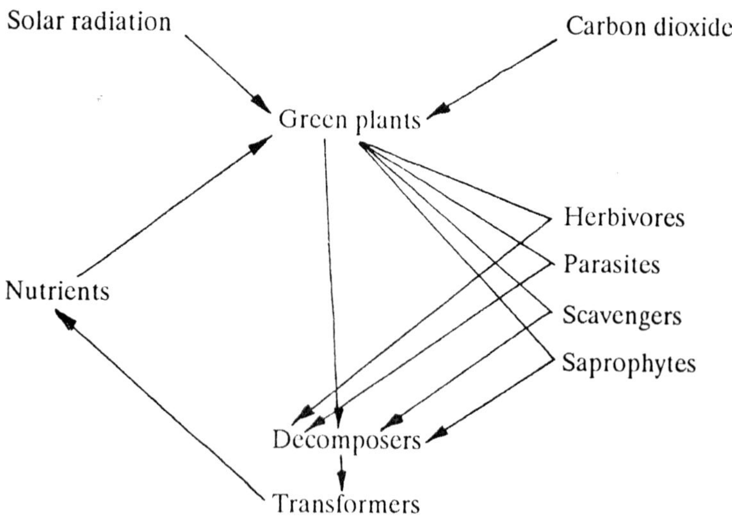

Fig. 54. Different components of ecosystem.

The producers and consumers in the ecosystem can be arranged into several feeding groups, each known as a trophic level. Each trophic level is composed of a number of kinds of organisms, which collectively makeup the species structure of the ecosystem.

The community is an aggregation of species. Among these only a few species are found in abundance either in numbers or in biomass; the majority are comparatively rare. In some communities these dominants may exert some influence over the other organisms. In the forest, for example, certain trees of the upper canopy influence the amount of light and moisture reaching the ground and the type of shelter offered to other plants and animals.

Occasionally an animal comprises the dominant mass of the community; this is the case for the coral reef; found in clear, warm waters of tropical and subtropical regions.

The diversity of species within a community reflects in part the diversity in the physical environment. The greater the variation in the environment, the greater are the species. Diversity is least in the simpler, less stable ecosystems such as Tundra and agricultural ecosystems. In tropical regions where climate is stable, many ecological communities have evolved to maturity. Within them have developed a large number of different species, non of them are really abundant. Within any community, diversity is greatest in small organisms. There are more kinds of herbs than there are trees, more insects than birds, more birds than mammals and so on. Because they are small, these organisms have become adopted to the conditions afforded by the small, diversified microclimates found within the community. Singh and Misra (1968) concluded that species diversity increase productive efficiency, while dominance makes the system stable, though less efficient for production. With an increasing species diversity at maturity of the ecosystem, the organic metabolites are also of different nature and their complexity too increases. Many of the organic substances are excreted into the environment as byproducts. Margalef (1957) has observed increase in the variety of plant pigment in aquatic plants, with advancing maturity of the system, and this situation leads to an increase in biochemical diversity, which is characteristic of mature ecosystems.

The relative abundance of various species within a community depends in part on the species themselves. Some are strictly opportunistic, common when conditions are highly favourable and scarce when this is not so. No equilibrium is established within the population.

All communities can be divided into more or less distinct layers, or strata, both on vertical and horizontal plane. Each vertical layer in the community is inhabited by its own more or less characteristic organisms. Although considerable interchange takes place between several strata, many highly mobile animals confine themselves to only a few layers, particularly during the breeding season. Occupants of vertical strata may change during the day or the season. Such changes reflects daily and seasonal variations of humidity, temperature, light, acidity, oxygen content of water and so on. Horizontal stratification or zonation is caused by differences in climate or edaphic conditions. This type of stratification is most common around ponds and lakes.

Stratification increases the number and variety of places in which the organisms can live in given area. Highly stratified communities generally contain a wider variety of animal life than those with only a few layers.

Ecosystem Function

More than just occupying space, each species and each individual in the community perform some function. What the organisms does — or its occupation in the community, is called its niche. Some species occupy a very broad ecological niche. They may feed on many kinds of food, part plant, part animals, or if herbivorous, they may feed on wider variety of plants. Since each kind feeds on preferred food plants, between them they utilize all forms of vegetation from ephemerals and annual herbs to perennial trees.

The function of the ecosystem is related to processes and events which either change the form of or move on energy and materials within biotic and abiotic components of the ecosystem. The processes may be categorised as under :

1. Metabolic processes of the organisms — assimilation, respiration, growth, reproduction, nutrient and water uptake, transpiration, nitrogen fixation, litter fall, decomposition and mineralization.
2. Energy flow at various trophic levels.
3. Biochemical cycle of materials and water in abiotic biotic-abiotic state

The *metabolic processes* are characteristic of living organisms. It involves *anabolism* **which** means chemical upbuilding of complex substances by protoplasm, and *catabolism* means disruptive process of chemical change in the organism. Some total of metabolic activities result in the ultimate function of the ecosystem, that is, to produce, or to reproduce, and this is measured quantitatively by the productivity of the functioning ecosystem. Environments are different in different areas. There are those which are conductive to high productivity and those which are conductive to low productivity. Some environments produce forests, others produce grasslands, plankton, still others produce nothing.

The green plants, photosynthetic and chemosynthetic bacteria synthesize inorganic materials and directly trap the light energy of sun and their biomass increases, the rate of energy storage is known as *primary production.* The other organisms like plant parasites, fungi, bacteria and animals do not have the power to fix the solar energy and they get their energy for the metabolic activities through the food. The rate of energy storage (increase in biomass) at the consumer level is known as *secondary production.*

Organic matter production on our planet through photosynthetic activity is approximately 155 x 10^9 tonnes/year (Lieth, 1972). A greater part of this gross production is respired, consumed or decomposed as litter and only a small amount of the organic matter remains as plant biomass. Terrestrial vegetation is more effective than the marine or aquatic vegetation in fixing the solar energy, and approximately 56 per cent of the total production in the biosphere is by land plants. Productivity of the terrestrial ecosystems too differs and the tropical forests have highest production while the deserts are the least productive (Table 52). Relative production values for different ecosystems are graphically indicated in Fig. 54.

Energy Flow

The structural characteristics of an ecosystem could be evaluated through a knowledge of trophic structure and ecological pyramids. Elton (1947) indicated that the food chain in any ecosystem cannot be more than five steps long, as when the green plants (step 1) is the primary trophic level which capture energy in

Table 52. Net primary productivity (Lieth, 1972)

Vegetation unit	Range g/m²/yr	Mean
Forest		
Tropical rain forest	1000 – 3500	2000
Rain green forest	600 – 3500	1500
Summer green forest	400 – 2500	1000
Woodland	200 – 1000	600
Tundra	100 – 400	140
Desert scrub	10 – 250	70
Grassland		
Tropical	200 – 2000	700
Temperate	100 – 1500	500
Desert		
Dry	0 – 10	3
Ice	0 – 1	0
Fresh Water		
Swamp and Marsh	800 – 4000	2000
Lake and stream	100 – 1500	500
Ocean		
Reef and estuaries	500 – 4000	2000
Open ocean	2 – 400	125
Total Earth		303

the functioning ecosystem. The captured energy at this trophic level and the rate at which it is captured and used for its maintenance, all determine the productivity of this trophic level. This trophic level or energy level is subjected to use or consumption by the next trophic level such as non-green plants which live on green plants, the herbivorous animals etc. (step 2), only 10 per cent of the food is assimilated in the biomass of the animals and the rest 90 per cent is either ejected as excreta or used up to provide energy for the metabolic activities of this trophic level. At step 3, when a herbivore is eaten by a carnivore only 10 per cent of the total biomass of the herbivore is ingested as food is assimilated in the biomass of the carnivore animal. The carnivore may be eaten by the top carnivore (step 4) when when dies becomes the food of the decomposers and transformers (step 5), which ultimately breakdown the food into its inorganic components. This kind of energy flow in a functioning ecosystem from one level to another is termed as the *food-chain.*

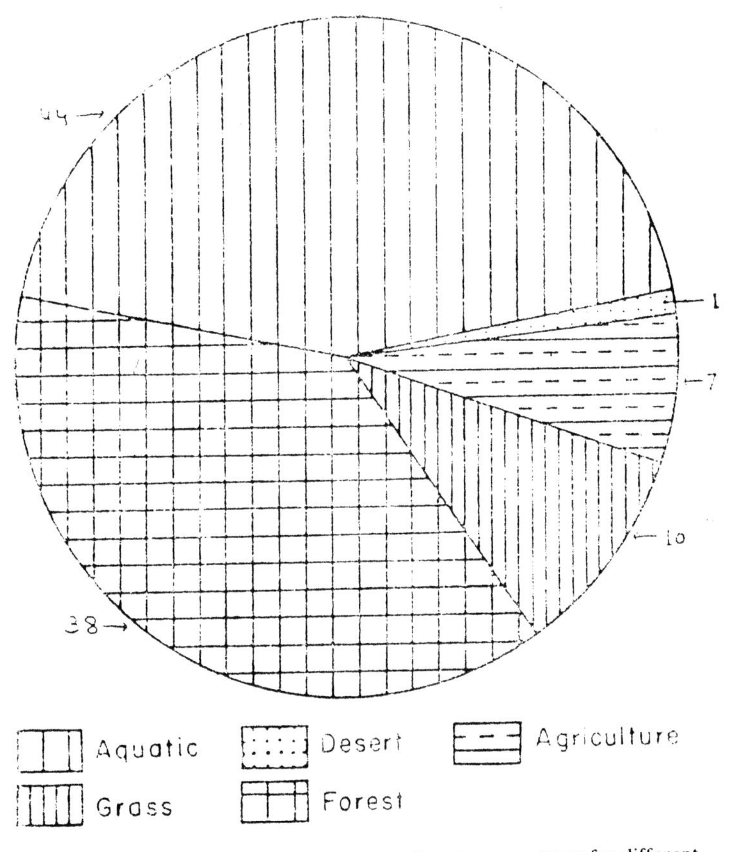

Fig. 54-A : Relative primary production values in percentages for different types of ecosystems (after K.C. Misra, 1980).

Trophic level one captures its energy from solar radiation and in the photosynthetic process the green plants expend some energy, further energy is expended by green plants in respiration. The energy thus expended in photosynthesis and respiration is getting out in the form of heat and therefore net plant production value of the trophic level one decreases. When we consider production in terms of energy — the expenditure of energy at otheretrophic levels is primarily in the form of respiration which is also getting out in the form of heat.

Thus we begin our flow of energy — originally from the sun — through the plants and animal communities. The energy flow in the ecosystem (Fig. 55) explains the energy capture and transfer as well as energy conversions that occur in an ecosystem.

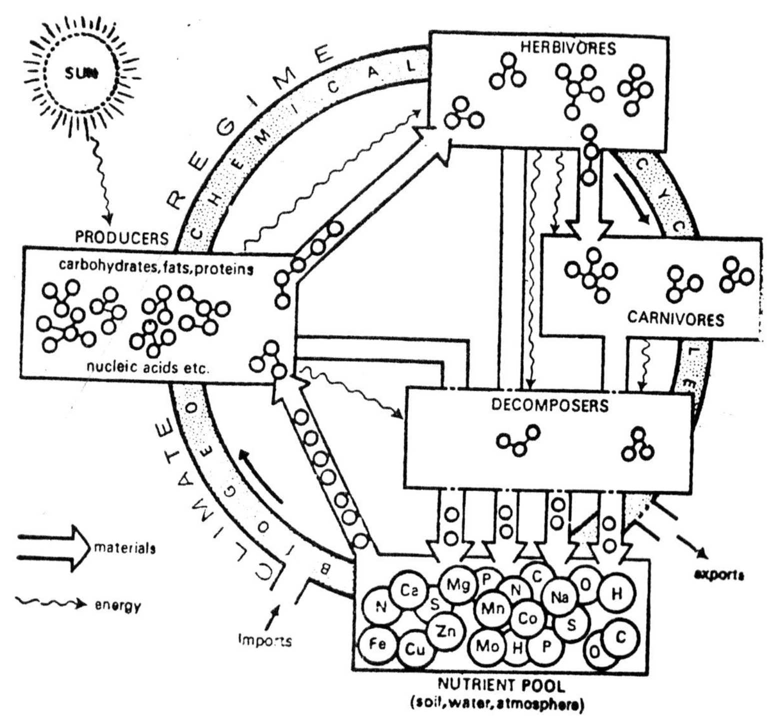

Fig. 55 : A generalised model of an ecosystem to show its structure and function. Note, the, biochemical cycles superimposed on the various components of the ecosystem; movement of materials in cyclic manner, and that of energy in non-cyclic (undirectional) manner. (after P.D. Sharma, 1985).

The energy manufactured by a plant and transformed into a herbivore and then into a carnivore — is sometimes described as a food-chain (Fig. 56). It does not take much imagination to recognise that this is an over simplified description of what happens in nature. Not only it is over simplified, but it is misleading — obviously one plant would not sustain the life of one herbivore animal, nor would one herbivore animal sustain the life of one carnivore animal. It is complex situation involving insects, birds, spiders, mice, bacteria, molds and many other living forms. However, the ecological principles of energy flow apply in all cases — it is just that nature adds many side branches and energy cycles which need to be discussed to illustrate the principle.

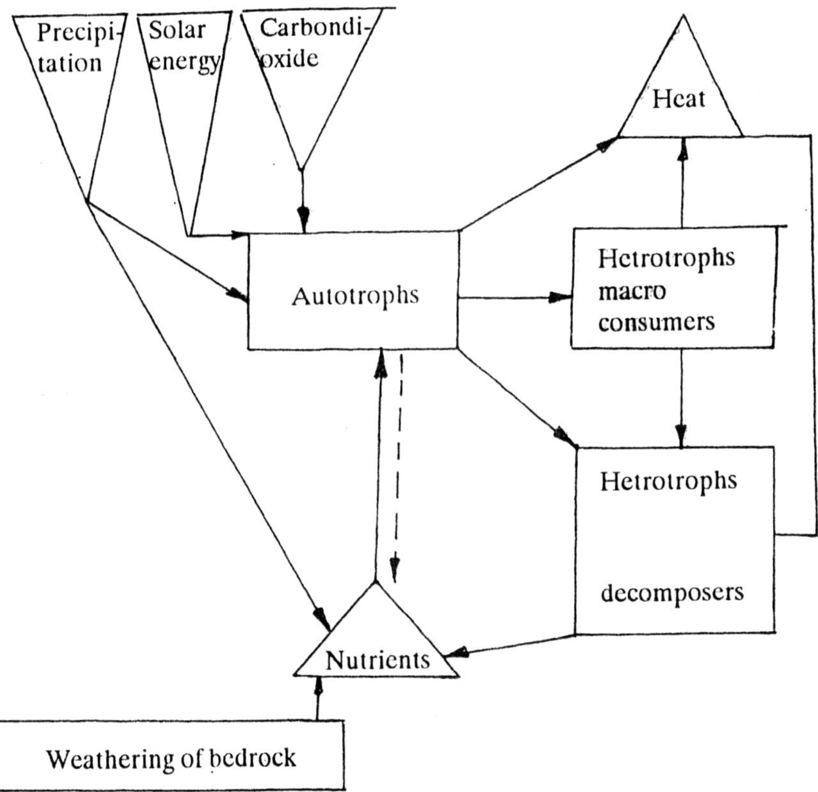

Fig. 56. Major linkages between the components of a simple terrestrial ecosystem (after Chorly and Kennedy, 1971).

Within the food-chain thus constituted in the natural ecosystem there is another complex situation known as the *food-web*. The flow of food may sometimes take forward direction, sometimes a backward, and at other times a number of side pathways, depending upon the actual number of individuals present at any trophic level in the functioning ecosystem. For example, in the food-web in a grassland ecosystem, at least three food-chains become interconnected (Fig. 57). The grasshoppers feed on grasses.

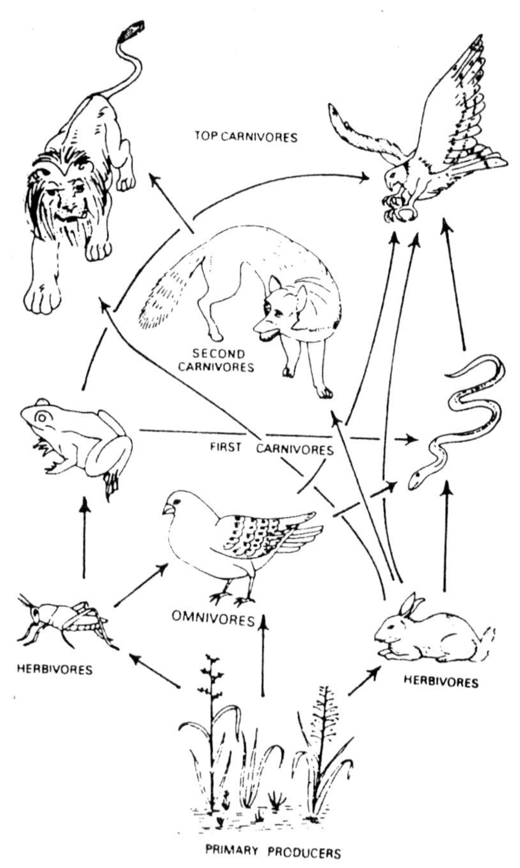

Fig. 57. Food web in a grassland ecosystem.

These grasshoppers are in turn eaten by frogs, which in turn are hunted by hawks and eaten. The frogs can also be swallowed by snakes, which in turn are eaten by hawks, which occupy the top of the food-web here. The insectivorous birds also feed on grasshoppers. These birds in turn can be hunted and eaten by hawks. The herbivorous birds directly feed on grasses. These birds are also captured and eaten by hawks. Rabbits feed on grasses. Snakes in turn feed on these rabbits. In turn these snakes are captured and eaten by hawks. Rabbits are also directly hunted and eaten by lion. Here lion and hawks are the top carnivores which occupy the top of the food-web. They have no enemies but for man.

The complexity of the food-web is at once apparent. Within an environment provided by climate and soil, the many plant and animal species will support and compete with each other in a variety of ways. Some of the interactions are direct : animals eating plants, or some animals eating other animals. Some are indirect : different species of plant compete for a common plant food supply.

Ecological Pyramids

There is some sort of relationship between the numbers, biomass and energy contents of the primary producers, consumers of the first, second order and so on to the top carnivores in any ecosystem. These relationships are represented in diagrammatic ways and are referred to an ecological pyramids (Fig. 58). The ecological pyramids are of three categories :

1. Pyramids of number
2. Pyramids of biomass
3. Pyramids of energy

Pyramid of numbers

Pyramid of numbers deal with the relationship between the numbers of primary producers and consumers of different orders. The base of such a figure is always the number of primary producers and the subsequent structures on this base are represented by the numbers of consumers of successive levels. The top represents the number of top carnivores in an ecosystem. In Fig. 58(a) the upright pyramid of number has been illustrated. The number of plants in a

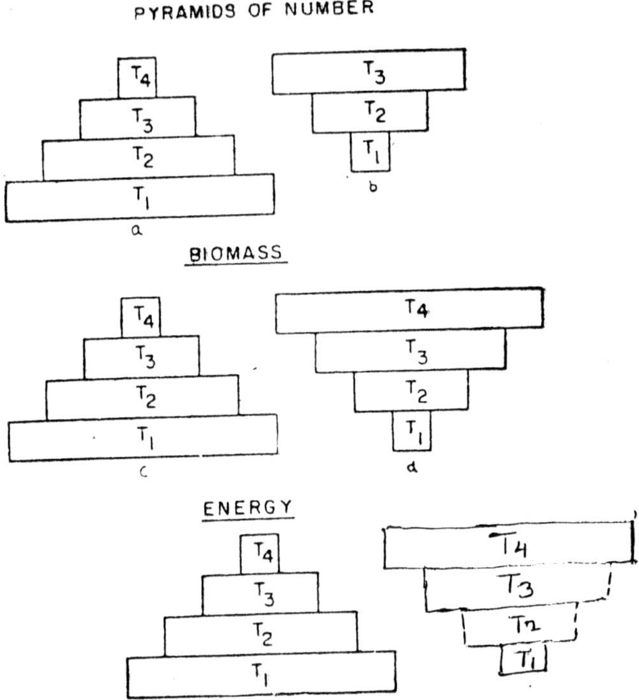

Fig. 58. Ecological pyramids ; pyramids of numbers (a) in grassland and cropland. (b) Forest; pyramids of biomass (c) forest and grassland. (d) oceans; pyramid of energy (e) forest and grassland. (f) oceans T_1, T_2, T_3, and T_4 are the various trophic levels.

grassland or cropland are very large. The number of rabbits in the grassland or the number of grasshoppers in the crop field are usually less than the number of green plants. The number of frogs are lesser than the number of grasshoppers. The number of carnivores is the least in the pyramid. However, there are instances, where the trend of decreasing numbers may not be true. Figure 58(b) illustrates an instance, where the number of primary producers (a tree) is less than the herbivore birds, feeding upon the tree fruits. The number of parasites like bugs and lice living and feeding upon birds body is still higher. Thus depending upon the size of producer organisms the pyramid of numbers may not be always upright, it may even be completely inverted in shape.

The pyramid of numbers ignores the biomass of the organisms. Although the numbers of certain organisms may be greater, their total weight or biomass may not be equal to that of the larger organisms. Neither does the pyramid of numbers indicate the energy transformed nor the use of energy by the groups involved and since the abundance of members varies so widely, it is difficult to show the whole community on the same numerical scale.

Pyramid of biomass

More informative is the pyramid of biomass (Fig. 58 C, D). These indicate the total bulk of organisms or fixed energy present at anyone time. Since some energy and material is lost in each successive link, the total mass supported at each level is limited by the rate at which energy is being stored below. The pyramid of biomass differs for terrestrial and aquatic ecosystems. In terrestrial ecosystems the biomass of all the primary producers is always maximum at any time and the top carnivores have minimum biomass (upright pyramid). In oceans the situation is different and the biomass of the consumers is always higher than that of the primary producers which are planktons and have a shorter life-span. The pyramid in the later type of ecosystem is inverted (Odum, 1963; and Phillipson, 1966).

The upright pyramid of biomass can represent a forest or a grassland ecosystem. The biomass of one tree is very high, and even the biomass of a number of birds feeding upon the tree is f less than that of the tree. The biomass of even a very large number of parasites in and on the body of birds is far less. Similarly, the biomass of grassland or cropland is higher than that of the herbivores feeding upon it and furthermore the biomass of carnivores is still lesser in comparison to that of the herbivores. The pyramid of biomass on land surface, therefore, becomes upright. Instances where the pyramid of biomass gets inverted are aquatic ecosystems. The biomass of phytoplanktons is quite negligible as compared to that of small fishes that feed on them. The biomass of large fishes living on small fishes is still higher.

One of the factors causing this inverse relation between the two habitats may be the difference in the energy flow pathway. For example, in marine community the grazing pathway is more efficient while in the forest community the detritus pathway is predominant (Odum, 1963), but this difference may not necessarily be inherent in aquatic and terrestrial ecosystems. In grassland, for example, the grazing pathway is very conspicuous and in certain cases as much as 88 per cent of the produce is removed for livestock consumption within a short period (Singh, J.S., 1967a) and yet the trophic biomass pyramid is not inverted.

The most significant difference in the aquatic and terrestrial ecosystems is that the variations in oxygen concentration of the gaseous medium on land are much less and being in abundance, it ordinarily never becomes a limiting factor for respiration (Meyer *et al.,* 1963) of the terrestrial communities. The much limited oxygen concentration of aquatic environments, on the other hand exhibits greater fluctuations both seasonally and diurnally.

According to Misra, Singh and Singh (1968) the rate of respiration is lower in case of aquatic animals than in terrestrial ones (Table 53).

Table 53. Rate of respiration of some aquatic and terrestrial animals at rest.

Animals (Aquatic)	Range of O_2 consumption (mg/g wt/hr)	Animals (Terrestrial)	Range of O_2 consumption (mg/g wt/hr)
Fishes	0.005 – 0.349	Mammals	0.124 – 13.70
Insects	0.192 – 0.381	Birds	1.50 – 10.70
Crustaceans	0.05 – 0.11	Insects	0.63 – 1.70
Molluscs	0.002 – 0.186	Arthropods	0.27 – 7.40
Worms	0.008 – 0.031		

Further, in case of terrestrial animals the respiration may increase many fold during periods of activity, for example, in man the rate of energy dissipation through respiration may go up 15-20 times than the resting value; and in insects, flight may cause 50-200 times increase. However, the active metabolism of fish may increase only four times the rest metabolism. Thus, both in rest and activity the energy loss by aquatic animals is smaller. Moreover, considerably greater species diversity of terrestrial ecosystems leads to more efficient loss of energy. Thus, the inverted biomass pyramid in the deep water ecosystems is primarily due to slower energy dissipation by animals and longer turnover period, which leads to greater buildup of consumer biomass.

Ecological Energetics

Energy is the basic force responsible for running the machine of life. In fact energy is the capacity to do work and all living organisms need energy to do work. Green plants capture solar energy through photosynthesis and convert it into chemical energy. In this stored form the energy is passed on from organism to organism as food. Even the coal and petroleum which we use to provide energy, is solar energy which was converted and stored by organisms in the geological past and became fossilised. Thus, sun is the source of all energy, either directly or indirectly, for all organisms of earth.

Different forms of energy interchange their forms under certain set of rules. These are defined in two laws of thermodynamics :

(1) The first law states that, "Energy cannot be created nor destroyed but may be converted from one form to another".

(2) The second law states that, "Processes involving energy transformations will not occur spontaneously unless there is a degradation of energy from a non-random to random form" (Phillipson, 1966).

In ecological energetics we are mainly interested in the (1) quantity of energy reaching an ecosystem per unit area per unit time, (2) quantity of energy trapped by green plants and converted to a chemical form, and (3) quantity and path of energy flow from green plants to organisms of different trophic levels over a period of time in a known area.

Efficiency of energy capture

The first step in the process of energy flow within an ecosystem is the capture of solar energy by green plants. The productivity potential of different ecosystems depends much on the efficiency with which vegetation accumulates this energy in the net primary production.

The quantity of solar energy entering the Earth's atmosphere is approximately 15.3 x 10^8 cal/m^2/year. Much of this is scattered by dust particles, or is used in evaporation of water. The average amount of radiant energy per unit area per unit time actually available to plants varies with geographical location. In Britain the figure is of the order of 2.5 x 10^8 cal/m^2/year; in Michigan, USA 4.7 x 10^8 cal/m^2/year; in Georgia, USA 6.4 x 10^8 cal/m^2/year; in Udaipur, India 3.39 x 10^5 cal/m^2/year. Figure 59 shows the results obtained for the utilization of solar energy by the herbaceous community at Udaipur. Of the solar energy, available over the growing season of 152 days (July 1 to November 30), 3.23 x 10^3 cal/m^2/year was directed to gross production, out of which 2.49 x 10^3 cal/m^2/year was retained in net primary production, while 0.74 x 10^3 cal/m^2/year was lost in respiration. Gross production does not represent the food potentially available to heterotrophs. Autotrophis in synthesizing organic matter must perform work, and the energy for this work is obtained from the breakdown (oxidation) of organic substances in the process of respiration. Gross production minus respiration therefore represents the food energy potentially available to heterotrophs. Of the solar energy available over the growing season in the grassland ecosystem of Udaipur, 50 per cent of the incident radiation during the growing season of 152 days (July 1 to November 30) was directed to the herbaceous community, 0.88 per cent was utilised by gross production, 0.65 per cent in the net primary production, while 0.23 per cent was lost in respiration. The efficiency of energy capture of the *Apluda* community was found to be 0.25 per cent (Vyas, 1975), while that of the herbaceous community at Udaipur to be 0.3075 per cent (Sharma, 1976).

Ovington and lawrence (1967) have reported 4525, 4827, 4817 and 4865 cal/g for maize field, prairie, savanna and oakwood ecosystem in Minnesota. In the alpine plant. Hadley and Bliss (1964) have reported calorific value for shoots (4557-5648 cal/g) as compared to underground parts (4405-4996 cal/g). The energy content of the grassland vegetation at Varanasi varied from 4018-4356.6 cal/g in the above ground parts, and from 4528.1-4770.4 cal/g in the underground parts, with an average value of 4150.23 and

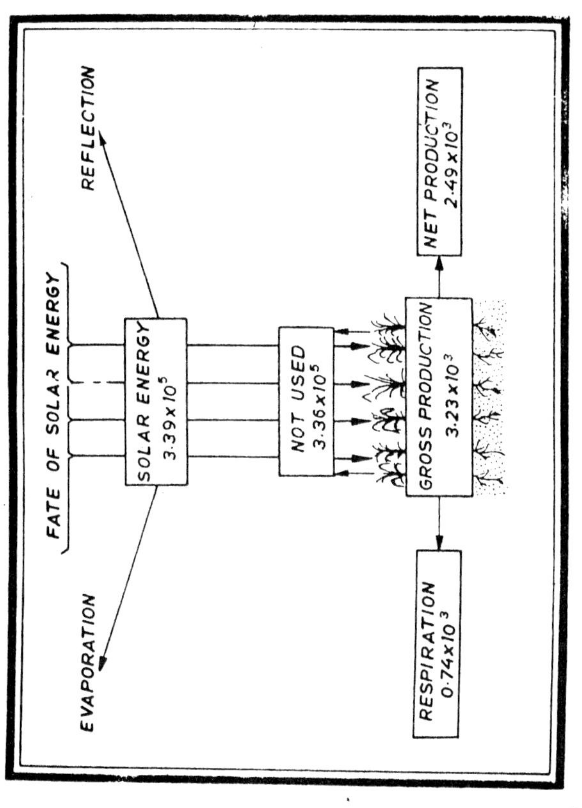

Fig. 59. Fate of solar energy in herbaceous community at Udaipur. All measurements in cal/m^2/year.

4648 cal/g for above ground and below ground parts respectively (Singh and Misra, 1968). Thus, energy content in the above ground parts of tropical grasslands seems to be little lower than that of temperate vegetation. However, the energy content in the underground parts seems to be reverse.

Energy accumulation mostly occur during the grand period of growth (23rd June to 30 September) as shown in Table 54. In the remaining part of the year, although there was some addition, yet the quantity added was appreciably lower.

Table 54. Net primary production in the grasslands at Varanasi in terms of energy (K cal/m^2).

Particulars/ Grassland	Period	Least disturbed	Moderately disturbed	Over disturbed
Aboveground	GPG	1449.72	1822.78	1958.91
	TA	1798.50	2105.41	2139.44
Underground	GPG	1436.11	1193.39	990.96
	TA	1436.11	1193.96	990.96
Total (per year)	GPG	2885.83	3016.83	2929.87
	TA	3234.41	3298.80	3130.40

GPG = Grand period of growth (June-September)

TA = Total annual

Source : Singh and Misra (1968a)

Purohit (1977) observed that energy content of *Helianthus annus* seeds was maximum (4.938 K cal/g), followed in a decreasing order by leaf (3.821 K cal/g), stem (3.734 K cal/g) and root (3.466 K cal/g).

Joshi and Aery (1989) observed that energy content of different parts (root, stem, leaf and seed) of *Crotalaria medicaginea* growing in various localities of Udaipur during vegetative, flowering and fruiting stages ranged rom 2.02 K cal/g (root) to 5.29 K cal/g (leaf) at the vegetative stage, 3.18 K cal/g (root) to 5.22 K cal/g (leaf) at the flowering stage, 2.38 K cal/g (root) to 6.42 K cal/g (leaf) at the fruiting stage. Energy content of seeds ranged from 3.14–4.16 k cal/g (Fig. 60). The old stem and leaves were found to be richer in energy content in comparison to the young organs. The higher values of energy content of pods may be explained on the basis of their fat and other reserve material content.

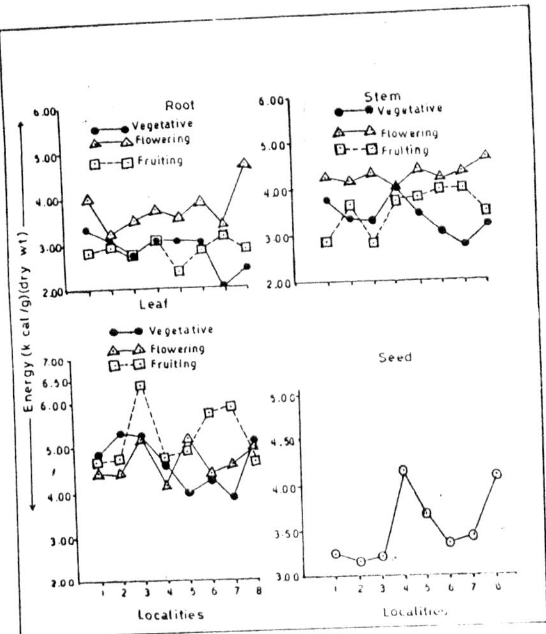

Fig. 60. Energy content (K cal/g dry weight) of Crotalaria medicaginea during different phases of growth in **root, stem,** leaf and seeds at Udaipur (Joshi and Aery, 1989).

A comparison of average energy values in different communities of various climatic regions (Table 55) suggests relatively lower values for the herbaceous community at Udaipur, than those of the other investigated areas. However, the values of the herbaceous community at Udaipur are slightly higher than that of *Apluda* grassland community at Kalighati, Udaipur. The lower values obtained in the grassland community of Udaipur may be viewed in the light of quality and quantity of food reserve and genetic and nutritive constitution of the plant material, as well as the life-history of plants (Golley, 1961). In addition to these facts, the lower values of the grasslands of Udaipur may be viewed in the light of the findings of McNair (1945) who suggested that increase in altitude and latitude and a decrease in temperature result in a corresponding increase in Iodine number (fat) and calorific values.

Table 55. **Average energy values (K cal/g dry weight) of dominant vegetation of different grasslands.**

Location	Communit	Average value	Reference
Georgia (USA)	*Andropogon*	3.905	Golley (1961)
Georgia (USA)	Herb old field	4.177	Golley (1961)
Varanasi (India)	*Dicanthium*	3.887	Choudhary (1967)
Sagar	*Heteropogon*	3.862	Jain (1971)
Ujjain	*Sehima* and *Dicanthium*	3.827	Singh (1969)
Ujjain	*Sehima*	3.389	Mall (1972)
Ratlam	*Sehima*	3.403	Billore (1973)
Ujjain	*Dicanthium*	3.330	Misra (1973)
Kalighati	*Apluda*	2.985	Vyas (1975)
Udaipur	Herbaceous community	3.098	Sharma (1976)

The amount of energy in the primary net production when expressed as percentage of half of the solar radiation received during the period represent the *efficiency of energy capture.* Here only half of the incident radiation is considered because approximately 50 per cent of the total radiation (that in the ultraviolet and infra-red portions of the spectrum) is not usable by plants in photosynthesis. The total incident solar radiation for the year to Varanasi was 1865150 K cal/m^2 and that for the grand growth period 352590 K cal/m^2. The efficiency of energy capture based on 50 per cent of these values (Table 56) indicate that progressive disturbance increases the net production and efficiency of above ground parts. It implied therefore, that in comparatively protected fields energy is stored in the underground parts with greater efficiency, while with increasing disturbance more energy is accumulated in the above ground parts. The efficiency is more than five times greater during grand growth period as compared to the average efficiency for the whole year.

Table 56. Efficiency of energy capture in the grassland at Varanasi as per cent half total solar incident radiation (Singh and Misra, 1968).

Particulars/ Grassland	Period	Least disturbed	Moderately disturbed	Over disturbed
Aboveground	GPG	0.82	1.03	1.11
	TA	0.19	0.22	0.23
Underground	GPG	0.81	0.67	0.56
	TA	0.15	0.12	0.10
Total (per year)	GPG	1.63	1.71	1.67
	TA	0.34	0.35	0.33

GPG = Grand period of growth (June to September)
TA = Total annual

A comparison of energy capture efficiencies values for different herbaceous communities (Table 57) suggests that though the efficiency of herbaceous community at Udaipur lies in the range as observed by Sims and Singh (1971), yet the values are minimum in comparison to that of the other locations. The cause of this decline may be related to the photosynthetic structure — chlorophyll and leaf area index, which play a key role in energy capture (Mall *et al.*, 1973). It is interesting to mention here that the peak of community leaf area index occur approximately during the same period (October) when the peak community chlorophyll, live (green) biomass, density and basal area of the herbaceous species was also observed. This clearly confirmed a close relationship of these structural parameters with the energy capture efficiencies. The decease in leaf area index during winter season, as well as its minimum value during the summer, may be correlated with unfavourable temperature and the level of soil moisture, reaching below the optimum requirement of the plants. The chlorophyll content of the investigated herbaceous community was found to range between 26.2 mg/m^2 (June) and 1253.1 mg/m^2 (October).

Table 57. Comparative energy capture efficiencies of herbaceous communities.

Location	Type of vegetation	Growing season (days)	Efficiency (%)	Reference
Minnesota	Corn and weeds	92	2.10	Ovington and Lawrence (1967)
Missouri	Tall grass prairie	180	1.20	Kucera *et al.* (1967)
Colorado	Tall grass prairie	140	1.20	Mori (1969)
West Siberia	Meadow steppe	100	3.40	Rodin and Bazilevich (1965)
Michigan	*Poa compressa*	180	1.10	Golley (1960)
Colorado	Short grass steppe	100	1.30	Klipple and Costello (1960)
North America	Mixed	145	0.16-0.97	Sims and Singh (1971)
Varanasi	Grassland -Least disturbed	365	0.34	Singh and Misra (1968)
	-moderately disturbed	365	0.34	-do-
	-over disturbed	365	0.33	-do-
Ratlam	*Sehima* community	365	0.33	Billore (1973)
Ujjain	*Dicanthium* community	365	0.38	Misra (1973)
Kalighati	*Apluda* community	365	0.25	Vyas (1975)
Udaipur	Herbaceous community	365	0.3075	Sharma (1976)

This extremely wide range itself is indicative of the effect of periodic variations and the environment, particularly temperature and soil moisture (Sharma, 1976).

Ambasht, Singh and Misra (1982) compared a gradient of grassland, savanna and forest stands under similar climatic and edaphic conditions. Herbaceous vegetation was found to be more efficient solar energy converter than the forests. Ambasht (1990) reported that low statured protected grass stands dominated by *Desmostachya-Heteropogon* grasses with 1.15 per cent energy capture efficiency were more efficient producers than shrub savanna with 0.9 per cent tree plantation with 0.74 per cent and natural mixed forest stand with 0.81 per cent energy capture efficiency. The 50 per cent of incident solar radiation in Chandraprabha sanctuary in Varanasi forest division in 1976-77 was 6557×10^6 K cal/ha/year out of this the grasslands showed net energy fixation of 75.32×10^6 K cal/ha/year. In savanna the shrub *Zizyphus jujuba* accounted for 10.62×10^6 K cal/ha/year and *Heteropogon-Bothriocloa* for 48.28×10^6 K cal/ha/year. In teak plantation of 16 year age only 19.99×10^6 K cal/ha/year and the ground layer dominated by *Oplismenus burmannii* accounted for as much as 28.89×10^6 K cal/ha/year. However, in near natural mixed forest stand protected against cattle grazing 47.4×10^6 K cal/ha/year goes into tree layer dominated by *Anogeissus-Diospyros-Buchanania* species followed by only 3.76×10^6 K cal/ha/year in the shrub layer dominated by *Holarrhena antidysentrica*.

Oceans are poorer energy fixing systems than the land. Therefore, man has to look upon the terrestrial systems as the principal source of energy for the increasing human needs.

System Transfer Functions of Energy

Vyas (1975) observed that in *Apluda* community at Udaipur the average energy content decreased from green to standing dead and to litter. The average energy content of the below ground biomass was found slightly less than that of the green compartment. The mean values observed were 3.165 (live green), 3.159 (roots), 2.941 (standing dead and 2.677 K cal/g (litter). These observations are in confirmity with the earlier reports of Golley (1965) who reported that as the death and composition of vegetation proceeds, the energy content decreases from green tissues through standing dead to litter compartment.

The system transfer functions of energy dynamics of *Apluda* community (Table 58) are comparable to those of Ujjain (Misra, 1973) and Ratlam (Billore, 1973). With slight divergence from that of Ujjain, the system transfer function of roots to roots disappearance was much higher in *Apluda* community at Udaipur than that of Ratlam. The system transfer function of *Apluda* community lies in the same range as reported by Golley (1965) from Broom sedge community of South Carolina.

Table 58. System transfer functions of energy dynamics in different grasslands.

	Compartment	South Carolina (Golley, 1965)	Ujjain (Misra, 1973)	Ratlam (Billore, 1973)	Udaipur (Sharma, 1976)
1.	Total net production to roots.	0.19-0.31	0.42	0.49	0.453
2.	Total net production to above ground net production	-	0.56	0.53	0.514
3.	Above ground net production to standing dead.	-	0.74	0.90	0.788
4.	Standing dead to litter	0.40-1.62	0.77	0.82	0.764
5.	Litter to litter disappearance	0.82-1.20	0.52	0.65	0.562
6.	Root to root disappearance	-	0.76	0.57	0.851

Energy Transfer Along Food Chains

All organisms require a source of potential energy in order to be able to survive and work. The ultimate source of energy is the solar radiation. This energy is trapped by the green plants, the producers, and is converted into chemical energy which is stored in the form of various organic substances including the growing tissues. The growth of the plant body and the interconversion of organic matter within use up a part of the energy is respiration which maintains the structure. The resulting living and non-living organic structure is utilised by the heterotrophs as their source of energy for growth and respiration and hence they are termed as food. The food energy thus originating from the solar energy is transferred from the autotrophs through a few steps or trophic levels of eating and being eaten until energy is dissipated as respiratory heat of the living organisms. The transfer of energy from organisms to organisms is termed as the food-chain (Misra, 1968a).

Energy Flow Models

Energy is needed by all organisms. The behaviour of energy in an ecosystem can be conveniently called as 'energy flow', because energy transformation are directional in contrast to the cyclic behaviour of materials (Odum, 1968). The annual energy production represents the starting point of ecosystem metabolism and functioning., Macfadyen (1963) suggested that energy flow provides a unifying concept for studying and comparing productivities of systems.

A number of models have been proposed to generalise the transfer of energy along food-chains. Based on the trophic levels involved, the energy flow models can be either upright or inverted pyramid of energy, or Y-shaped energy flow model, or schematic energy flow model in an ecosystem.

Upright energy flow model

In a more general sense, in the transformation of energy through the ecosystem the energy is reduced in magnitude by 100 from primary producers to plant consumers and by 10 for each step thereafter. Thus, if an average of 100 K cal of light energy per square meter per day were fixed by green plants, about 10 K cal would be converted to herbivore tissue, 1.0 K cal as the first level carnivore tissue, and 0.1 K cal for the second level carnivore and the rest dissipated as heat. This is confirmity with the second law of thermodynamics which states that when energy is converted from one form to another, some energy is dissipated as heat. Thus, at every food level, there is a constant decrease in the amount of energy available for the next food level. The energy relations can be represented by an upright pyramid (Fig. 61).

The Lindeman (1942) model of energy transfer in an ecosystem envisages the presence of at least three trophic levels (there may be more), which are (a) the producers, (b) the herbivores, and (c) the carnivores. Energy dynamics at each trophic level consists of the following components: the energy entering into the trophic level in the form of food or sunlight, the energy found in the bodies of individuals which make up the trophic level, the energy lost to the trophic level through decomposers and the energy lost to the trophic level from consumption by next trophic level. Briefly, the Lindeman's model expresses that the rate

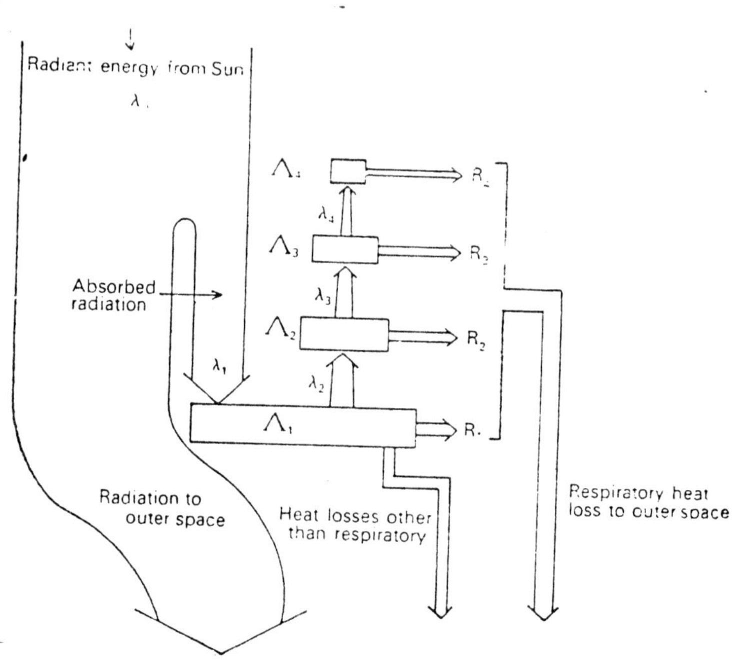

Fig. 61. Diagrammatic representation of flow of energy (after Lindeman).

of change in the energy content of a standing crop is equal to the rate at which energy is absorbed by that standing crop minus the rate at which energy is lost from it. Hence, it is apparent that each succeeding trophic level will have lower calories than the preceding one and there will be a finite number of trophic levels (Misra, 1968a).

The upright energy flow model are always sloping, since less energy is transferred from each level than was transferred into it. In instances (Figure 61), where producers have less bulk than consumers, particularly in open water communities, the energy they store and pass on, must be greater than that of the next level. Otherwise the biomass that producers support could not be greater than the producers themselves. The high energy flow is maintained by a rapid turnover of individual plankton, rather than an increase of total biomass.

Y-shaped energy flow model

In an ecosystem besides primary producers, herbivores, carnivores, there exists saprophytic micro-organisms which consume the dead bodies of the producers and the consumers, for meeting their energy requirements. Thus the energy flow channel in an ecosystem is Y-shaped as shown in Fig. 62.

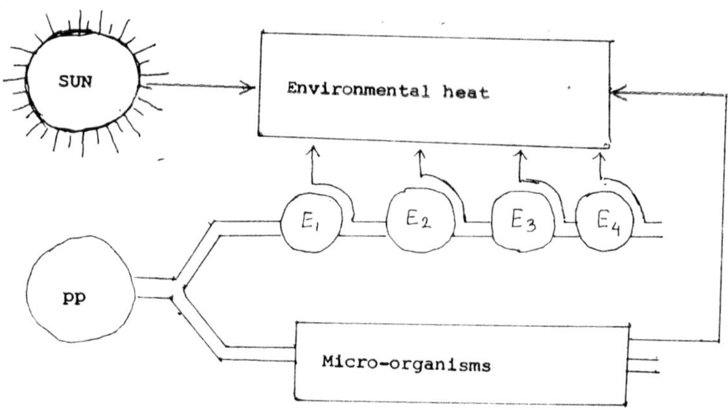

Fig. 62. Y-shaped energy flow model in an ecosystem.

Schematic model

Patten (1959) proposed a schematic model (Fig. 63), with the help of modern information theory, based on the following principles. The balance between information gained (or negative entropy) and entropy generated in a unit of time determines the dynamic state of the ecosystem. If more information is accumulated than used, the excess is converted to biomass. Such a condition of positive balance represents a growing or successional stage. If more entropy is generated than information gained, the system suffers a net loss of order and declines, as is senescence. When an exact balance between negentropy gains and losses is achieved, the ecosystem has reached a stable equilibrium with respect to order-disorder. Such a condition, sustained, markes climax.

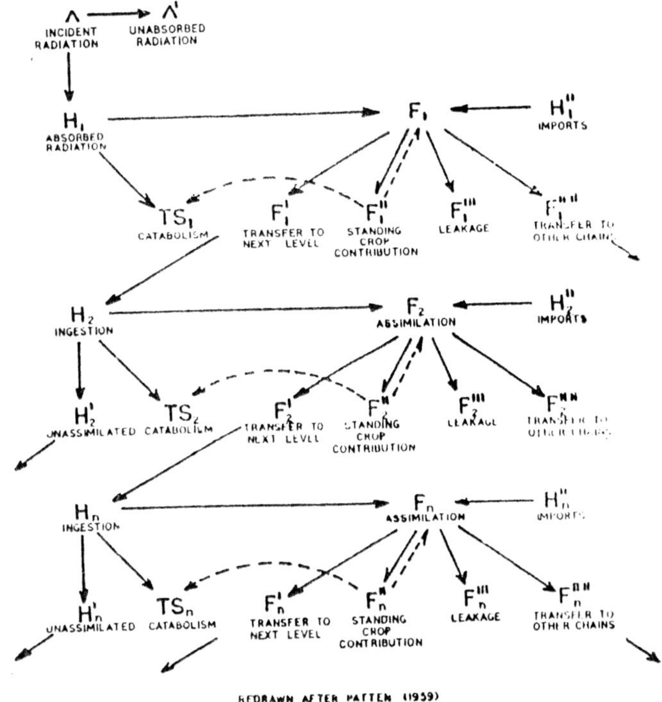

Fig. 63. Patten's (1959) schematic model of energy flow based on the relationship H F T S. H is analogous to enthalpy, F to free energy, and TS to entropy. Ingested energy (Hn) must be assimilated (Fn) before it is respired (TSn).

There are two intrinsic limitations in the proposed models. Obviously they are designed for systems which are in equilibium and as such their usefulness is limited in situation of imbalance. Secondly, there are seldom such simple food-chains in ecosystems as depicted. In fact, the food-chains are enmashed in what we may call a food web and presentation of energy flow scheme has to be based on much assumptions and approximations. Nevertheless, these models are finding interesting use in ecological literature. Gathering and consolidation of more data may perhaps, indicate points where these models may be suitably amended (Misra, 1968a).

It should be remembered that the energy captured by the autotrophs does not revert back to the herbivores, does not pass back to the autotrophs etc. As it moves progressively through the various trophic levels, is no longer available to the previous levels. This un-directional flow of energy, however, is subjected to progressive decrease at each each trophic level. Moreover, the amount of energy available to second level and third level carnivore is so small that only few organisms can be supported if they depend on that source alone.

Progressive Energy Efficiency

The efficiency with which the energy is utilized between any two trophic levels has drawn the attention of ecologists who have proposed different models. Lindeman (1942) proposed the term "progressive efficiency" as the rations :

$$\text{Progressive efficiency} = \frac{\lambda_n}{\lambda_{n-1}} \times 100$$

Where λn is the number of calories ingested by the organisms feeding on λ_{n-1} producers and λ_{n-1} is the number of calories ingested by λ_{n-1}.

This ratio is synonymous with "ecological efficiency" (Slobodkin, 1960), "Efficiency of transfer of ingested energy" (Patten, 1959), utilization efficiency (Davis and Golley, 1963), and gross efficiency of yield divided by ingestion (Wiegert, 1964). According to Phillipson (1966) the gross ecological efficiency is represented by the calories of prey consumed by predator divided by calories of food consumed by prey times 100 and the food-chain efficiency represents the calories of prey consumed by predators divided by calories of food supplied to prey times 100.

Quantitative description of energy transfer through any food-chain are rather difficult to obtain. Golley (1960) studied a food-chain in the oldfield habitat, involving a grass (*Poa pratensis*) a herbivore (*Microtus*) and a carnivore (*Mustella*). The net production of vegetation varied between 44.3 x 10^6 cal/ha/yr which represented 1.1 per cent of the incident solar radiation. *Microtus* used 1.6 per cent of the total net production of the grass, while *Mustella* consumed 31 per cent of the net production of the *Microtus* population. Although not calculated Golley, ecological efficiency for *Mustella* and *Microtus* would be 2.3 per cent (Engelmann, 1966).

Menhinick (1967) studied energy flow through the herb stratum of a stand of lespedeza (*Lespedeza cuneata*). The grazing food-chain in this stand was mainly comprised of Arthropods. The gross and net energy production by producers was about 7300 and 2550 K cal/m^2/year respectively. Approximately 1 per cent of the total net production was ingested by herbivores (23 K cal/m^2/year). Carnivores ingested about 1.6 Kcal/m^2/year which was approximately equal to herbivore production. Thus, in general, primary consumers utilize only a small fraction of net primary production in the lespedeza community, secondary consumers of carnivores, exerted greater pressure on their food supply. The ecological efficiency in this case works out to be 7 per cent.

The degradation of energy through successive trophic levels has broad implications for man, because man is as much a part of the energy flow through the ecosystem as the mouse and the weasel or the herrring and the sharks. Energy in food-chain diverted through man is lost in about the same ratio as it passes from plant → beef → man as it does when passing from plant → mouse → weasel. Man being carnivorous can support a much higher population if he becomes also a herbivore → this is possible through cooking of plants to make them digestible.

16. *Production Ecology*

Introduction

The rapid increase in the numbers and needs of the human population of the world and their demands on the natural environment has greately increased the need for biological research. The international Council for Scientific Unions (ICSU) has therefore, initiated an International biological Programme (IBP) entitled, "The biological basis of productivity and human welfare", with the objective of ensuring the world-wide study of :

(*a*) Organic production on land, fresh water and the seas, and the potentialities and uses of new as well as of existing natural resources.

(*b*) Human adaptability to changing conditions.

A primary production process is the very support for any ecosystem, it is therefore necessary for us to gain complete insight into, and probably even control of, this crucial process (Golley and Lieth, 1972). The primary production is one aspect of the sustained carrying capacity of the earth for man. Many problems of energy and nutrient flow and their relation to communities and potential harvests, make primary production of scientific interest. Determination of dry matter production by plants always constitute the basis for production ecology. Comparative data on rates of dry matter production by plants are required to enable the assessment of limits to potential production in different climates.

Desert communities have fewer component populations, and consequently a simpler organisation. They offer a special opportunity for the comprehension of the bioenergetics of a complete community. The prosperity of a plant mainly depends upon its efficiency in dry matter production. It is a function of carbon gain due to photosynthesis minus its losses in respiration (Figure 64).

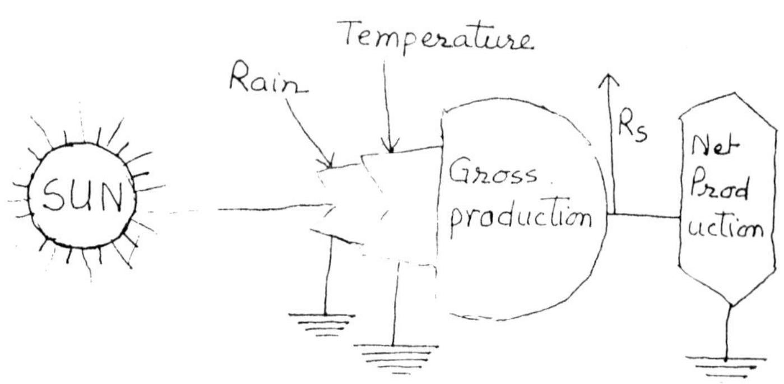

Fig. 64 : Primary production model.

Terminology

The total weight of dry matter present in the ecosystem at any-one time is termed as *biomass*, while the amount of dry matter accumulated within a specified period is the *production*. Rate of production is usually expressed as g/m^2/day and is termed as *productivity*. Since it is the first and basic form of energy stored by the green plants, it is more specifically known as *primary production*. It is a general term that refers to all or any part of the energy fixed by plants. Thus, the garin harvested from a wheat field is primary production, so are the chaffs, the atalks and the roots, as well as the weeds of the same field. The rate at which the radient energy is stored by green plants in the form of organic substances which can be used as food material is known as *primary productivity*. The total rate of energy assimilation including the organic matter which is used up in respiration is known as gross *primary production*. It is mass of carbon fixed per unit area per unit time by the producer photosynthetic organisms. Since plants, like other organisms must overcome the tendency of energy to disperse; free energy must be expended for production as well as for reproduction and maintenance. The energy required for this is provided by respiration. The rate of storage of organic matter in plant tissues in excess of the respiratory utilization by the plants is known as the *net primary productivity*. It is the mass of carbon fixed per unit area per unit time corrected for the losses due to respiration of the authotrophs and also heterotrophs. The rate at which energy is stored at consumer level is referred to as *secondary productivity*. Periodic estimations of biomass over a unit time have been referred sometimes as *turn over*, for example estimations of biomass over a certain period have been done for many consecutive years in grasslands. Sometimes turn over rate is also used for certain elements like phosphorus returning from plankton to water in fractions per hour.

Clearly total primary production or gross production exceeds that which can be measured. The relationship can be summerised most simply by the formula :

$$NPP = GPP - [R_S\ (A) + R_S\ (H)]$$

Where NPP = Net Primary Production, GPP = Gross Primary Production, Rs (A) = Respiration of autotrophs, Rs (H) = Respiration of Hetrotrophs (Woodwell and Whittaker, 1968).

Since, the primary production of organic matter in plants depends mainly upon the photosynthesis by foliage, a close relation of plant growth to the leaf is undoubtedly significant (Tedeki, 1966). The production of a forest is commonly expressed by foresters as the volume of wood produced per unit area of land per unit time. The rate of production of stem wood dry matter depends upon the rate of production of total dry matter of the whole forest. The accumulation of dry matter in turn depend upon the product of the mean net assimilation rate of the leaves and the leaf area of the forest per unit land area (Attiwill, 1962). In the forest ecosystem studies, it is logical to identify biologcal production from commercial production of timber. However, timber is only a part of the products of trees as a living entity. For a better understanding of the processes involved in timber production, studies on dry matter production of forest ecosystem as its basic process is necessary (Satoo, 1966).

Roots are necessary for the intake of water and minerals and so forth, but they are less accessible part of the standing crop, and because of the difficulty of sampling underground parts, data are limited. Therefore, in the study of biomass production, we find available reports of leaf, branch and root biomass scare. Recent studies provide greately increased amount of information on below ground production, especially of herbaceous plants. A possible explanation of this increased study may be the discarding of the previous reluctance to sample below ground systems because of their assumed great depths (Bray, 1963).

A large number of factors, both internal and external affect the primary production process in the arid and semi-arid parts of Rajasthan, which abounds in rainy season annuals and decidous perennials, where there is no scarcity of sunlight and carbondioxide, except for water, such a study becomes more interesting and rewarding. The water stress has a role on the regulation of stomatal opening, thus influence the biomass production. This paper reviews data on biomass production by grassland, shrubland and forests.

Grassland Production

Net primary production by grasses and forbs in trophical grasslands of arid regions have been studied by Gupta, Saxena and Sharma (1971, 1972), Kumar and Joshi (1972), and Shankar and Dadhich (1972). Whereas that of semi-arid regions of Kota by Agarwal and Kasat (1979), and that of Udaipur by Vyas, Garg and Agarwal (1972), Shrimal and Vyas (1975), Vyas an Vyas (1978), and Tiagi and Katewa (1987).

In these grasslands a number of grass species with several species of leguminous and non-leguminous forbs constitute the natural herbaceous vegetation. All these species are mixed together in various proportions on different physiographic units. A significant difference occurs in the phenological behaviour of these species, on account of the marked difference in the climatic conditions prevalent in different seasons (Vyas and Vyas, 1978). These grasslands initiate growth activity soon after the first premonsoon shower, when the anuals become conspicuous, and new crop of foliage appears in the perenial species, but as the dry weather approaches, these plants complete their life-cycle and show a gradual decline in standing biomass. The increase in biomass production of legumes and forbs, and a corresponding decrease in that of grasses may be due to severe competition from legumes and forbs (Vyas *et al.*, 1972; Agarwal, 1979).

Table 59 gives a comparative idea about production of various types of grasslands in arid and semi-arid climates. It is apparent that precipitation is a limiting factor in arid areas where the biomass production has been reported to be feeble in comparison to the semi-arid areas, where comparatively higher annual precipitation leads to higher values of biomass production.

Table 59. : Net primary production in arid and semi-arid grasslands of Rajastan.

Site Grassland	ANP $g/m^{-3}/yr$	BNP $g/m^{-3}/yr$	TNP $g/m^{-3}/yr$	Rainfall mm	Author
Arid					
Jodhpur Mixed	164	—	—	311	Gupta *et al.* (1972)
Pilani Mixed	217	61	278	391	Kumar and Joshi (1972)
Semi-arid					
Kota *Botherioclova*	632.5	—	—	846	Agarwal and Kasat (1979)
Udaipur Mixed	184	—	—	627	Vyas *et al.* (1972)
Udaipur *Apluda*	255.9	255.0	510.9	603	Vyas & Vyas (1978)
Udaipur *Sahima*	705.2	256.6	961.8	555.8	Tiagi & Ketewa (1987)
Udaipur *Heteropogon*	1168.9	160.5	1329.5	555.8	-do-
Udaipur *Apluda*	353.1	206.8	559.9	555.8	-do-

ANP = Above ground net primary production
BNP = Below ground net primary production
TNP = Total net primary production

Shrimal and Vyas (1975) observed that in Udaipur the grassland vegetation was composed of 17 species of flowring plants, including 8 species of grasses, 4 species of legumes and 5 species of other herbs. This vegetation was dominated by *Heteropogon contortus* and *Dicanthium annulatum*. In the unprotected and protected stands of this vegetation above ground biomass increased from June to September. In September, most of the species reached maturity and started dying, which caused a decline in the above ground biomass from the month of October. The underground biomass increased with the above ground biomass during June and

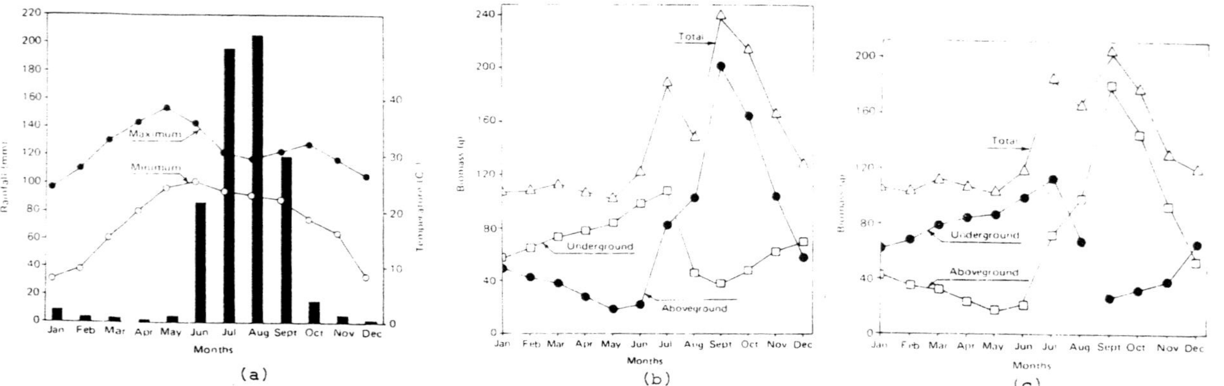

Fig. 65 : Climate and grassland biomass production at Udaipur (a) Ombothermic diagram for Udaipur on the average of five year data. Bars represent rainfall, (b) Monthly variation in biomass in a unprotected grassland, (c) Monthly production in biomass in a protected grassland.

July, however, it continued to increase up to May. In contrast to the above ground biomass, underground biomass was slightly higher in most of the months in the protected stand (Figure 65). The pattern of rainfall seems to be correlated with biomass production. Negiligible rainfall during winter and summer months limits biomass production. It has been concluded that protection leads to community stability but reduces biomass production.

Table 60. : Net above ground biomass production (g/m²/yr) on the two stands : Stand 1 (unprotected) and Stand 2 (Protected for 3 years) after Shrimal and Vyas (1975).

Species	Maximum green standing crop (g/m²) Stand 1	Stand 2
Grasses		
Apluda aristata	9.2	14.1
Aristida funiculata	5.4	3.2
Chloris virgata	6.7	4.0
Dicanthium annulatum	35.1	29.3
Eragrostis tenella	2.3	4.3
Heteropogon contortus	45.8	39.7
Themeda quadrivalvis	7.4	11.3
Tragus biflorus	3.0	—
Legumes		
Alisicarpus hamosus	6.3	8.4
Desmodium triflorum	8.6	6.1
Indigofera cordifolia	24.5	19.6
I. linnaei	13.0	9.2
Other Species		
Boerhaavia diffusa	11.4	—
Evolulus alsinoides	4.8	2.8
Glossocardia bosvallea	9.2	—
Justicia procumbens	4.7	11.2
Lepidogathis cristata	5.3	—
Net above ground production (g/m²/yr)	202.7	163.2
Rate of production (g/m²/day)	0.55	0.44
Number of dominant species	17	13

In these grasslands about 40 per cent of the community biomass was shared by *Heteropogon contortus* and *Dicanthium*

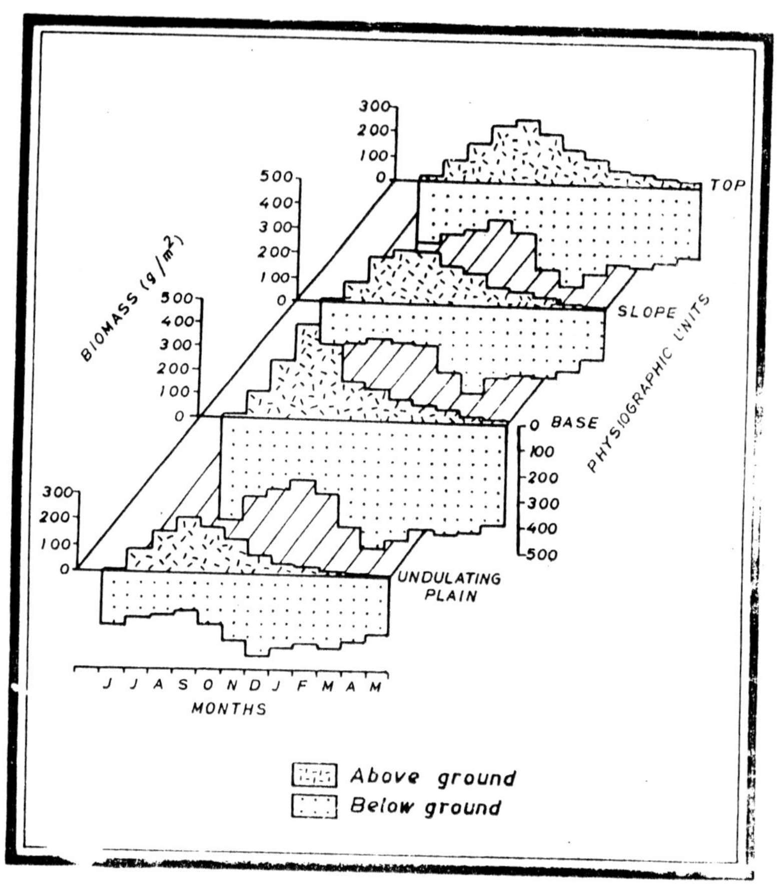

Fig. 66 : Monthly fluctuations in net above ground and below ground plant biomass in the herbaceous community at different physiographic units.

annulatum. The community net primary produtivity was between 163.2 and 202.7 g/m^3/yr in unprotected and protected stands respectively (Table 60). At Varanasi the productivity figures for *Dicanthium* grassland has been reported to be 400 - 5009 g/m^3/yr (Singh, 1968) and at New Delhi *Heteropogon* produces 800 g/m^3/yr (Varshney, 1972). These differences illustrate the unfavourable effects of semi-arid climate of Mota Magra forest block of Udaipur.

Vyas (1975) observed that in the grassland community of *Apluda* at Kalighati, Udaipur (Figure 66), the standing crop showed distinct periodic variations. The above ground biomass increased from June till it attained the peak during September, except for the hill top where the peak was attained in October. The above ground biomass was found to vary from 6.5 - 207.1 gm/m^2 (undulating plains), 19.0 - 394.0 gm/m^2 (hill base), 15.0 - 226.5 gm/m^2 (hill slope), and 19.6 - 256.3 gm/m^2 (hill top). The below ground biomass behaved quite similar in all the physiographic units. The maximum values were observed in December, while the minimum values were obtained in September. The below ground biomass was found to vary from 159.0 346.0 gm/m^2 (undulating plains), 186.0 - 495.0 gm/m^2 (hill base), 156.0-379.0 gm/m (hill slope), and 146.0-410.0 gm/m^2 (hill top).

The net above ground primary production at the hill top, hill slope, hill base and the undulating plains were assessed at 236.8, 211.4, 375.0 and 200.6 gm/m^2/yr respectively (Table 61). Similarly the net below ground biomass production recorded was 270.0. 227.0, 331.0 and 192.0 gm/m^2/yr for the hill top, hill slope, hill base, and the undulating plains respectively. Thus on the basis of net production the various physiographic units may be arranged in the order : hill base > hill top > hill slope > undulating plain. The observations further suggested that at the hill top and hill slope the net below ground production was more than the above ground production, while at the hill abuse the position was just reverse. In the undulating plain the net above ground as well as below ground production were observed to be almost equal (Vyas, 1975).

Sharma (1976) reported that on the hills around Udaipur the maximum biomass of mixed grassland community was attained during September (hill top), October (hill slope and hill base). The values recorded being 397.46 gm/m^2 (hill base), 314.34 gm/m^2 hill slope), and 262.10 gm/m^2 (hill top). Further, at the protected site

Table 61. : Monthly net positive increase ($gm.m^{-2}$) in above ground and below ground biomass for *Apluda* community at different physiographic units of Kali Ghali forest block, Udaipur (Vyas and Vyas, 1978).

Months	Above ground				Below ground			
	Top	Slope plain	Base	Undulating	Top	Slope	Base	Udulating plain
June to July	74.9	74.4	87.5	80.3	-42.0	-16.0	-86.0	-31.0
July to Aug.	70.0	95.9	130.3	74.2	-16.0	-21.0	-36.0	-17.0
Aug. to Sept.	71.2	41.2	157.2	46.1	-48.0	1.0	-30.0	-19.0
Sept. to Oct.	20.6	-8.1	-81.5	-28.0	56.0	21.0	55.0	58.0
Oct. to Nov.	-57.4	-53.8	-140.5	-43.2	138.0	119.0	174.0	77.0
Nov. to Dec.	-53.7	-34.8	-37.2	-55.7	70.0	82.0	80.0	52.0
Dec. to Jan.	-47.4	-49.7	-42.2	-31.2	-54.0	-72.0	-30.0	-39.0
Jan. to Feb.	-34.8	-20.5	26.8	-15.7	-36.0	-13.0	-50.0	-11.0
Feb. to March	-7.0	-22.2	-10.1	-12.3	6.0	4.0	22.0	5.0
March to April	-30.6	-20.0	-2.0	-2.0	-12.0	-30.0	-15.0	-25.0
April to May	-6.0	-1.1	-7.0	-2.2	-28.0	-50.0	-30.0	-28.0
Total positive inerement	236.8	211.4	375.0	200.6	270.0	227.0	331.0	192.0

the net above ground biomass productiona at the hill base, hill slope, and hill top was observed to 351.12, 274.18 and 238.90 gm/m^2yr respectively. Similarly, the net below ground biomass production was 340.0, 316.0 and 277.0 $gm/m^2/yr$ at the hill base, hill slope and hill top respectively. At the unprotected site net above ground biomass production at the hill base, hill slope, and hill top was 210.02, 222.01 and 224.24 $gm/m^2/yr$ respectively. While net below ground biomass production was 237.00, 216..00 and 226.00 gm^2/yr at the hill base, hill slope and hill top respectively. Thus, on the basis of net production at this site, it is very interesting to record that both hill base and hill slope have similar values, while the hill top showed a slightly higher values. The observations further suggest that net below ground biomass production was more than net above ground biomass production at the hill base and hill top, while at hill slope the net above ground biomass production was higher than the net below ground biomass production.

Net primary production by grasslands in relation to grazing intensity was studied by Gupta, Saxena and Sharma (1972) for the arid regions. They selected three stands - protected, medium grazed, and over grazed areas at Jodhpur (Table 62). At site 1 and 2 *Elusine compressa* and *Dicanthium sindicum* were highest contributors, while at site 3 *Tragus biflorus* an annual grass was the main

Table 62. : Net primary above ground biomass production (g/m^2) by grassland at Jodhpur in 1968-69 (Gupta, Saexna and Sharma, 1972).

Period	rainfall mm	Site 1 Protected	Site 2 Medium grazed	Site 3 Over grazed
July	169.8	97.90	41.06	—
August	8.8	137.30	67.55	46.83
September	—	145.80	98.71	33.49
October	—	126.16	82.45	19.27
November	—	101.64	77.65	18.54
December	—	91.88	-57.68	22.76
January	6.2	91.78	58.85	19.98
February	0.2	86.45	52.22	18.92
March	0.4	71.78	51.76	19.90
April	—	79.66	61.36	26.12
May	—	63.23	44.74	20.00
June	0.5	67.66	60.92	23.28

contributor. At site 1 and 2 the maximum standing crop was recorded during September and Minimum in May. While at site 3 the maximum was recorded in August and minimum in November.

Shrubland Production

There have been very meagre studies on the biomass production by shrublands in Rajasthan. Agarwal (1980, 1983) initiated such studies in the semi-arid parts of Darrah Game Sanctuary, which was dominated by *Grewia flavescens* Juss and *Lantana indica* Roxb. The disitribution of dry matter production varied in the two species. In general, the photosynthetic, non-photosynthetic and total above ground biomass production increased gradually (Table 63) with increase of plant dimensions. The distribution of dry matter was maximum in stem and minimum in the leaves. The ratio between non-photosynthetic and photosynthetic biomass was highest in the young plant of *G. flevescens* followed by a decline. However, *L. indica* showed a reverse trend, where this ratio gradually increased with an increase in plant dimensions. The cause of such a differential behaviour has yet to be explored.

Table 63. : Net above ground biomass production by two shrub species at Kota (Agarwal, 1983).

Species/ S. No.	gm/plant			non-photo-synthetic/ Photosynthetic ratio
	Photo synthetic	Non-photo-synthetic	Total	
Grewia flavescens Juss.				
1	45.66	1510.41	1556.07	33.07
2	45.91	1644.26	1690.17	35.81
3	293.47	1133.33	2126.80	6.24
4	321.42	1625.34	1946.76	5.06
5	1181.81	6788.93	7970.74	5.74
Lantana indica Roxb.				
1	225.66	281.25	506.91	1.24
2	410.52	1152.34	1562.86	2.80
3	588.23	2405.68	2993.91	4.08
4	666.66	3024.46	3691.12	4.53
5	792.45	4226.78	5019.23	5.33

Table 64. : Above ground biomass (dry wt Kg./ha) of *Calotropis procera* L at Kota (Agarwal and D'souja, 1980)

Sample No.	Density /ha	Stem	Leaves	Flowers	Fruits	Total
1	1000	337.5	247.3	32.8	8.3	625.9
2	1700	168.7	71.3	17.0	6.6	263.8
3	1200	281.2	118.4	43.7	16.5	459.8
4	1300	253.1	224.7	21.9	3.3	503.0
5	1500	253.1	165.7	15.3	3.3	437.4
6	1200	450.0	390.7	37.5	16.5	894.7
%		54.7	38.2	5.3	1.7	

Net primary production of *Calotropis procera* community in Chambal ravines near Kota was studied by Agarwal and D'souja (1980). Observations (Table 64) reveals that the biomass production was independent of densdity of plants. Areas with high density had low organic production and *vice-versa.* This might be due to competition and excessive grazing. On area basis, the above ground biomass of *C. procera* was 530.7 Kg/ha based on the assumption that the shrubland was composed of an average of 1313 plants/ha. Out of this dry matter 54.7, 38.2, 5.3 and 1.7 per cent was contributed by stem, leaves, flowers and fruits respectively. The quite high percentage in stem and leaves reflect the high photosynthetic efficiency and energy accumulation.

Zizyphus nummularia is one of the important shrub species of the arid and semi-arid tracts which continue to supply foliage for a large period of time. The dry matter production of the shrub and its distribution into stem and leaves have been studied by Saxena and Sharma (1981).

The leaf biomass was observed to be minimum in Summer cut and maximum in winter cut bushes. The shoot biomass production and productivity followed a similar trend. Even the number of stumps were significantly higher in winter plants than summer and spring season plants. Plants gain highly significant height and their basal area by the end of November (start of winter) than summer and spring season's plants (Table 65).

Table 65. : Growth attributes of *Zizyphus nummularia* in flat alluvial plains of Jodhpur (Saxena and Sharma, 1981).

Season	Above Leaf	ground biomass Shoot	(gm/bush) Total	Productivity g/bush/day	Plant height cm	Basal area cm^2	No. of stumps
Spring	73.26	348.47	421.73	1.15	141.7	3.53	3.17
Summer	40.03	338.22	378.25	1.04	138.3	4.18	3.35
Monsoon	152.08	248.64	400.72	1.09	146.9	6.02	3.85
Winter	193.09	574.85	767.94	2.10	171.3	7.42	4.07
mean	114.61	377.54	492.16	1.35	149.6	5.29	3.60
SE $\pm$	34.80	69.87	—	—	5.67	0.61	0.23
F Test	sig	sig	—	—	H sig	H S	sig
CD at 5 %	100.7	202.15	—	—	16.40	1.76	0.66

Forest Production

It has been observed that in forests, tree dimensions such as girth, height of bole, branch volume, crown area, root spread, above ground biomass and below ground biomass production increase with age of plants. Ovington and Madgwick (1959) observed that in *Pinus sylvestris*, the trees with larger bole girth classes than those of smaller girth classes have extensive root system. The roots of the largest tree, spread about twice as far laterally and downwards as the roots of the smallest tree, and similar difference was reported to occur in the spread and depth of the living crown. Trees of larger girth tend to be heavier in all respects having greater weights of leaves, cones, living branches and roots (Figure 67).

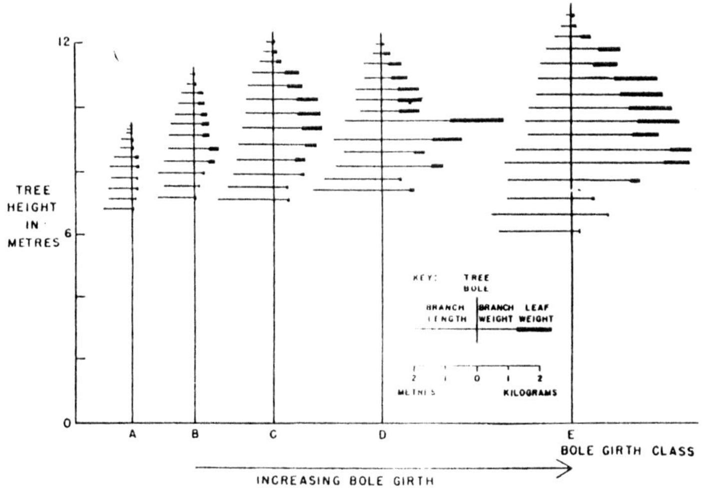

Fig. 67 : The average whorl heights, branch length and fresh weights of branches and leaves for trees of different girth classes (after Ovington and Madgwick, 1959)

The extent of information available on the biomass production by forest trees and forest communities in Rajasthan, suffers from paucity of data for arid regions, while for the semi-arid regions much informaion is one record. Tree biomass production have been studied by Vyas, Agarwal and Ranawat (1971), Vyas, Garg and Ranawat (1971), Agarwal (1972), Vyas, Agarwal and Garg (1972),, Garg, Ranawat and Vyas (1972), Vyas, Garg and Ranawat (1973), Vyas *et al.* (1973), Vyas, Ranawat and Garg (1973), Vyas, Garg and Ranawat (1974), Vyas Garg and Vyas (1976), Vyas, Garg and Vyas, (1977), Agarwal (1978), Vyas *et al.* (1978), Vyas, Garg and Jindal (1979), Agarwal (1980), Vyas *et al.* (1980).

Primary production is dependent on leaf area, leaf dry weight and life-span of the leaves. Leaves of the four species studied by Agarwal (1980) in the forests of Udaipur showed a differential behaviour as regards their life-span. The maximum life-span of leaves (10 months) was observed in *B. monosperma*, while in *L. coromendelica* it was minimum (5 months), *T. grandis* and *B. serrata* had intact leaves for 7 months. The dry matter in the leaves of all the four species was observed to increase with maturity and it was maximum at the time of leaf fall (Table 66). Among the four species studied *T. grandis* showed the maximum rate of organic matter production (74.62 ± 8.36%) while *L. coromendelica* the minimum (36.25 ± 0.57%).

Table 66. : Pattern of foliar dry weight production (in % ± S.D.) in deciduous trees (Agarwal, 1980).

Months	*B. monosphema*	*T. grandis*	*B. serrata*	*L. coromendelica*
April	19.80 ± 1.95	—	—	—
May	27.54 ± 1.71	—	—	—
June	26.35 ± 3.38	—	—	—
July	30.05 ± 4.24	22.21 ± 3.20	30.57 ± 3.34	29.54 ± 0.89
August	28.99 ± 2.07	31.27 ± 6.21	29.27 ± 4.09	37.00 ± 6.85
September	29.64 ± 2.82	37.49 ± 1.22	35.30 ± 4.10	35.58 ± 0.35
October	25.72 ± 4.59	51.53 ± 4.09	37.66 ± 3.39	36.25 ± 0.57
November	26.97 ± 1.04	56.24 ± 9.41	34.76 ± 5.67	—
December	28.57 ± 1.73	54.12 ± 2.72	41.00 ± 3.68	—
January	37.10 ± 2.09	74.62 ± 8.36	46.16 ± 6.16	—
February	—	—	—	—
March	—	—	—	—

Table 67. :Average above ground biomass (kg/tree) production of tree species in the semi-arid regions of Rajasthan.

Species	Bole kg/tree	Branch kg/tree	Leaf kg/tr	Total kg/tr	Leaf area m^2/tree	Life span of leaves months	Source
Erythrina suberosa	6.93	6.12	1.14	14.19	29.76	6	Vyas *et al*. (1971 a)
Anona squamosa	7.85	1.14	0.49	9.48	19.00	5	Vyas *et al*. (1971 b)
Wrightia tinctoria	11.06	2.26	1.94	15.26	84.00	6	-do-
Boswellia serrata	28.69	32.15	1.43	62.27	45.24	7	Agarwal (1980)
Butea monosperma	80.53	13.63	6.99	101.15	107.22	10	-do-
Lannea coromendelica	58.00	60.91	4.03	122.94	55.17	4	-do-
Tectona grandis	32.13	13.76	2.73	48.62	47.54	7	-do-

Observations on net above ground biomass production by tree species growing in the semi-arid region have been presented in Table 67. It seems likely, that the relation between wood biomass (stem + branch) and leaf area was responsible for limiting the tree size and also for increase in bole and branch dry weight. The high biomass production in *Butea monosperma* have been explained on the basis of leaf area, leaf biomass as well as life-span of leaves, all of which were maximum for it. While the maximum non-photosynthetic biomass production in *Lannea coromendelica* in spite of its lower leaf biomass and lowest life-span have been accounted for its highest photosynthetic efficiency.

It has been observed that net primary production of different plant parts increases with an increase in circumference at breast height of the bole. The regression equations obtained indicate a clear positive correlation between number of growth rings (Ngr) and their circumference at breast height (CBH) for different species :

For Bole

Adina cordifolia	Y = 4.2 + 1.17 X
Anogeissus latifolia	Y = 5.4 + 1.3 X
Boswellia serrata	Y = 5.9 + 1.04 X
Butea monosperma	Y = 35.3 + 0.37 X
Diospyros melanoxylon	Y = 4.95 + 1.14 X
Mitragyna parvifolia	Y = −34.81 + 2.67 X
Schrebera swietenioides	Y = 14.5 + 0.84 X
Terminalia tomentosa	Y = 7.55 + 0.93 X

For Branch

Anogeissus latifolia	Y = −0.9 + 2.3 X
Butea monosperma	Y = −9.5 + 1.7 X
Mitragyna parvifolia	Y = −1.41 + 2.4 X
Schrebera swietenioides	Y = −8.2 + 1.3 X

(Where Y = CBH and X = number of growth rungs)

Vyas, Garg and Jindal (1979) calculated linear correlations between leaf area and plant biomass. The regressions equations obtained are as follows :

Table 68. : Above ground biomass (mt/ha) production in different forest stands of the semi-arid parts of Rajasthan (Vyas *et al.*, 1980).

Stand	Forest community	Density	Bole	Branch	Leaf	Total
Kewara Nal	*Boswellia-Lannea*	1008	17.10	12.20	2.15	31.45
Chotta Ghata	*Aegle-Boswellia*	1396	18.67	6.57	1.94	27.18
kali Ghati	*Lannea-Butea*	706	13.87	8.25	1.86	25.96
Borimalan	*Boswellia-Anogeissus*	1081	22.68	12.21	2.41	37.32
Prasad	*Tectona-Lannea-Boswellia*	1310	17.91	7.34	3.12	28.36
Chitari Mata	Tectona-Boswellia	996	14.83	6.50	2.59	23.92

Table 69. : Average increment of non-photosynthetic biomass in deciduous trees in semi-arid parts of Rajasthan.

Species	Growth rings	Biomass kg/tree/yr	Author	Species	Growth rings	Biomasa kg/tree/yr	Author
Adina cordifolia	0-8	0.93	Vuyas *et al.* (1973 a)	*Mitragyna parvifolia*	0-14	1.5	Vyas *et al.* (1971 b)
	8-9	1.38			14-30	1.87	
	9-26	2.15	-do-		30-35	8.85	-do-
	26-27	6.92	-do-		35-40	3.84	-do-
	27-30	0.75	-do-	*Soymida febrifuga*	00-9	0.80	Vyas *et al.* (1973)
	30-32	1.17	-do-		9-18	0.56	
	32-40	0.80	-do-		18-20	0.80	-do-
	40-41	1.60	-do-		20-23	1.33	-do-
Boswellia serrata	0-42	1.84	Vyas *et al.* (1978)		23.36	3.33	-do-
	42-47	2.53			36-39	4.30	-do-
	47-52	5.34	-do-		39-47	1.76	-do-
	52-58	6.10	-do-				
	58-65	15.24	-do-		47-61	1.63	-do-
	65-71	4.34	-do-	*Terminalia tomentosa*	12-21	1.71	Vyas *et al.* (1976)
	71-81	3.97	-do-		21-33	1.70	
	81-85	1.08	-do-		33-38	5.0	-do-
					38-55	3.8	-do-
Erythrina suberosa	0-6	2.17	Vyas *et al.* (1972)		55.61	4.8	-do-
	6-10	2.83			61-62	4.4	-do-
	10-18	4.5	-do-				
	18-22	6.34	-do-	*Diospyros melanoxylon*	16-18	0.35	Vyas *et al.* (1977)
					18-21	0.70	
					21-23	1.31	-do-
					23-28	2.56	-do-
					28-36	1.26	-do-
					36-49	1.68	-do-
					49-76	2.17	-do-

Adina cordifolia	Y = −8.45 + 1.15 X
Anogeissus latifolia	Y = 1.37 + 1.66 X
Anogeissus pendula	Y = 0.11 + 1.88 X
Schrebera swietenioides	Y = −13.87 + 2.33 X
Soymida febrifuga	Y = 2.39 + 2.01 X

(Where Y = leaf area in m^2, and X = plant biomass in kg).

Table 68 presents various forest stands of Udaipur, their forest communities and above ground biomass (mt/ha). It is apparent from the data that the above ground biomass was maximum (37.32 mt/ha) at Borimalan forest stand and minimum (23.92 mt/ha) at Chitari Mata stand. it may be further observed that in teak dominated forest stands leaf biomass consstituted 10 to 11 per cent of the total above ground biomass, while in non teak forest stands it constituted only 6 to 7 per cent of the total above ground biomass.

The average increment of non-photosynthetic (stem + branch) above ground biomass with respect to number of growth rings (in the absence of any record about the age of the trees) presented in Table 69 suggest that high rate of biomass increment occured in the younger trees which may be due to higher leaf area, greater crown exposure or longer photosynthetic efficiency of the leaves in the younger plants - features which Watson (1958), Peterken and Newbould (1966) and Misra *et al.* (1967) invoked to explain the higher productivity. These values showed a gradual rise with age followed by a decrease. The age range of *Erythrina suberosa* was perhaps not sufficiently extensive to include the period of decline. However, the possibility of poor growth conditions in the micro-environments of the area and also their influence cannot be ruled out.

Litter Production in Forest

The study of quantitative aspects of litter fall remains an important path of forest ecology, dealing with a major pathway for both energy and nutrient transfer (Bray and Goham, 1964). The litter is the fuel for nutrient cycles and is particularly important in the nutrition of woodlands on soils of low nutrient status. If forest production is to be used with maximum efficiency it is essential to understand the organic fraction of forest litter. In India, estimation

Table 70. : Contribution of leaf litter by different species in Koriyat forest community (expressed in guns/m^2) from October 1971 to Match 1972 (Ranawat and Vyas, 1975).

S. No.	Plant species	Months	October 71	Nov. 71	Dec. 71	Jan. 72	Feb. 72	March 72	Total
1.	*Acacoa catecji*		—	—	1.200	0.040	—	—	1.240
2.	*Acgle marmelos*		—	0.340	19.200	60.000	12.200	2.000	-93.740
3.	*Albizzia odoratissima*		—	0.208	0.720	1.200	0.840	—	2.968
4.	*Anogeissus latifolia*		2.480	5.360	26.800	20.800	17.300	0.200	72.940
5.	*Boswellia serrata*		22.180	46.200	13.200	3.600	0.200	—	85.380
6.	*Erythrina suberosa*		1.040	17.200	14.400	1.800	—	—	34.440
7.	*Emblica officinalis*		1.480	8.000	17.000	10.800	—	—	37.280
8.	*Grevia tiliaefolia*		—	0.080	—	—	—	—	0.080
9.	*Lannea coromandelica*		9.360	33.600	4.200	—	—	—	47.160
10.	*Mitragyna parvifolia*		—	—	1.200	1.200	0.600	1.400	4.400
11.	*Schrebera swietenioides*		—	—	2.400	6.400	12.400	—	21.200
12.	*Soymida febrifuga*		1.440	0.040	0.040	—	0.400	0.900	2.820
13.	*Sterculia urens*		12.000	3.600	—	—	—	—	15.600
14.	*Tamarindus indica*		—	0.020	1.080	2.200	10.800	10.000	24.100
15.	*Wrightia tinctoria*		—	2.780	10.800	1.200	—	—	14.780
16.	*Wrightia tomentosa*		2.000	6.400	6.400	0.480	—	—	15.280
17.	*Zizyphis jujuba*		0.600	0.600	3.000	2.800	0.020	0.080	7.100
			52.580	124.428	121.640	112.520	54.760	14.580	480.508

of annual litter production in some evergreen and deciduous forests have been attempted by Puri (1953), Upadhyaya (1955), Singh (1968), Seth *et al.* (1963), Subba Rao *et al.* (1972) etc. The only contribution on litter production in semi-arid regions of Rajasthan is that by Ranawat and Vyas (1975).

Observations on the pattern of litter production in deciduous forests of Koriyat, Udaipur (Table 70) suggest that in all 17 plant species have been recorded to contribute to the litter produced in the area. Further, duration of leaf-fall varied from species to species, being only one month (*Grewia tiliafolia*) ; two months (*Acacia catechu, Sterculia urens*) three months (*Lannea coromendelica, Schrebera swietenioides, Wrightia tinctoria*) four months (*Albizzia odoratissima, Erythrina suberosa, Emblica officinalis*; five months (*Aegle marmelos, Boswellia serrata, Tamarindus indica*). In *Anogeissus latifolia* and *Zizyphus jujuba* leaf fall has been recorded for six months.

The amount of litter fall was quite low (52.58 gm/m^2) in October, when litter fall started, became maximum (124.428 gm/m^2) in November and declined thereafter to a minimum (14.58 gm/m^2) in March. It was further observed that the number of species contributing litter varied from month-to-month and followed a similar trend as that of the amount of litter except that amount of litter fall was maximum (124.428 gm/m^2) in November while the number of species contributing litter was maximum (15 out of 17 recorded) in December. Among various species studied *Aegle marmelos* has been observed to contribute the maximum (19.508%) while *Grewia tiliaefolia* the minimum (0.016%) of the total litter.

Table 71 presents average annual leaf litter production in different climatic zones of the world. The data suggests that leaf litter production in tropical deciduous forests is comparatively more (1½ or 2 times) than that of cool temperate or warm temperate forests. The litter production in Koriyat forests, Udaipur was 4.80 m ton/ha/yr. The lower values obtained in comparison to that of equatorial zones and other forests of the Indian subcontinent, may be explained on the basis of unfavourable climatic conditions prevailing in this semi-arid zone. Singh (1968) attributed differences in litter production to temperature, length of growing season and the amount of insolation.

Table 71. : Annual leaf litter production

Region	Location	Leaf litter m ton/ha/yr	Author
Alpine	67°N	0.7	Bray & Gorham (1964)
Cool temperate	37-62°N	2.5	-do-
Warm temperate	30-40°N	3.6	-do-
Equatorial	0-10°N/S	6.8	-do-
Indian Subcontinent			
Shorea robusta	30.19°N	5.0	Seth *et al.* (1963)
Tectona grandis	30.19°N	5.3	-do-
Deciduous forests (Sagar)	23.50°N	2.6-9.3	Upadhyaya (1955)
Tectona grandis (Varanasi)	24-25°N	1.01-6.21	Singh *et al.* (1967)
Deciduous forests (Udaipur)	24.35°N	4.80	Ranawat and Vyas (1975).

Foliar Pigment Estimates

Foliar pigment estimates have been used as an index of poductivity of natural communities. A direct correlation between the amount of chlorophyll and dry matter production have been reported by Bray (1960), Lieth (1965), Madiana and Lieth (1964), and Lieth *et al.* (1965). Vertical distribution pattern of chlorophyll in the canopy is a relevant characteristic of a photosynthetic system (Newbould, 1967). Chlorophyll content in various foliage strata have been invariably correlated to the density of irradiation (Rabinowitch, 1945; Odum and Odum, 1959). In *Rhododendron maximum* where most of the leaves were shaded, chlorophyll content was higher in the upper leaves, while a reverse trend was observed in heathbalds, the leaves of which are well exposed to radiation (Whittaker and Garfine, 1962). The obsevations on the deciduous trees of semi-arid regions of Rajasthan (Table 72) fall in line with these observations. Of the trees, *Butea monosperma* grows in light exposed conditions on the tectonic hills and consequently, the chlorophyll content was higher in its lower canopy strata as well. Of the remaining four trees minimum height was attained

Table 72. : Canopy characteristics and pigment estimates in dry deciduous trees of Udaipur (Vyas and Vyas, 1975).

Species	Canopy stratum above ground (m)	Total leaf biomass (gm/m²)	Total chlorophyll (mg/m²)	Total carotenoid (mg/m²)
Soymida fabrifuga	3 - 4	5.36	13.97	7.42
	4 - 5	8.64	21.68	17.21
	5 - 6	16.95	39.15	40.01
	6 - 7	24.68	55.26	51.56
	7 - 8	28.85	74.88	57.62
	8 - 9	38.46	92.16	67.61
	> 9	34.16	78.24	51.24
	average	22.44	53.62	41.81
Diospyros melanoxylon	3 - 4	3.14	9.31	3.76
	4 - 5	5.06	19.52	6.63
	> 5	4.16	17.22	5.28
	average	4.13	15.35	5.22
Butea monosperma	2 - 3	4.16	14.97	4.62
	3 - 4	9.24	35.34	14.72
	> 4	2.74	7.02	2.43
	average	5.38	19.11	7.26
Wrightia tinctoria	3 - 4	1.86	2.70	1.14
	4 - 5	5.42	16.20	8.10
	5 - 6	4.36	8.81	3.96
	> 6	3.92	10.24	3.12
	average	3.89	9.48	4.08
Anogeissus latifolia	3 - 4	1.64	3.42	1.44
	4 - 5	9.47	56.41	15.04
	5 - 6	18.74	56.12	16.83
	6 - 7	21.05	54.62	18.91
	> 7	24.16	48.42	21.12
	average	15.01	43.79	14.07
Total contribution to the forest stand		276.24	735.66	419.77

by *Diospyros melanoxylon*, its leaves remain shaded and consequently chlorophyll content was higher in the upper canopy strata. The presence of miaximum amount of chlorophyll and carotenoid pigment per unit leaf area and leaf dry weight in the middle strata of the canopy of these deciduous trees in the semi-arid regions may be explained on the basis of unfavourable effects of intense irradiation and deep shading effect on the pigment accumulation (Gopal and Bandhu, 1972).

17. *Biogeochemical Cycles*

With the origin of life on earth, there came a change in the physical and chemical processes which normally lead to a random dissipation of free energy. In a non-living system, with no external energy source, the general trend is for ordered arrangements of molecules to become disordered by thermal agitation, the loss of orderly structure being defined by the increasing *entropy* of the system. The first *chemolithotrophs* were able to harness the free energy of naturally occurring chemical compounds to create localized accumulations of organic matter with an extremely complex, ordered structure, while the evolution of the photosynthetic process further sophisticated the impact of life on the geochemical environment as the exploitation of solar energy allowed much more extensive synthesis of living material. The utilization of solar energy permitted the apparent reversal of the universal entropy stream at the earth's surface and thus, the evolution, firstly of the chemolithotrophs and *photolithotrophs* and then of more complex *heterotrophic* organisms imposed a directional force on virtually all surface geochemical processes and caused an immense alteration in the history of the development of planet earth. The most dramatic change was the institution of photochemical oxygen production in an atmosphere, which was, primevally, in the reducing condition. The first chlorophyll bearing plants appeared some three to three and a half thousand million years ago in the Pre-Cambrian. Today, the photosynthetic plants on the earth's surface inhabit an atmosphere which they, themselves have produced and a soil system which is a reflection of the dynamic balance between the input of plant material and the catabolism of saprophytes (Ethrington, 1963).

The life on earth depends upon the availability of energy and circulation of some 33 to 40 elements, which plants and animals require for their normal growth and development. These elements or nutrients are termed as *biogenic salts* and are of two types—the *micronutrients* and the *macronutrients*. The macronutrients include elements and their compounds that play a key role in the protoplasm and are required in large quantities such as carbon, nitrogen, oxygen, potassium, calcium, magnesium, phosphorus etc. The micronutrients are very essential for various metabolic and ketabolic activities of plants and animals, but they are required in trace amounts, such as iron, zinc, copper, sodium, molybdenum, cobalt, strontium, boron etc. The elements flow from non-living to the living, and back to the non-living. The more or less circular path of the chemical elements passing back and forth between the ortanisms and the environment are known as *biogeochemical cycles* (Odum, 1963). *Bio* refers to living organisms and *geo* to the rocks, soil, air and water of the earth. *Geochemistry* is an important physical science, concerned with the chemical composition of the earth and the exchange of elements between different parts of the earth's crust and its oceans, rivers etc. *Biogeochemistry* is thus the study of the exchange of materials between living and non-living components of the biosphere.

Some of the elements are returned as fast as they are removed. They are stored in *short term nutrient pools*. Others may be tied chemically or burried deep in the earth for long period in *Long term nutrient pools*.

The consequence of these relationships is the establishment of a number of cycles, some of which are essentially *local* and some *global* in scale. The local cycles involve the less mobile elements in which there is no mechanism for long distance transfer but the global cycles have a gaseous component which links all the world's living organisms to form the giant ecosystem which we know as the biosphere (Ethrington, 1973). Chemically these cycles may be divided into three types : (1) the ggaseous cycles of carbon, oxygen, hydrogen and nitrogen, (2) the sedimentary cycles of phosphorus, potassium, calcium, magnesium, copper, zink, barium, molybdenum, manganese, iron, and silicon, (3) the gaseo-sedimentary cycle of sulphur.

The global cycles of carbon, nitrogen, phosphorus and sulphur have been studied extensively in recent years. The main aim of these studied have been to account for the overall balance of the material fluxes due to motions of air and water and the determination of the biological and chemical processes that play a key role in this context. Even though the present level of our knowledge about many of these processes is still inadequate, we understand the dynamics of the major biogeochemical cycles in their broad outline.

In Figure 68 is shown a simplified biogeochemical cycle. *Abiotic pool* is composed of atmosphere, hydrosphere and lithosphere. *Biotic pool* is composed of autotrophic, heterotrophic and carnivorus and saprophytic organisms. There exists varying degrees of exchange in every pool and among both the pools. An atom in the cycling pool is instantly available to the organism, while an atom in the reservoir pool may or may not be permanently unavailable to the organisms. Almost always there is a slow movement of atoms between the unavailable and the available pools (Odum, 1963).

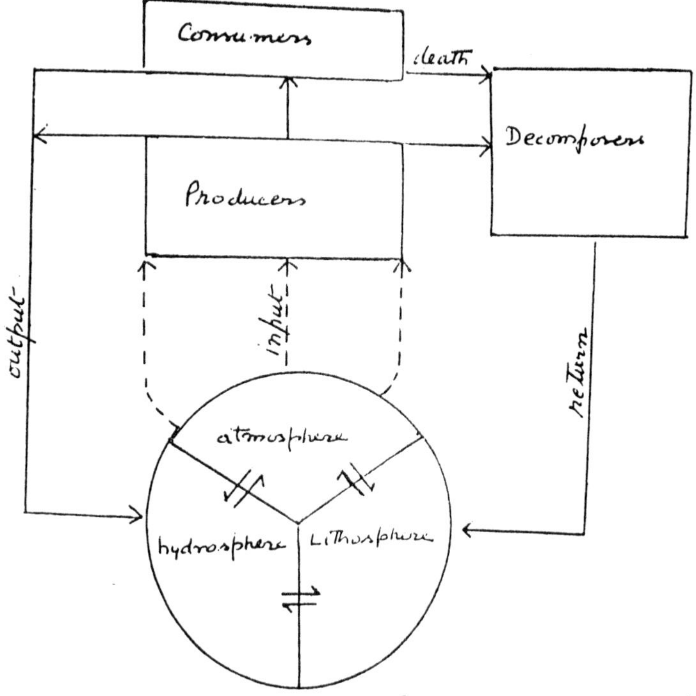

Fig. 68 : A model of the transport of a chemical element within the biosphere.

Carbon Cycle

Carbon is a basic component of all organic compounds, the building material of which all living things are constructed. Green plants get their carbon supply from the atmospheric carbon dioxide through the process of photosynthesis. The primary consequence of the photosynthetic process is the storage of energy in reduced carbon compounds. These may be then explited by the consumers and decomposers of the ecosystem so that the throughput of energy is accompanied by a cyclic exchangee of carbon dioxide with the atmosphere. This exchange takes place over the wholel and and ocean surface of the earth wherever temperature is high enough, or water supply adequate, to support life.

The global distribution of carbon shown in Table 73 indicate that it clearly involves the annual exchange of significant amounts of carbon between the atmosphere, terrestrial biosphere and its oceans.

Table 73 : Global distribution of carbon (Singer, 1975; and Baes *et al.*, 1977).

Particulars	g^C
Atmosphere	7.00 X 10^{17}
Terrestrial biosphere	1.76 X 10^{18}
Oceans	3.90 X 10^{19}
Fossil. fuel	1.20 X 10^{19}
Ocean life sediments	2.05 X 10^{18}
	Fluxes of carbon
	g^C/Yr
CO_2 consumed in phtosynthesis	3.1 X 10^{16}
CO_2 released from fuel consumption	4.1 X 10^{15}
CO_2 equilibrium flux with oceans	9.0 X 10^{16}
CO_2 increase in atmosphere	1.4 X 10^{15} (1.0 ppm/yr)

On a geological time scale it may be seen that photosynthetic carbon consumption would rapidly exhaust available resources if it were not for the respiratory return of carbon dioxide to the atmosphere. The carbon locked up in the animal wastes and in the protoplasm of the plants and animals is released by the activities of decomposers.

The atmospheric carbon dioxide concentration is globally rather constant, though a small seasonal variation over the earth's surface is correlated with vegetational cover and photosynthesis (Leith, 1970), and local environment derives from urban-industrial burning of fossil fuel. This constancy of the atmoispheric carbon dioxide concentration is due to the buffering effect of its solubility in ocean water, coupled with the mixing caused by large- and small-scale are turbulence. The interchange between the two phase occurs through diffusion. The direction of diffusion is dependent on relative concentrations in the two environments. The dissolved carbondioxide reacts with water in the soil or in aquatic ecosystems to form carbonic acid. This reaction is reversible and may proceed both forward and backward. In turn the carbonic acid dissociates in a reversible reaction into hydrogen and bicarbonate ions. The bicarbonate ion, in another reversible reaction, gives hydrogen and carbon ions. The various reactions may be summerized as follows :

$$CO_2 + H_2O \rightleftharpoons H^2OCO^3 \rightleftharpoons H^+ \; HCO^-_3 \rightleftharpoons H^+ + CO_3^{-2}$$

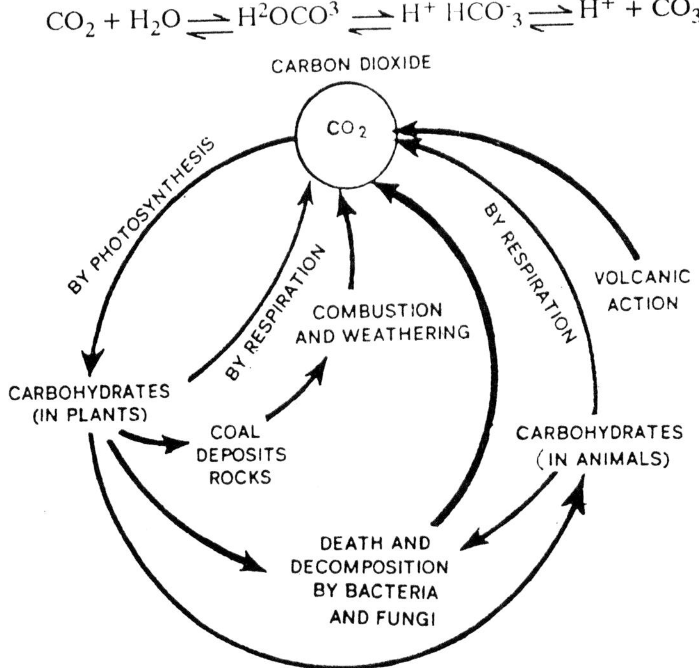

Fig. 69: The global cycle of carbon.

Since all these reactions are reversible, therefore, if the concentration of carbondioxide into the atmosphere is less there will be movement of carbondioxide into the atmosphere. However, when there is deficit of carbondioxide into the atmosphere. However, when there is deficit of carbondioxide in the sea, the movement will be from atmosphere towards the sea.

Despite the large amount of fossil carbon which has been returned to the atmosphere since the industrial revolution, the ocean buffer has so far prevented an excessive increase in concentration, but it has been estimated that, by the 2000, there will be a 25 per cent increase in its concentration in the atmosphere. The consequence of such an increase will be alteration of the energy budget of the earth's surface by the 'green house effect'. The obvious consequences would be an increase in atmospheric temperature and global cloud cover (Revelle *et al.*, 1965).

Nitrogen Cycle

Gaseous nitrogen is the most abundant element in the atmosphere and its global circulation (Figure 70) provides an indexhaustible reservioir for the nitrogen fixing organisms which supply almost all of the nitrogen utilized by plants. The quantity of nitrogen combined in living and dead organic matter is small in comparison to the total capacity of the atmospheric reservoir, a characteristic in which nitrogen differs from carbon cycling. Besides the gaseous reservoir of enormous magnitude nitrogen also occurs as a complex soil based reservoir which is localised and of small magnitude.

Though the atmosphere contains 79 per cent nitrogen, yet most of the plants and animals cannot make use of this gaseous nitrogen. Plants can make use of nitrogen salts like nitrates and animals must have their nitrogen in the form of aminoacids.

The global emission rate of ammonia from soils and from plants and animals has been estimated at 75×10^6 tons/year (NRC< 1979). Pollution sources primarily from ammonia production plants and from fertilizer applications account for 0.32×10^6 tons/year in the United States (Lamb, 1984). According to Hutchinson (1944) biologically 140-700 mg/m^2/year nitrogen is fixed whereas a small amount 35 mg/m^2/year is fixed by electrification and photochemically. But, according to Delwiche (1965, 1970) the amount of nitrogen fixed biochemically is 1 gm/m²/year. The amount of nitrogen fixed in fertile areas may range up to 20 gm/m²/year.

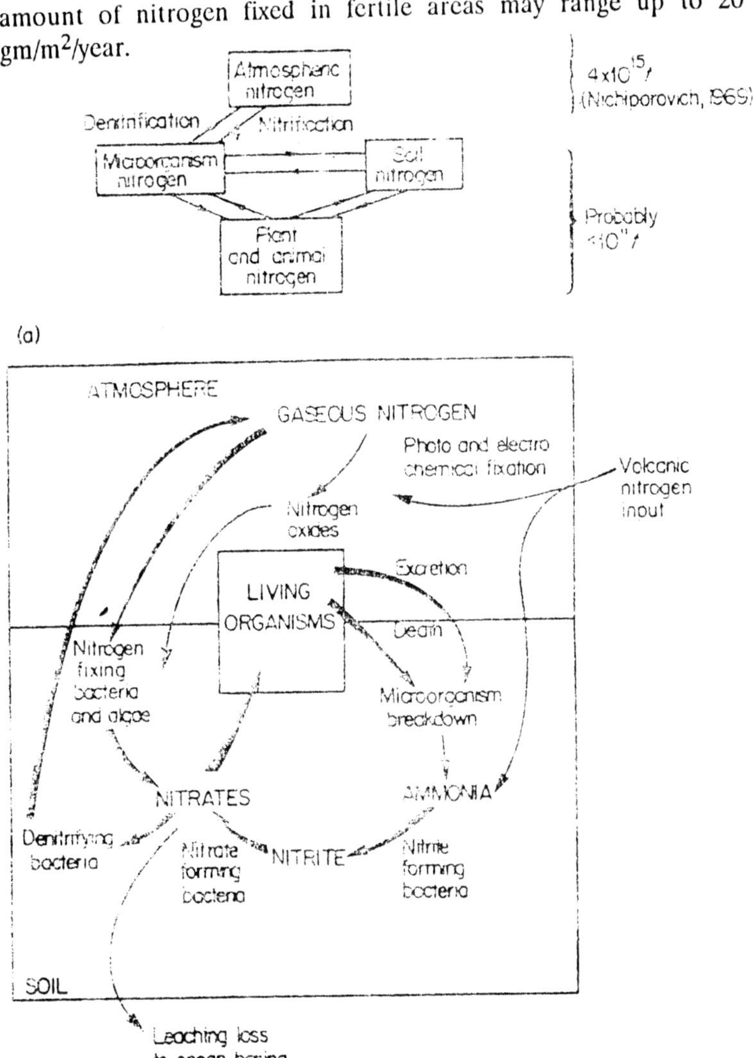

Fig. 70 : The nitrogen cycle. (a) Major relationship between the very large atmosphere pool of gaseous nitrogen and the biosphere. (b) The complex interrelationships of the soil-based portion of the cycle. (after Ethrington, 1975).

The nitrifying bacteria like *Nitrosomonas* convert ammonia into nitrate, whereas others like *Nitrobactor* act on nitrates and convert them into nitrates. Considerable quantities of nitrogen is added to the soil by free living nitrogen fixing bacteria—*Azotobactor* and *Closteridium*. Other nitrogen fixing bacteria like *Rhizobium* grow in the root nodules (perticularly legumes) and fix nitrogen into nitrates for use of the plants. Among blue-green algae *Anabaena*, *Nostoc*, *Gleotrichia echinulata*, *Trichosdesmium* etc. are reported to play a significant role in the fixation of nitrogen in aquatic situations.

Nitrogen exists in the biosphere primarily in protein and nucleic acids. Decomposition of organic matter leads to the enzyme catalyzed release of ammonia. For example, an amino acid may undergo the following reaction :

$$R-\underset{\substack{|\\ NH_3}}{CH}\text{-}CH\text{-}COOH + \frac{1}{2}O_2 \xrightarrow{\text{enzyme}} R-\underset{\substack{||\\ O}}{C}\text{-}C\text{-}COOH + NH_3$$

Similarly urea, which is found in high concentration in sewage, can be broken down catalytically to release ammonia :

$$CO(NH_2)_2 + H_2O \xrightarrow{\text{Urease}} 2NH_3 + CO_2$$

Denitrification carried out by anaerobic organisms release nitrogen and nitrogen oxide :

$$NO_3^- + HCHO \longrightarrow \frac{1}{2}N_2O + \frac{1}{2}H_2O + CO_2 + OH^-$$

This type of oxidation and release can also occur with fats, fatty acids, amino acids and methane. Nitrogen oxide (Nitrous oxide) formation typically amounts to approximately 10 per cent of the nitrogen returned to the atmosphere. Some nitrogen is also returned to the atmosphere by denitrifying bacteria. Most of the denitrifying bacteria reduce nitrate to nitrite, and still others to ammonia. Denitrification to molecular nitrogen occurs under anaerobic or partially anaerobic conditions.

A large part of nitrogen from the atmosphere is fixed up by chemosynthetic, symbiotic and mycorrhizal activities and stored in the plant biomass. Nitrogen that leaves the system is the same amount as is taken in from the air or water.

Sulphur Cycle

Sulphur cycle links air, water and soil. In the atmosphere it occurs in the form of sulphur dioxide which is formed during combustion of fossil fuel or due to volcanic activities. Hydrogen sulphide is another component of the atmosphere formed during decomposition of organic sulphur under anaerobic conditions. The gaseous phase does not, however, form a large reservoir as it does with nitrogen.

Sulphur dioxide is converted to sulphuric acid vapour in the presence of moisture. These acid vapours remain in the atmosphere at high temperature. As the temperature begin to decrease, the vapours also begin to condense slowly. The resulting condensate takes the form of aerosol droplets which owing to the presence of unburnt carbon particles is black, acidic and carbonaceous in nature. This matter is popularly called "acid must", has been found to spot and/or blister and corrode the substrates. Problems associated with acid must take on an even more serious nature as it encounters rain or snow in the atmosphere and return to the earth's surface in the form of "acid rain" torrents. Depending upon the acidity of the precipitation, acid rain can cause serious ecological imbalances by (1) running fresh water lakes, streams and rivers, consequently killing fish and other marine life, (2) causing essential plant and crop nutrients to leach out of the soil, inhabiting the process of nitrogen fixation by micro-organisms, affecting soil fertility and consequently damaging crops, and (3) causing serious health hazards by contaminating drinking water, since toxic sediments of metals like cadmium, lead and aluminium are liberated from the affected fresh water lake, river bed by the process of acid extraction.

The soil based part of the cycle (Figure 71) involves a microbial oxidation-reduction interchange between sulphate, element sulphur and sulphide. Under aerobic conditions a variety of heterotrophic organisms such as *Aspergillus, Neurospora, Escherichia, Proteus* etc., decompose organic sulphur compounds and release sulphates. But under anaerobic conditions *Desulphovibrio desulphuricans* converts sulphates into sulphide.

When a waterlogged soil drives, and becomes aerobic, bacteria such as *Theobacilus thiooxidans* oxidise H_2S to sulphate, and as a result extreme soil acidity may arise. Problems have occurred when metal or concrete structures are exposed to the corresive effects of such soils.

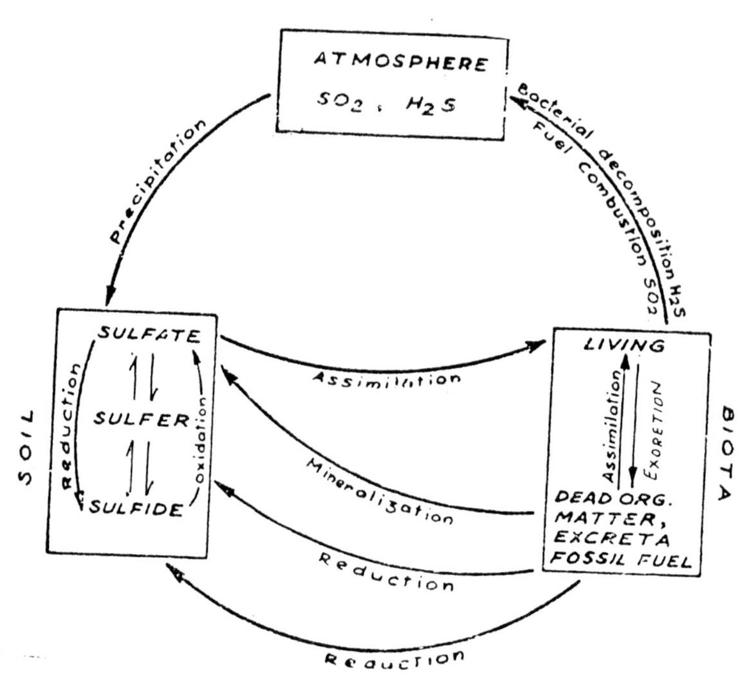

Fig. 71 : Sulphur cycle involving land, soil and organisms.

Most fossil fuel contain a considerable amount of sulphide or elemental sulphur, as they were deposited under anaerobic conditions conductive to sulphide accumulation. The burning of these fossil fuels has added a sulphur load to the atmosphere. Sulphur dioxide causes acute phytotoxicity ranging from blistering, bleaching and defoliation to complete destruction of surrounding vegetation.

Phosphorus Cycle

Phosphorus cycle is simpler than nitrogen and sulphur cycles. It has fewer pathways and chemical conversions but, lacking a large and mobile reservoir, it is more easily interrupted by both natural processes and human interference. Figure 72 shows that the cycle has two main storage pools of insoluble organic and inorganic phosphorus while a much smaller pool of dissolved phosphorus permits movements between them. Mobilization of phosphorus from the organic form is entirely a biological process while micro-organisms and root surfaces probably play a large part in dissolving inorganic phosphorus.

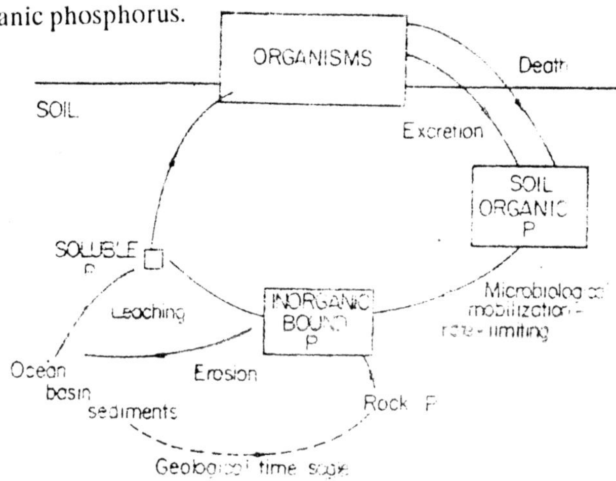

Fig. 72 : The local cycle of phosphorus. Both the organic and inorganic pools of soil phosphorus are essentially rather immobile with the result that the mobilization of organic P into the plant-available phosphate form is rate-limiting to the cycle, and also that the leaching loss of soil P is fairly slow. In many cases it is physical loss by soil erosion which removes P from natural ecosystems and agricultureal land. Loss by these two mechanisms is only counteracted on a geological time-scale when ocean basin sediments become elevated to form new land surfaces. Soil P is normally very insoluble but the addition of fertilizer P causes a great increase in the rate of leaching to ground water or water courses.

The removal of phosphorus to ocean basins by sedimentary processes enormously increase the time scale of the cycle which becomes complete only when geological processes expose sediments or newly formed land surfaces. The loss has now been accelerated by the mining of phosphate rock fertilizers, a form in which it is easily leached to the ground water and thence to rivers. Though the present resources of phosphate rock are adequate, the irreversible loss may ultimately be disastrous for agriculture unless some means of recycling the surplus phosphorus can be found. At present the impact of this loss is felt in phosphorus shortages but in the superbundance of phosphorus which is entering natural waters, so forming part of the pollution—eutrophication syndrome (Ethrington, 1976).

A perusal of the biogeochemical cycling processes within the ecosystem make it amply clear that the abiotic components of the environment are transformed into biotic structure through metabolic activities and are locked up in the biomass for some time depending on the turnover rate. In lower plants with short life-span and soft tissues the turn-over rate is quicker than the higher plants having long life and with hard tissues.

In nature sometimes the biogeochemical circulation process is held up for a long time, may be millions of years, as in the case of fossil fuel deposits which are organic derivatives and are locked up in the lithosphere. Man is now harnessing energy from fossil fuel and hastening the biogeochemical cycling processes. Fire is also an important factor which hastens the biogeochemical processes on the terrestrial ecosystems. The decomposers activities also accelerate the cycling processes, faster decomposition rate in humid tropics results in rapid release of materials, held up in the organic debris, to the environment (Misra, 1974).

Cycling of Radioactive Materials (Strontium-90)

In recent years, man has been adding radioactive materials into the biogeochemical cycles due to some of his activities. These radioactive materials result from chain reactions in nuclear weapon testing, nuclear warfares and in nuclear reactors. One of the most important radioactive material released into the environment is Strontium-90. Later these return to the earth along with rain, dust and other materials as atomic fallout. It is taken up by plants, more readily, both directly by foliage absorption and via soil and water. Strontium-90 enters the animal tissue most easily through the grazing food-chain, especially in the region with high rainfall and low level of calcium and other mineral content of soil. It has been found that amount of Strontium-90 is many times greater in bones of sheep grazing on the English moors, than sheep grazing in drier and more fertile areas. When taken in with food by man and vertibrates, it becomes concentrated and is retained by bones in close proximity to blood making tissue in the bone marrow.

18 *Systems Ecology and Ecosystem Modelling*

In trying to describe, what systems ecology is or means ? One is in the somewhat same position, as the five blind men who were trying to describe an elephant by touching his different parts. The only agreement on a definition seems to be that it is an approach to problem solving and focuses on the entire problem or system, rather than on each of its separate components. It is an approach that tries to quantitatively define alternate courses of action, and proovide a means of evaluating them in an orderly fashion with the aim of selecting those that are better than others. The idea of comprehensive analysis is not new, but it is only in recent years that the mathematical, statistical and computational tools have become available in the field of ecology, that permit systematic storage, retrieval and analysis of large amount of data than have heretoforce been possible (Deininger, 1980). With the systems involved and with superfine interacting links, the problem appeared and solutions were searched. The result was the birth of the so-called 'system ecology'—where simple mathematical and statistical concepts are involved in an analysis. Systems ecology is thus, not only a mathematical technique, but rather is a broad research strategy to offer best choice. The whole process is nothing but logical and orderly organisation of data, to permit the analysis of many more alternatives. It is hoped, by being able to compare many alternatives, the decision—makers should be able to define those combinations of system component that best meet the stated objectives.

The aim of systems ecology and ecosystem modelling is mainly to formulate and desscribe the structure and functions of the ecosystems through verbal, graphic and mathematical models by scientists so that diverse sets of data collected for different ecological systems could be brought on comparable grounds, through systems ecology and modelling. This kind of inter-comparison is extremely useful in management of ecosystems. Models particularly play an important role in synthesis by providing mechanisms for combining diverse experimental results and integrate them with the literature (Van Dyne *et al.*, 1978).

Ecological systems are complex systems and often difficult to define in space and time (Odum, 1971). Models incorporating various *components* or *holons* of the system that can be quantified (biomass, nutrient, energy etc.) help in simplifying a unified picture of the system. Webster's dictionary defines it "*as regulatory interacting and interdependent components forming a unified whole*". Patten (1971) definessystems *as "a group of physical components connected or related in such a manner as to form and/or act as an entire unit*". These component parts or observable attributes are termed as state variables, which then give us the state or condition of the system.

Systems change with time. They can either be *deterministic* (wherein the previous state determines the succeeding one and that everything is determined by some cause), or *stochastic* (where behaviour is random and probabilistic). System can be either *static* (in a steady state) or *dynamic* (a transient state). However classified, a system should be definable, observable, measurable and a recognisable whole which has an inbuilt concept of size, levels and hierarchy.

Levels of Hierarchy

A system could be visualized as a three level hierarchy :

(1) System (at reference level)

(2) Subsystem (at part level)

(3) Supersystem (at environment level).

Levels of hierarchy are time and space dependent. The strength (energy) of interactions decrease with increase in distance. Hence, the levels of organisation are invariably proportional to the interaction strength. These levels of hierarchy move downwards and inwards, that is from a whole to a part. A generalized hierarchial model has been presented in Table 74.

Table 74 : Hierarchy in ecological units.

Level	Time-scale	Size-scale
Ecosystem	Years, centuries	Square kilometer
Habitat	Months, decades	Hectares
Transitional habitats	Months, years	Square meter/hectare
Within habitats	Hours, months	Square meter

Elements of Systems Ecology

Basically there are four elements of systems ecology as indicated in Figure 73. The information generated can either be used for prediction and explanation; or it can also be fed back to the systems for refining the model.

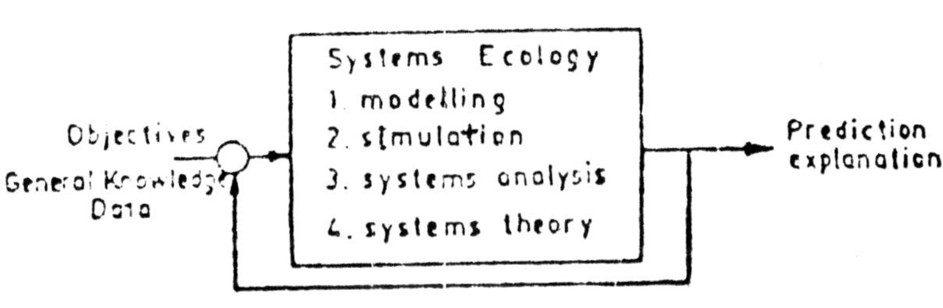

Fig. 73 : Elements of systems ecology.

(a) Modelling

The model is an abstraction of serial system, and one of the major reason for modelling is to provide a frame-work for bringing together the parts of the system which is too complex to organise and comprehand otherwise. The second major reason for having model is testing an experimentation (Pandeya, 1977). In terms of Odum (1971), a model is a formulation dynamics, a real world phenomenon, and by means of which prediction can be made. In the simplest form, models may be verbal or graphic, but ultimately statistical and mathematical. He further defines that the world 'statement' is in fact word 'model' of ecological principles in question. The modelling involves recognition, definition of goals and objectives, generation of solutions, courses of action and prediction about any system (Dubey, 1990).

Models may be evaluated in terms of three basic properties or goals : *realism*, *precision* and *generality*. Realism refers to the degree to which the mathematical statement of the models, when translated into words, correspond to the biological concepts that it intends to represent. Precision is the ability of the model to predict numerical change and to mimic the data on which it is based. Generality refers to the breadth of applicability of the model. The number of different situations in which it can be applied. Indeed the question asked of ecological models are often complex. Because ecosystems have random inputs such as weather, it has been unreasonable to construct model of high predictive power when basic inputs frequently cannot be measured or predicted. Thus, ecological models are often judged in terms of ability and generality to guide research efforts rather than on numerical predictive power (precision).

Modelling is very useful in understanding the systems behaviour. Models can never be exact replicas of the system itself. They are infact abstractions. Nevertheless, a model is informative. Model is a sequential and repetitive process which helps in conceptualisation, synthesis, simulation and analysis of synthesis. The key to effective modelling is to strike a proper balance between realism and abstraction. Ashby (1956) defines a model as homomorphism of some real system which it represents. Models have ability to describe and predict behaviour of ecological systems, governed by the physical, chemical and biological principles.

Modelling is the process of making a simplified representation of complex reality for some specific purpose; to show the relationship between entities; possibly in a mathematical form, to illustrate what we know and do not know, for a specific predictive purpose. Before we go ahead and construct a model let us examine the elements that are essential for any model. Walter (1971) listed some of these elements as :

1. *State variables* - These are sets of numbers which are used to represent the state or condition of the system at any time.

2. *Transfer function or functional relationships*—These are the equations which represent the flows or interactions between components or compartments of the system.

3. *Forcing functions*—These are the equations which represent the inputs to the system, or factors affecting but affected by the components of the system.

4. *Parameters*—These are the constants of mathematical equations.

In practice, there are two kinds of approaches in ecological modelling: (*i*) the deterministic, and (*ii*) the stochastic approach.

The deterministic approach

It involves simple constants, and emphasizes the quantities of materials and energy in ecosystem compartments. Such models are usually well developed as systems of fairly differential equation like :

$$\frac{dy}{dt} = ry \qquad (1)$$

Where y = density of population at time t; y = a constant. Thus, if applied to bacterial colony, may reveal that colony may increase and exhaust the resources and thus we may need two constants. The equation of the model will be thus :

$$\frac{dy}{dt} = ay - by^{2} \qquad (2)$$

where a and b are constants.

The stochastic approach

In this approach the equation (2) will be :

$$\frac{dy}{dt} = (a + y(t))\,y \qquad (3)$$

where yt is a random variable with mean zero, that is, the value of it will differ from occasion to occasion. Thus, stochastic models involve many mathematical equations mostly of poisson series, or distributions, or binomial distributions. A frequency distribution thus will be presented as $n(q + p)k$. Where n = number of sampling units,

p = probability of any point in the sampling unit being occupied by the individual, $q = 1 - p$, and k = maximum possible number of individuals a sampling unit could contain.

Among the main models we have the following at use :

1. *Multivariate*—A common model with a large number of forms, involving more variables.
2. *Descriptive*—Consisting of principle component analysis and cluster analysis.
3. *Predictive*—Involving multiple regressions.
4. *Game theory*—Closely related to mathematical programming models based on theory of games.
5. *Catastrophy theory*—An elegant example of mathematical topology.

The development and use of model

In essence, hypothesis form the key of modelling. Three classes of hypothesis may be distinguished :

1. Hypothesis of relevance—identifying and defining the variables, species and subssystems which are relevant to the problem.
2. Hypothesis of processes—linking the subsystem within the problem and defining the impacts imposed upon the system.
3. Hypothesis of relationship—and the formal representations of these relationships by linear, non-linear and interactive mathematical expressions.

The development of models often follows a somewhat standardized pattern in which there is a progressively detailed breakdown of each component into modules which can be more easily translated into research activities. The following conclusions can usually be drawn :

(a) The quality of the available data and the understanding of causal pathways, especially as they relate to ecology are generally poor.

(b) Systems analysis and data collection must develop a mutual feedback from which the decision-makers can draw maximum benefit.

(c) Training in systems analysis is valuable for stressing a broad, interdisciplinary, problem oriented philosophy of research.

(d) Systems modes can be improved by building them and striving to correct their weakness.

(e) Systems analysis teams must be broadly interdisciplinary.

(f) Research which does not use systems analysis may demand large quantities of high quality data and may consequently be expensive.

What is needed are models that consider the country or the city as one integrated and mutualistic life-supporting system. The common practice of 'sub-optimising' a model for only part of the system, as for example, considering only food production without taking into consideration the environmental stress, economic and social consequences of increased production can be extremely misleading and dangerous if such models are taken to the 'whole truth' (Dubey, 1990).

The Construction of an Ecosystem Model

The first step in modelling is to construct a base or a conceptual model, with the following information :

1. Verbal description of subsystem level holons.
2. Verbal description of environmental inputs and outputs.
3. Adjacency matrix.
4. Pictorial energy models for component sub-system and interactions. Two kinds of symbols are used, (*i*) Forrester diagrams, and (*ii*) H.T. Odum's modules for energy circuit language.
5. Verbal description of control factors.

The construction of an initial-output model (Fig. 74) allows comparisons between different ecosystems. Because of the spatial and temporal differences in heterogeneity between systems, the study of the processes within such systems must be adjusted to appropriate scales. Comparisons are then possible by scaling the system down to a common unit, say an area of 1 m^2 or whatever suitable.

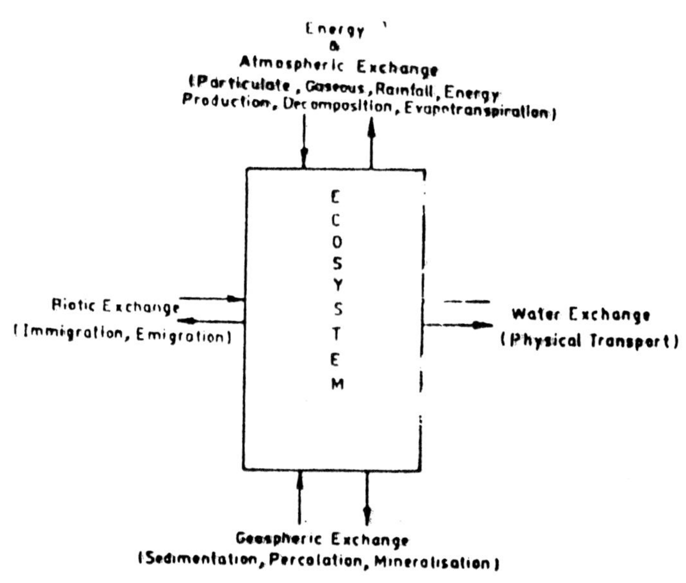

Fig. 74 : Initial input-output model of an ecosystem.

Assigning biotic compartments within the above generalised model is the next stage of model evolution. The various biotic components of different strata of a forest ecosystem are shown in Figure 75, a compartmental conceptual model along different trophic lines. Also the flows to and from the system are compartmentalised with respect to the source of the material within the system.

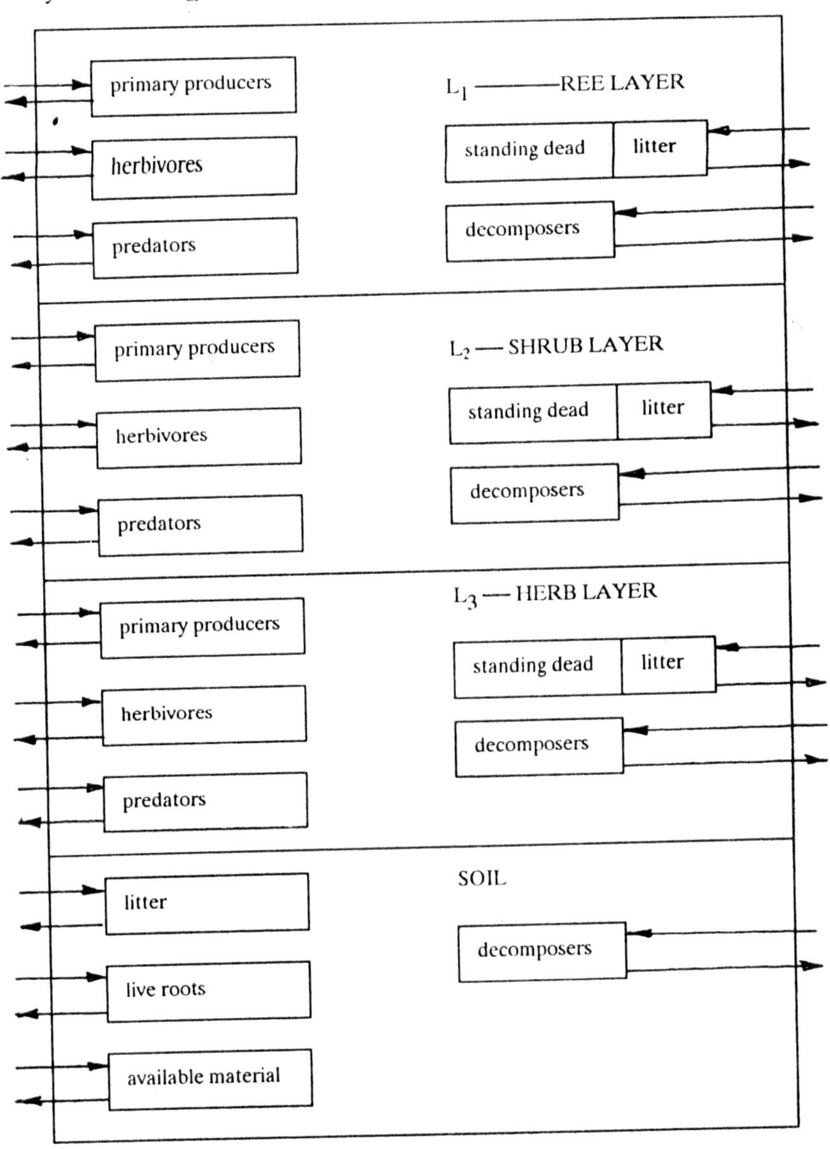

Fig. 75 : A compartmental conceptual model of a forest ecosystem. The biotic components of different levels of orghanisation (tree, shrub, herb layers, soil etc.) make the four, small black boxes of the larger black box — the forest ecosystem.

Now the individual compartment is taken for detailed study. The details of one compartment have been shown in Fig. 76. The adjacency matrix of each compartment is now conveniently represented in the form of an adjacency matrix. An adjacency matrix is a set of number or symbols organised into rows and columns. The various abbreviations used in the figure and text are as follows :

leaf - Lf, Branch - Br, Bole - Bo, Underground - Ug, Standing dead - Sd, Fresh litter - Lt, Decomposed litter - Ld, Soil - So, Mycoflora - Mf, Fauna - Fa, Leaching - Lh.

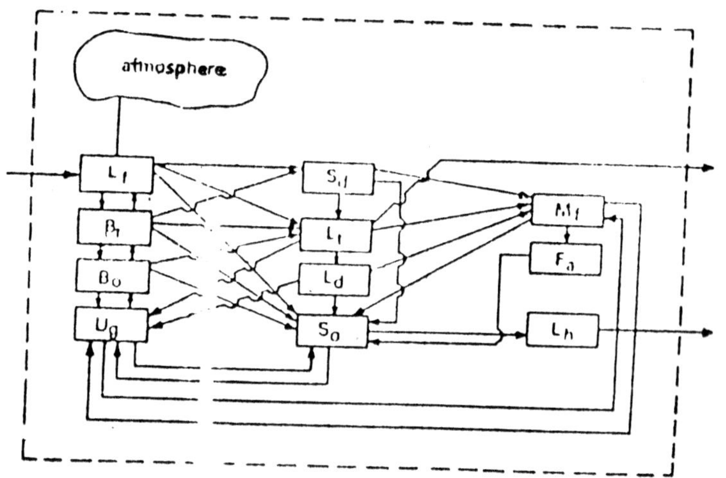

Fig. 76 : A single compartment (smaller black box) of ecosystem model snowing the details of flow of materials through its biotic and abiotic components.

A detailed working model for nitrogen cycling in a teak forest of India have been proposed by Pandey (1979) based on individual compartment detailed study. Values within boxes show storage of nitrogen and values on arrows as annual rate of nitrogen (Fig. 77).

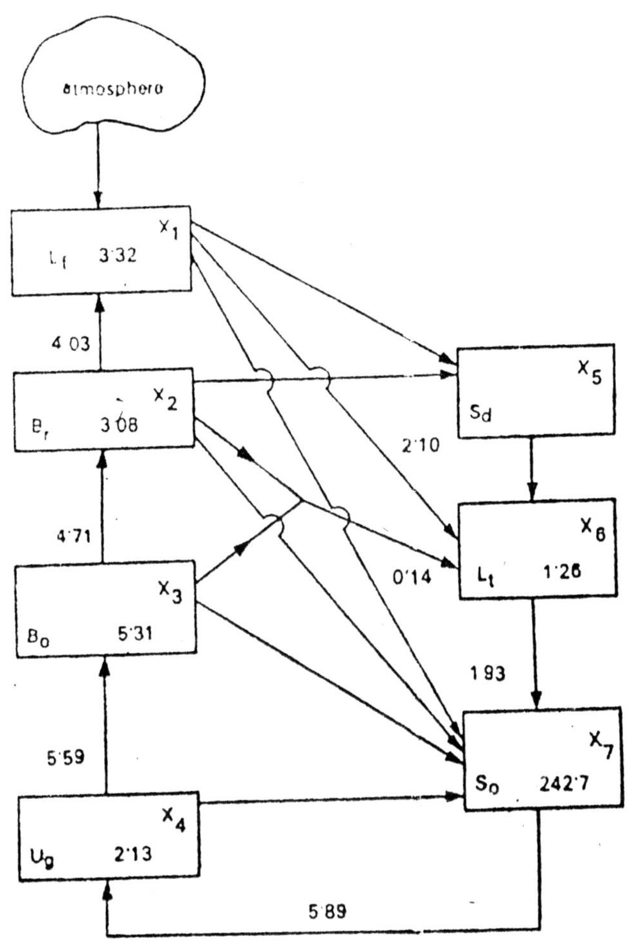

Fig. 77 : A detailed working model for nitrogen cycling in a teak forest of India. Values within boxes show storage of nitrogen (g/m²) and values on arrows as annual rates of nitrogen (g/m²/yr) (After V.N. Pandey, 1979).

Compartments or holons are now represented by specific symbols in a specific language (Forrester diagrams - Fig. 78; or Odum's energy circuits language - Fig. 78) *i.e.*, symbol for energy sources, primary producers, consumers, heat sinks, passive storage, work gates etc. Energy symbol has a mathematical definition and therefore synthesizes the functions and relationships of the reference system. Although the energy diagrams network tends to get rather complex at times, it has an added advantage over the Forrester diagrams in that a lot of verbal description is reduced by use of specialized symbols.

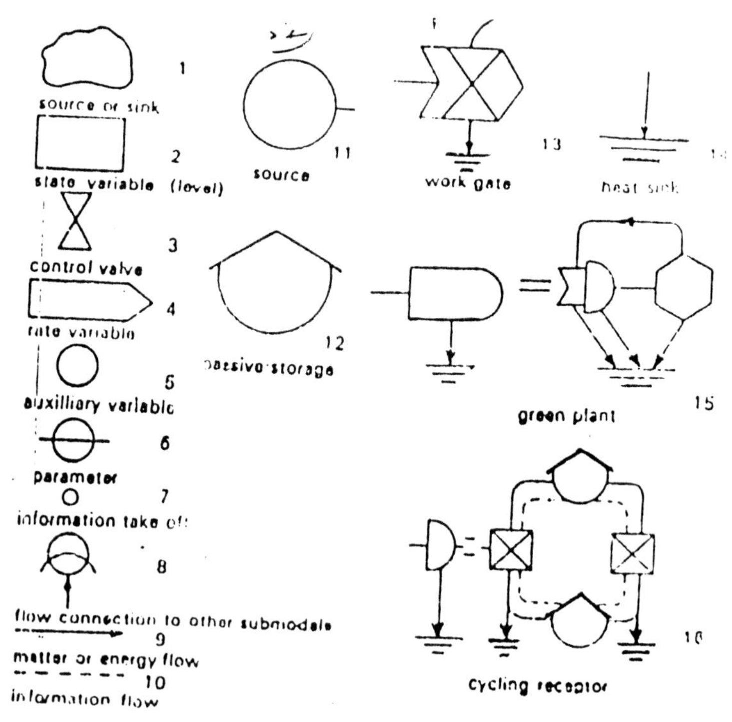

Fig. 78 : Symbols used to show various processes in ecosystem models. 1-10, as used by Forrester (1968) with Rykiel's (1977) modifications and 11-16, the modules of the energy circuits language, by H.T. Odum (1971).

Use of Compartmental Study

A study of four compartments (live green, standing dead, litter, and root) was attempted in the herbaceous vegetation of Kalighati, Udaipur (Vyas, 1967). The significant observations are presented in Fig. 79.

Live green biomass was maximum (393.99 g/m^2) in September, and was contributed by 30 species (13 grasses and 17 forbs). The contribution of grasses alone was 280.17 g/m^2, while the remaining 113.82 g/m^2 was contributed by forbs. The minimum value (19.00 g/m^2) of live green biomass was observed in June, when only 3 perennial species were found to survive.

Standing dead biomass varied throughout the year. The value 22.00 g/m^2 recorded in June represent a carry over from the previous growing season. The amount of standing dead biomass was found to increase gradually from June till it attained a peak (346.8 g/m^2) in December. The observations further suggest a continuous fall in the standing dead biomass up to the month of May. Among the grasses the total contribution towards standing dead biomass was found to be maximum by *Apluda mutica* followed by *Sehima nervosum*. Amongst the forbs maximum contribution was by *Cassia tora* followed by *Bidens biternata*.

Litter biomass (198.00 g/m^2) recorded in June represent a carry over from the previous growing season. The amount of litter was observed to drop gradually after June until a minimum value (47.8 g/m^2) was attained in August. The maximum value (315.0 g/m^2) of litter biomass was observed in May.

Below ground compartment biomass (338.0 g/m^2) recorded in June decreased at the advent of monsoon showers and attained a minimum of 186.0 g/m^2 in September. The observations further indicate that from October onwards the below ground biomass gradually increased and was observed to be maximum (495.0 g/m^2) in December, followed by a gradual decrease.

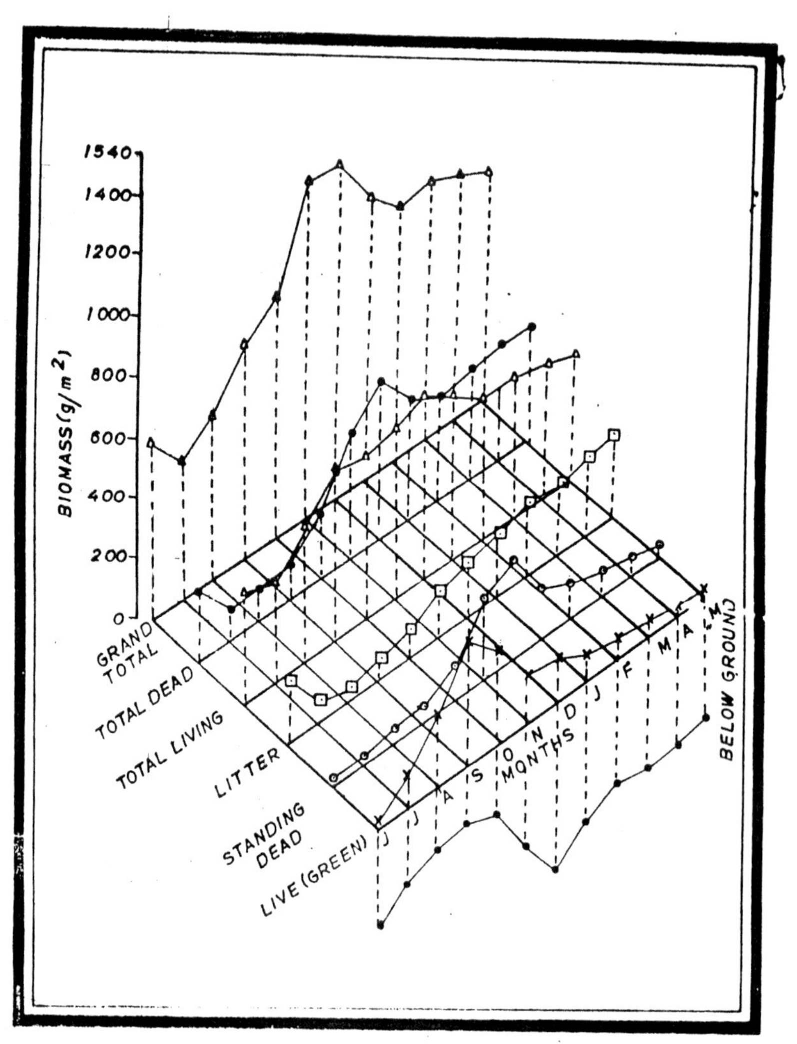

Fig. 79 : Monthly fluctuations in live green, standing dead, litter, below ground, total living, total dead and total plant biomass in the herbaceous community at Udaipur (after Vyas, 1967).

Dry Matter Dynamics and Budget

The flow of dry matter through different compartments presented in Fig. 80 suggests that the gross primary production by grassland vegetation at Udaipur was 971.47 g/m^2/yr, of which 224.18 g/m^2/yr was lost in respiration, and 791.74 g/m^2/yr was available for total storage, while only 747.29 g/m^2/yr was observed in net primary production. Out of which 331.0 g/m^2/yr was contributed by roots and 381.75 g/m^2/yr was contributed by above ground litter. The biomass lost from roots and litter was 282.00 and 150.20 g/m^2/yr, making a total loss of 432.20 g/m^2/yr.

System transfer function between different compartments indicate the intrinsic behaviour of each compartment. Observation presented in Fig. 81 suggest that higher production was directed towards above ground (STF = 0.501) as compared to below ground (STF = 0.442) biomass production. The rate of transfer from above ground net primary production to standing dead, from standing dead to litter, from litter to litter disappearance was found to decrease following a system transfer function of 0.849, 0.839 and 0.562 respectively, as compared to the rate of below ground net primary production to root disappearance (STF = 0.851).

Tripathi *et al.* (1991) studied single tree influence on primary productivity and transfer dynamics on a stabilized sand dune at Ajmer. Biomass flow under *Acacia leucophloea* and *Eucalyptus camaldulensis* have been shown in Fig. 82 and 83 respectively. Data on above ground biomass production except for litter (338.9) and litter disappearance (328.0) which were significantly higher under *E. camaldulensis,* indicate that litter had little effect on the above ground net production and shot disappearance because in some cases single tree influence was not detected due to soil homogenization and litter redistribution. However, reasonably high production of *E. camaldulensis* leaf litter and decomposition influence the soil chemical properties adversely by way of releasing some organic compounds from leaves or roots. This may be a possible reason of reduced below ground production under *E. camaldulensis.* Therefore, differences in production potential under different tree plantations may be attributed to the quantity and type of litter production at the site.

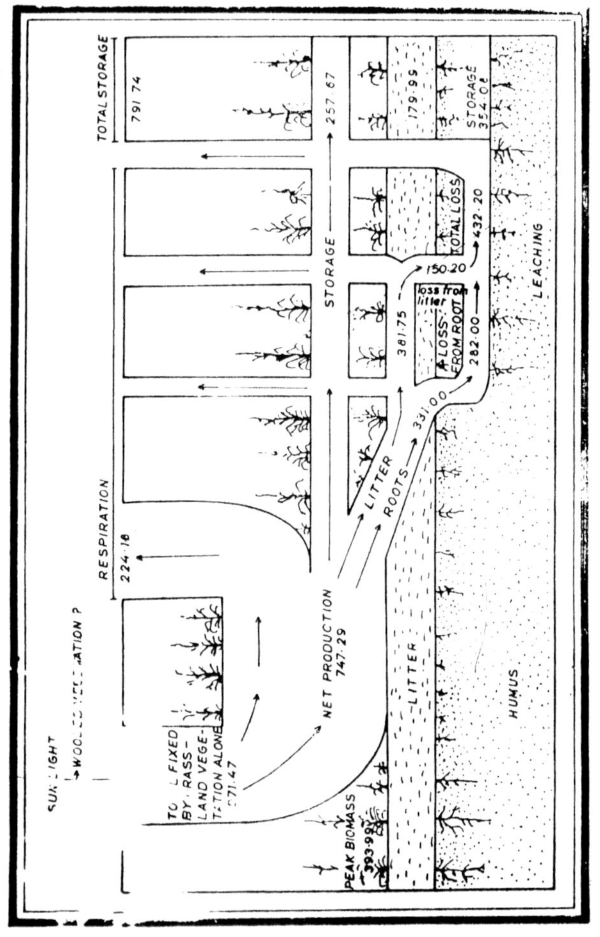

Fig. 80 : Dry matter relationships among different compartments for Apluda community.

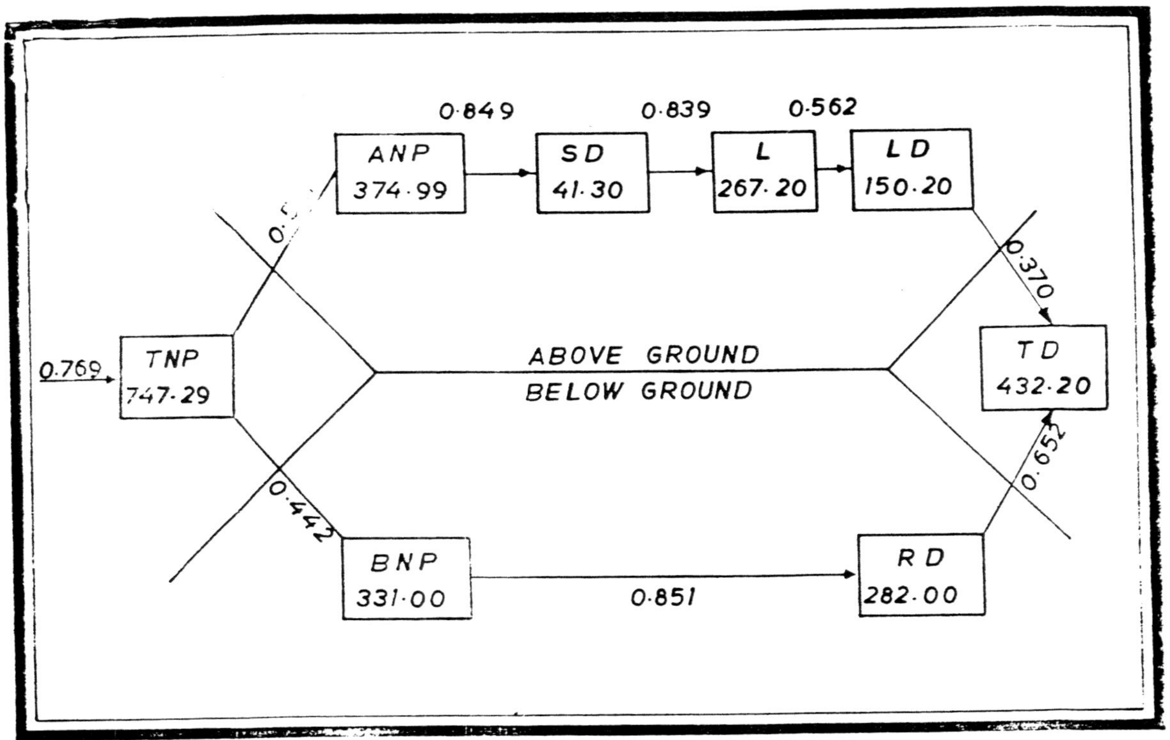

Fig. 81 : Dry matter transfer and functional rates in different compartments of Apluda community at Udaipur.

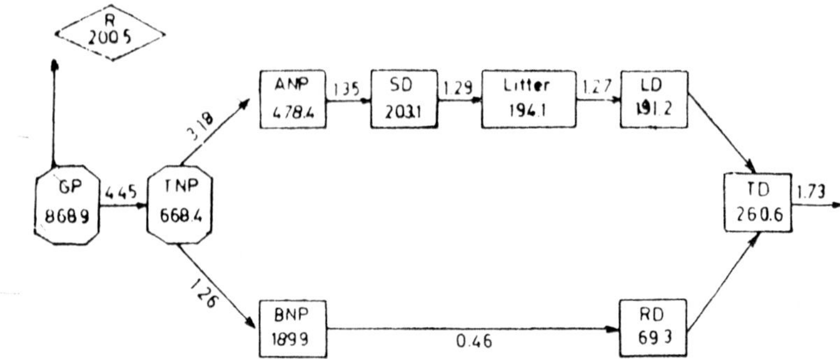

Fig. 82 : Flow of biomass in Various producer compartments under *Acacia leucophloca* plantation. GP = Gross productivity, R = Respiration, TNP = Total net productivity, ANP = Above ground net productivity, BNP = Below ground net productivity, SD = Standing dead, LD = Litter disappearance. Box indicates accumulation gm^{-2}, Arrow indicates flow rate gm^{-2} day^{-1}

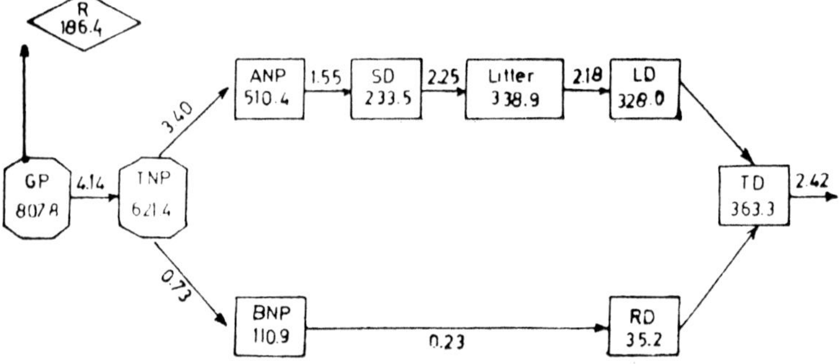

Fig. 83 : Flow of biomass in Various producer compartments under *Eucalyptus camaldulensis* plantations. GP = Gross productivity, R = Respiration, TNP = Total net productivity, ANP = Above ground net productivity, SD = Standing dead, LD = Litter disappearance. Box indicates accumulation gm^{-2}, Arrow indicates flow rate gm^{-2} day^{-1}

Simulation

The simulation models represent a very high level of abstraction. Such models reproduce some features of the real models and it generates the time domain behaviour to the observer. Such models are mostly mathematical and computerised. They are used for productive purposes and help the observer in guiding or controlling the interactions of causal agencies like grazing, fire, floods etc. They are of two types : (1) Linear, and (2) non-linear.

Linear simulation model

A system is linear if it is : (*a*) Zero state linear, (*b*) Zero input linear, and (*c*) Satisfies decomposition property. Let us consider a holon (H) where the input (Z) gives an output (Y) as shown in Fig. 84.

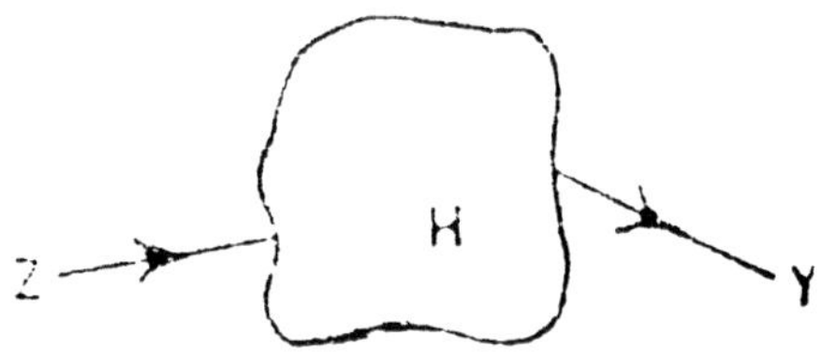

Fig. 84 : A holon (H) with input (Z) and output (Y).

In general the mathematical representation of this process can be written as :

$$Y_p = fp\ (Z_p) \tag{1}$$

where p is an integer having values 1 .... n, describes the various events of the system, fp represents the functional dependence of Yp or Zp for a particular processes.

In a linear system the event is described as follows :

$$Y_1 = a_1\, Z_1 \tag{2}$$

Where a is a constant

Hence, for a input Z_1' we shall get

$$Y_1' = a_1\, Z_1' \tag{3}$$

For a input Z_1'' we get an output

$$Y_1'' = a_1\, Z_1'' \tag{4}$$

If $Z_1'' = Z_1 + Z_1'$, we get from (4) $Y_1'' = a_1\, (Z_1 + Z_1')$

using (2) and (3) $Y_1'' = Y_1 + Y_1'$ (5)

or a linear combination of the input result in a linear combination of the output. Such systems are said to be linear systems. We shall further illustrate the concept of a linear system graphically :

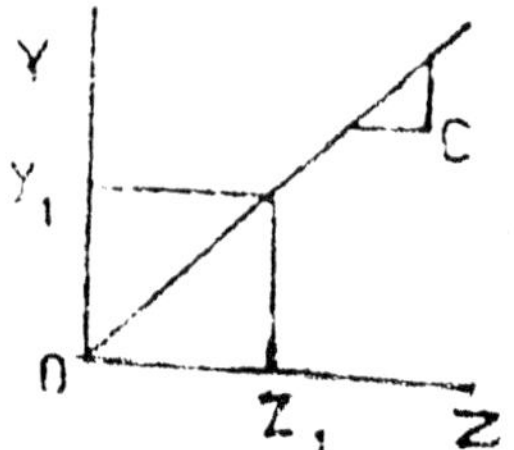

Fig. 85 : Linear system

$Y = CZ + O$ (6)

where C = slope of the line. If we force the first element of input Z, that is, Z_1', we get the corresponding output Y_1 then equation (6) will change to $Y_1 = CZ$ (7)

Further if we force the input Z_1 by a coefficient a in (6) we get $Y = a\,CZ_1$, But $CZ_1 = Y_1$ from (7)

Therefore $Y = a\,Y_1$

Similarly if $Z = Z_1 + Z_2$, then $Y = Y_1 + Y_2$

In other words again a linear combination of inputs yields a linear combination of outputs.

Non-linear simulation model

Let us consider a case where the output Y is dependent on the quadrature of the input Z. In this case the equation becomes

$Y_1 = a_1 Z_1^2$ (8)

for a different input Z_1' and Z_1'' equation (8) gives

$Y' = a_1\ (Z_1')^2$ (9)

and $Y_1'' = a_1\,(Z_1'')^2$ (10)

If we assume that $Z_1'' = Z_1 + Z_1'$ equation (1) gives

$Y_1'' = a_1\ (Z_1 + Z_1')^2$

$Y_1 + y_1'$

Therefore $a_1(Z_1 + Z_1')^2 = a_1\,[Z_1^2 + (Z_1')^2 + 2Z_1Z_1']$

$a_1Z_1^2 + a_1(Z_1')^2$

In this equation we see that the linear combination of the input does not yield a linear combination of the output. Such a system are said to be non-linear systems.

If there is a close relationship in the simulated and measured values that is, the behaviour of the validation compartment is fairly predictable, the model is a workable one. The structure of the system is important and will give reasonably sure data sets, generating dynamics which are reasonably correct. As far as possible, the validation compartment should be close to the measured inputs, or forcing functions. The analog computers are one way of generating simulation models by solving differential equations. The analog computers consists of a set of basic components which are capable of (1) summation, (2) integration, (3) multiplication, and (4) arbitrary function generation (Patten, 1971). Another computer that can be used for generating dynamics is the digital computer. It differs from the above in that it processes numerical information and generates numerical outputs. While analog computers give us instantaneous solutions leading to better understanding of the systems behaviour, the digital computers give greater numerical accuracy. Computers need an instruction language to perform functions in a definite order. Some of languages more commonly used are—Fortran IV, CSMP (continuous system modelling programme), ILM (Intermediate Level Model).

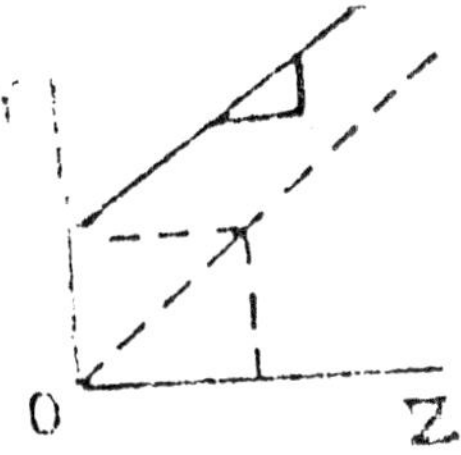

Fig. 86 : Non-linear system.

Systems Analysis

Systems analysis involves a sequence of representation which progresses from a physical domain to logical domain to mimic logical but simple behaviour; to deduce or infer characteristics of a real system or lower domain model from properties of a higher order hierarchial model for explanatory purposes in guiding or controlling further interactions with reality through understanding. Systems analysis permit a division of holons together with the environments created by each. Total flows can therefore be assessed from the total inputs and outputs. Systems analysis confirms to the forms of system theory or theory of the environment.

There are a number of more or less standard steps in a systems analysis study. These are

1. Formulation of the problem.
2. Construction of a mathematical model that describes the most important variables of the system.
3. Definition of a criterion function or measure of merit.
4. Collection of data to allow an estimation of the various parameters of the model.
5. Derivation of optimal solutions through formal algorithms.
6. Testing of the model, the solutions and the sensitivity of the parameters.
7. Implementation of the best solutions.

The first step involves the formulation of the problem and a definition of the objectives of the study. Alternate courses of action must be determined and the relation between the courses of action and the major variables of the problem under study must be delineated. This leads to step 2, namely the construction of a mathematical model. The quantitative relationships between the variables of the problems must be established. The third step is the definition of a criterion function, or measure of merit. Since the function under investigation may be in many different states, or the system to be constructed may have many alternate solutions, it is necessary to have a criterion which compares one system against another. Such criteria might be a minimum cost system or one having the greatest benefit — cost ratio. The fourth step is the collection of data to allow an estimate of the parameters of the model. This may the most time-consuming step, and depending on its outcome the entire model may have to be modified due to lack of data. Step 5 then seeks to determine an optimal solution to the problem, if it is possible at all, through the application of a number of formal algorithms. The most important step is a testing of the model, the solution, and the sensitivity of the parameters.

The major techniques and tools of the systems analysis are the following :

Linear programming

Non-linear programming

Dynamic programming

Integer programming

Stochastic programming

Network analysis, PERT and CPM

Simulation techniques

The technique of linear and non-linear programming deal with optimisation problems where the maximum or minimum of linear or non-linear functions is to be found subject to equality or inequality constraints, which may be linear or non-linear. Dynamic programming deals with the optimization of multistate decision processes, where as integer programming attempts to find solutions to problems where the variabels may take on only integer values. Stochastic programming is concerned with problems in which some of the variables and parameters are random variables. Network techniques, PERT which stands for programme evaluation and review technique, and CPM which stands for critical pathway methods are management techniques for control of large-scale research and development projects or for large construction jobs. Simulation techniques are used for the simulation or large systems without the explicit attempt to find the optimum but rather just to observe what happens when certain basic parameters are changes.

In the following paragraphs the use of systems analysis for water and air pollution control has been described. Two small numerical examples have also been given.

Use of Systems Analysis in Water Pollution Control Models

Probably because water pollution was recognised early, the use of systems analysis is more developed in this area. Models have been constructed that start with the collection of waste waters, their treatment, and their final disposal in a body of water and the quality variations in this body of water.

Despite the fact that the expenditure for waste water collection systems are very large, relatively little work has been conducted on these aspects. Only recently have efforts been undertaken to develop general mathematical models for simulating the entire rainfall-runoff process from the onset of the precipitation, through collection, conveyance, storage, treatment and final disposal to the water course. Basically these models aim at describing the entire, unsteady flow in the canal network to study its functioning and behaviour; to see if it is over- or under-disignated, and to get traces of the water quantity and quality at the various outlet points as a function of time.

Other efforts have been to determine the least costly collection systems for sanitary wastes alone under steady-state conditions. These models try to determine networks that will collect all the wastes at the required points. Convey them to a central collection point, and minimise the cost of the system, consisting of the cost of the pipes and the costs of excavation. From a strict mathematical point of view, many of these problems are unsolve; from a practical point of view, these models allow the designer to find very good solutions, even if they may not be optimal in the strict mathematical sense.

The design of waste water treatment plants has found more attention. Typically, the models view the treatment plant consisting of a number of unit processes and the objective is to select that combination of processes that will attain the required effluent quality at minimum overall cost. Both linear and dynamic programming models have been formulated and solved.

The planning of regional and inter-community waste water collection and treatment systems has also been given attention. The aim of these models is to determine the most economical location of treatment plants and the interconnecting sewarage system so that the regional costs are minimised. Expansion of capacity and possibly changing degrees of treatment over time have been given consideration.

By far the most attention has been given to models of entire river basins. Under the assumption that a regional water authority exists, models have been constructed for the most economical location of treatment plants and their required efficiencies. The earlier models considered in essence only the degree of treatment as a variable, whereas the later models also include effluent charges, storing of waste waters in lagoons, and artificial stream aeration. Newer models consider the effects of low flow augmentation from reservoirs and the problems of heat discharges.

Use of Systems Analysis in Air Pollution Control Models

The use of mathematical models in air pollution control is of more recent origin. It may also be said that air pollution is more difficult to model. Water pollution is in most cases (except estuaries and lakes) a one-dimensional problem—along a river; solid-waste disposal is two-dimensional-location of facilities in a plane; but air pollution is a three-dimensional problem, namely modelling the distribution and ultimate deposition of pollutants in all three dimensions.

Thus, at the beginning of modelling the focus was primarily on obtaining the data and coefficients of the pollutants in the atmosphere for a diffussion model. There are several of these models available and while there is some difference in the mathematics, the greatest uncertainties lie with the coefficients and parameters that enter the equations. The selection of one model over the other is based on the analyst's truth in the various parameters.

With a basic dispersion model available, the first step again were simple simulation models. Given a number of line, area or point sources, the task was to simulate how the pollutant would distribute over an area of interest. The simulation was usually limited to two parameters, sulphur dioxide and particulates. Now some models have appeared which aim at finding the combination of control processes and energy sources that will attain the air-quality standards and minimise the economic costs for the region.

Two examples have been given here, one from the area of water pollution and one from air pollution.

(*a*) *Water pollution problem*

On a small tributary to a main river three sources of pollution are situated, which are discharging 1000, 3000 and 2000 kg BOD per day. The flow in the river is 1.158 m^3/sec. The velocity of the flow in the river is such that the travel time from location 1 to 3 is 0.5 days, from 2 to 3 about 1 day, and from 3 to the mouth of the tributary about 0.5 days. The BOD is the only pollutant of concern and the rate of BOD excretion is 0.2/day. To remove 1 kg of BOD at the three locations costs 5, 3 and 4 cost units respectively. The maximum achievable degree of treatment at anyone location is 90 per cent. Given that the water quality standards require that the BOD in the river should not exceed 10 mg/l at anyone place, what is the required degree of treatment at each of the three sources of pollution.

Fig. 87 summarizes the information and also shows a load plan of the tributary if no treatment would be provided. For this the assumption was made that the discharge mixes completely and instantaneously with the flow in the river, and that the actual flow of water from the three sources is negligible. Prior to discharge from the first source, there is no pollution in the river. The load plan was drawn as follows: at point 1, 1,000 kg BOD are discharged into a flow of 1.158 cm^2/sec. This leads to a BOD concentration of—

$$L = \frac{100 \times 10^6}{1.158 \times 24 \times 3600 \times 10^3} = 10 \text{ mg/l}$$

(1000 kg BOD per day result in a concentration of 10 mg/l). After a travel time of 0.5 days the BOD in the river at point 2 due to discharge at point 1 will have been reduced through self purification to

$$L_{10} = 10 \times 0.9 = 9 \text{ mg/l}$$

But at point 2 a new source of 3000 kg will lead to a concentration of 30 mg/l, and thus the total load in the river at point 2 is 9 + 30 = 39 mg/l. And so on. Let X_1 stand for the unknown fraction of BOD removed at location 1 ($\leq X_1 \leq 0.9$).

X_2 and X_3 are similarly defined. Then we may write for location 1 an inequality of the form

$$10 (a - X_1) \leq 10 \tag{1}$$

(Note: $(1 - X_1)$ is the fraction of BOD discharged) at location 2 the BOD in the river must also be below 10 mg/l, so one writes:

$$9 (1 - X_1) + 30 (1 - X_2) = \leq 10 \tag{2}$$

and similarly for location 3 one writes :

$$7.4 (1 - X_1) + 24.6 (1 - X_2) + 20 (1 - X_3) \leq 10 \tag{3}$$

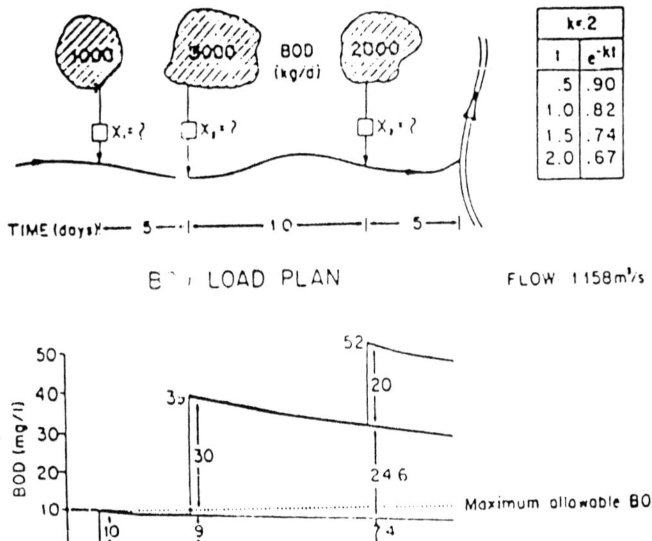

Fig. 87 : A simple river basin with the three sources of pollution (Deininger, 1980).

Please note that it is sufficient to write these inequalities for the 3 points to guarantee that the BOD at every point in the tributary is below 10 mg/l.

The first equation one might wish to ask is : Is there a feasible solution ? If at every location, the maximum possible degree of treatment would be used, can the inequalities (1) through (3) be satisfied ? Setting X_1, X_2 and X_3 equal to 0.9, one would obtain :

$$10\,(0.1) \leq 10 \tag{1'}$$

$$9(0.1) + 30\,(0.1) \leq 10 \tag{2'}$$

$$7.4\,(0.1) + 24.6\,(0.1) + 30\,(0.1) \leq 10 \tag{3'}$$

All three inequalities would be satisfied. The cost of this solution would be

$1000 \times 5 \times 0.9 + 3000 \times 3 \times 0.9 + 2000 \times 4 \times 0.9 = 19800$ cost units

In other words, for 19800 cost units one would obtain the maximum possible reduction in pollution and one would assure that at no point in the river the BOD is more than 10 mg/l. Let's call this possible solution "Solution A".

Consider now another policy of pollution control which may be stated as follows : Each polluter may discharge as much pollutants as he wishes as long as the river just below his outfall does meet the water quality standards.

Inspecting inequality (1) it becomes clear that no treatment would be required at location 1 ($X_1 = 0$). Using $X_1 = 0$ in equality (2) leads to

$$9\,(1 - 0) + 30\,(1 - X_2) = \leq 10$$

$$1 - X_2 = 1/30$$

$$X_2 = 29/30 = 0.97$$

In other words, the required fraction of BOD to be removed at 2 would be 0.97 (or 97 per cent). Using $X_1 = 0$ and $X_2 = 0.97$ in inequality (3) would lead to a treatment requirement at 3 of $X_3 =$ 0.911.

The necessary treatment at both location 2 and 3 exceeds the postulated maximum degree of treatment possible. This policy of maximum upstream discharge leads therefore to an infeasible solution, and we note that "Solution B" is impossible.

Suppose we postulate that all 3 pollutors should provide the same degree of treatment. That would mean $X_1 = X_2 = X_3$. Inequality (1) leads to

$$X > 0$$

Inequality (2) leads to

$$9\,(1 - X) + 30\,(1 - X) \leq 10$$

$$39\,(1 - X) \leq 10$$

$$X > 28/29$$

Inequality (3) leads to

$$7.4\,(1 - X) + 24.6\,(1 - X) + 30\,(1 - X) \leq 10$$

$$52\,(1 - X) \leq 10$$

$$X = 42/52$$

The maximum of these three determines the necessary degree of treatment, which happens to be determined by inequality (3). Thus, the necessary degree of treatment would be 42/52, or roughly 81 per cent.

The cost associated with this "equal treatment" policy would be "Solution C". $(1000 \times 5 + 3000 \times 3 + 2000 \times 4) \times 0.81 = 17000$ cost units. Let us now ask the question : what would be the least costly solution (Solution D) ? Well, the total costs are :

$$1000 \times 5 \times X_1 + 3000 \times 3 \times X_2 + 2000 \times 4 \times X_3$$

$$= 5000\,X_1 + 9000\,X_2 + 8000\,X_3$$

$$= 1000\,(5X_1 + 9X_2 + 8X_3)$$

This function is called the objective function, and it is to be minimised. The common factor of 1000 can be eliminated. And thus the problem can be stated as :

minimise : $5X_1 + 9X_2 + 8X_3$ (4)

subject to : $X_1 \geq 0$ (5)

$9X_1 + 30X_2 \geq 29$ (6)

$7.4X_1 + 24.6X_2 + 20X_3 \geq 42$ (7)

$X_1 \leq 0.9$ (8)

$X_2 \leq 0.9$ (9)

$X_3 \leq 0.9$ (10)

and $X_1, X_2, X_3 \geq 0$ (11)

The above is a typical linear programming problem. A linear function (equation 4) is to be minimised subject to the constraints (5) – (7), which are just the constraints (1) – (3) written in a different form. Constraints (8) – (11) represents the so-called non-negetivity requirements. This problem may be solved using any of the alogrithms for solving linear programming problems, such as for example the simplex method, and the solution would be

$$X_1 = 0.25,\ X_2 = 0.9,\ X_3 = 0.9$$

The total cost of this solution (Solution D) would be 16550 cost units.

To summarise, the 4 different solutions discussed are shown in Table 75. In this small example, the advantage of the least cost solution over an equal treatment policy is not very large. But in actual situation, the least cost solution was less than 50 per cent of the cost of an equal treatment policy.

Table 75. Policy and cost relationship in waste water treatment.

Policy	Cost units	Comment
A	19800	Maximum pollution control
B	—	Maximum allowable discharge
C	17800	Equal treatment policy
D	16550	Least cost solution

(B) *Air pollution problem*

A refinery (source 1), a power plant (source 2) and a municipal incinerator (source 3) are the only major sources of air pollution in an area. The three sources emit particulates and sulphur dioxide. Only the latter pollutant will be considered. The diffusion equation by Pasquill-Gifford will be used.

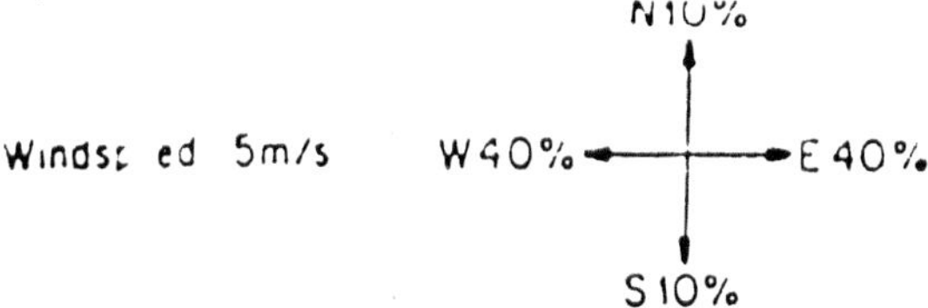

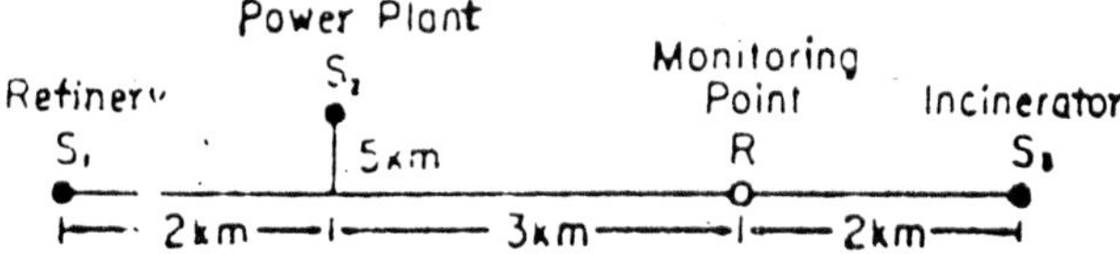

Fig. 88 : Simple Air Pollution Problem with 3 sources emitting SO_2.

$$C(x, y,) = \frac{Q}{\pi \sigma_y \sigma_z \mu} \exp(-0.5(y/\sigma_y)^2) \exp(-0.5(h/\sigma_2)^2)$$

Where Q = emission rate (g/sec)

μ = wind speed (m/sec)

$\sigma_r{}'\sigma_z$ = dispersion coefficients in y and z direction

h = effective height of source

x = downwind distance

y = crosswing distance

C(x, y) = concentration of pollutants at downwind distance x, crosswind distance y, elevation zero (g/m^2)

σ_x and σ_y must be estimated from the standard graph. The relevant data for the three sources are summarised in Table 76.

Table 76. Particulars of sources of sulphur dioxide

Source	S_1	S_2	S_3
x(m)	5000	3000	2000
y(m)	—	500	—
h(m)	200	100	50
Q(g/sec)	300	200	100
σ_y (m)	450	290	200
$\sigma_z(m)$	270	170	120

The ground level concentration at point R due to source 1 when the wind blows to the east is then (in ug/m^3)

$$C_{R_1} = \frac{300 \times 10^6}{3.14 \times 450 \times 270 \times 5} \exp(-0.5(200/270)^2$$

$$= 300 \times 10^6 \times 0.42 \times 10^{-6}$$

$$= 126\,\mu g/m^3$$

At the point same point the concentration from source 2 would be

$$C_{R_2} = \frac{200 \times 10^6}{3.14 \times 290 \times 170 \times 5} \exp(-0.5(500/290)^2$$

$$\exp(-0.5(100/170)^2$$

$$= 200 \times 10^6 \times 0.32 \times 10^{-6}$$

$$= 64\,\mu g/m^3$$

When the wind blows to the west, the ground level concentration at point R due to source 3 will be

$$C_{R_3} = \frac{100 \times 10^6}{3.14 \times 200 \times 120 \times 5} \exp(-0.5(100/120)^2$$

$$= 100 \times 10^6 \times 1.99\ 10^{-6}$$

$$= 199\,\mu g/m^3$$

Ground level concentration of R would therefore be

Wind to the east $126 + 64 = 190\,\mu g/m^3$

Wind to the west $190\,\mu g/m^3$

Wind to the north or south $0\,\mu g/m^3$

The average concentration at point R would then be

$$190 \times 0.4 + 199 \times 0.4 + 0 \times 0.2 = 156\,\mu g/m^3$$

Suppose that one would like to lower this ambient concentration to a maximum of $80\,\mu g/m^3$. If one completely disregards the location of the sources and the dispersion, one could argue as follows : into the airshed of concern, a total of $300 + 200 + 100 = 600$ g/sec of sulphur dioxide are emitted resulting in a ground level concentration of 156

$\mu g/m^3$. If this ambient concentration is to be reduced to $80 \mu g/m^3$, then the emission must be reduced to

$$x = 600 \times 80/156$$

$$= 308\, g/m^3$$

or roughly speaking, all the sources would have to cut their emissions into one half (exactly; 49 per cent reduction). Assume now that the costs for removal of sulphur dioxide are 0.6, 0.7 and 0.8 cost units at sources 1, 2 and 3 respectively. For a 49 per cent reduction the total costs would be

$$(300 \times 0.6 + 200 \times 0.7 + 100 \times 0.8) \times 0.49 = 196 \text{ cost units}$$

A strategy to minimise the costs of reducing the emission would lead to the solution that at source 1 only 8 g/sec could be discharged, and 2 and 3 no control would be necessary. The cost of such a solution would be

$$292 \times 0.6 = 175 \text{ cost units}$$

(control at source 1 was chosen because it has the lowest cost units)

An ambient least cost model could be formulated in the following way. The ambient concentration at point R must be below 80 $\mu g/m^3$. Letting X_1, X_2 and X_3 stand for the unknown fraction reduction necessary at the sources, one could write :

$$0.4(126(1X_1) + 64(1X_2) + 0.4 \times 199(1X_3) \quad 80$$

After simplifying, this leads to

$$50.4X_1 + 25.6X_2 + 79.6X_3 \quad 75.6 \quad (12)$$

The objective function — minimise the total costs — would be

$$0.6X_1 + 0.72X_2 + 0.8X_3 \quad (13)$$

Generally speaking, a least cost ambient model will always lead to the smallest total cost units for a region.

Systems Theory

There are three propositions to this theory :

1. Every eoject (holon, organism) has two environments—(*a*) input environment, and (*b*) output environment.
2. Internal cause or the propagating structure of objects or organisms cannot be determined to be consistent without reference to an external environment, that is, organism and environment go together as unity, and therefore cannot be separated logically or mathematically.
3. Ecosystem can be constructed as a union of disjointed input or output environment of all objects present. The environment of proposition -1 foms an ecosystem partition.

19. *Biogeography*

The word 'biogeography' has originated from the Greek *bios* = life; *ge* = earth; *graphein* = to write. Life has established within the frame-work of physical geography — relief, rocks, soils, glaciers, volcanoes and climate. No organism occurs uniformly throughout the world. Specific organisms are restricted to specific areas of the world. Three aspects of the distribution of an organism are generally recognised : *geographic range,* or the extent where the organism normally occurs; the *geologic range,* or the distribution in time — past and present; and the *ecological distribution,* or the major biotic communities. Certain biologists have also made distinction between geographic distribution (horizontal distribution) and *bathymetric distribution* (vertical distribution). Bathymetric distribution includes the following three realm : (1) *Holobiotic,* or vertical distribution of organisms in marine habitat, (2) *Limnobiotic,* or vertical distribution of organisms in fresh water habitat, and (3) *Geobiotic,* or altitudinal distribution of organisms on land. The study of distribution of plants is known as *phytogeography,* and that of animals as *zoogeography.*

Biogeographical Regions of the World

The whole world is divided into 8 biogeographical regions, each more or less embracing a major continental land mass and each separated by oceans, mountains ranges or deserts (Fig. 89). These are (1) Nearctic, (2) Palaearctic, (3) Africo-Tropical (4) Indomalayan, (5) Oceanian, (6) Australian, (7) Antarctic, and (8) Neotropical. Some geographers consider the Neotropical and the Australian regions to be so different from the rest of the world, that these two regions often considered as regions equal to the other rest combined. Accordingly, they are classified as (1) Neogea (Neotropical), (2) Notogea (Australian), and (3) Metagea (main part of the world).

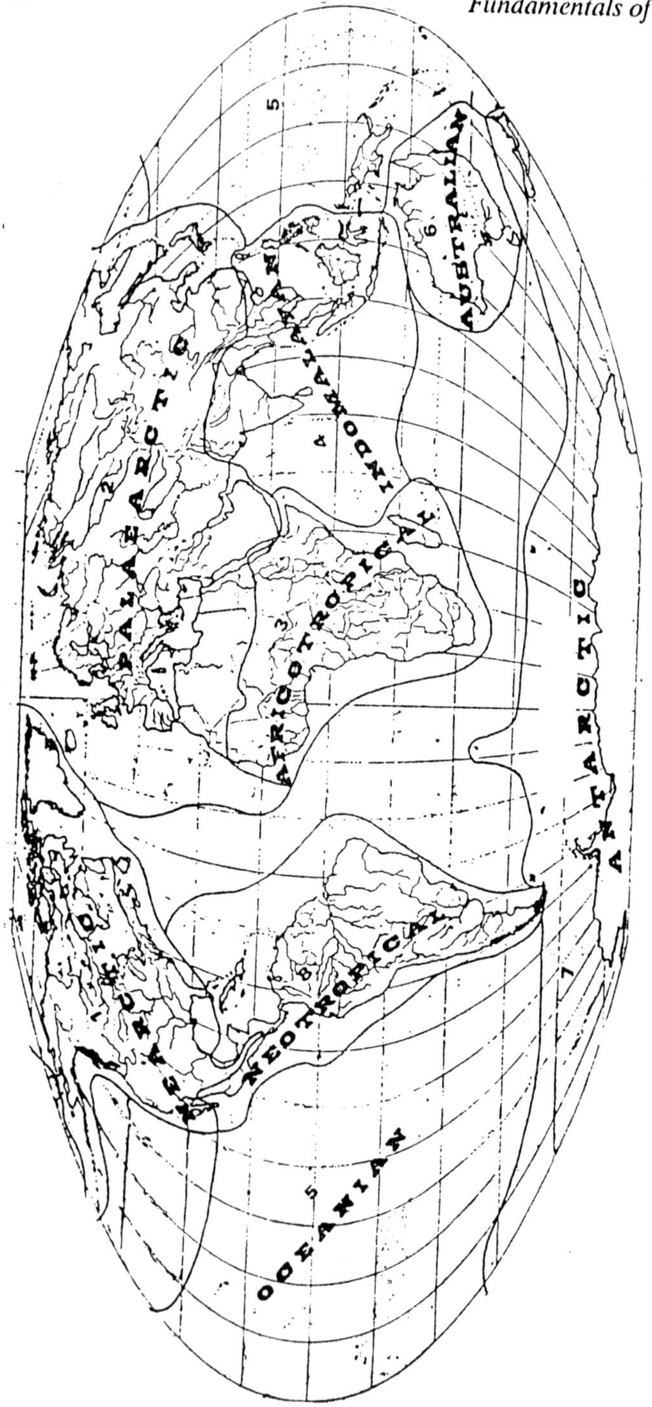

Fig. 89 : Polar projection map of the world showing major biogeographical regions.

Except for Australian region, each at one time or the other has had some land connection with another, across which animals and plants would pass. Palaearctic and the Nearctic are quite closely related, and often considerd as one, the Holarctic. Palaearctic contains the whole of Europe, all of Asia, North Himalayas, north Arabia, and a narrow strip of coastal north Africa. Both regions are similar in climate and vegetation; both possess coniferous and deciduous forest biomass, tundra desert, grassland and chaparrel. Neotropical region includes South America, part of Mexico and West Indies. The Africo-tropical was earlier called as Ethopian containing the continent of Africa, south of Atlas mountains and Sahara desert, and the south corner of Arabaia. It embraces tropical forests in central Africa and in the mountains of east Africa, savanna, grasslands and deserts. During the miocene and the pliocene Africa, Arabia and India shared a moist climate and a continuous land bridge, which allowed the animals to move freely among them. This explains the similarity in biotic composition. Tropical savannas are the world's best, but they are fast disappearing. Indomalayan (Oriental) region includes India, Indochina, South China, Malaya and the western islands of the Malay Archipelgo. It is bound on the north by the Himalayas and on the other side by the Indian and Pacific oceans. An imaginary boundary — the Wallace's line separated Oriental from the Australian regions.

Patterns of Distribution

The plants and animals exhibits the following specific patterns of distribution over the globe :

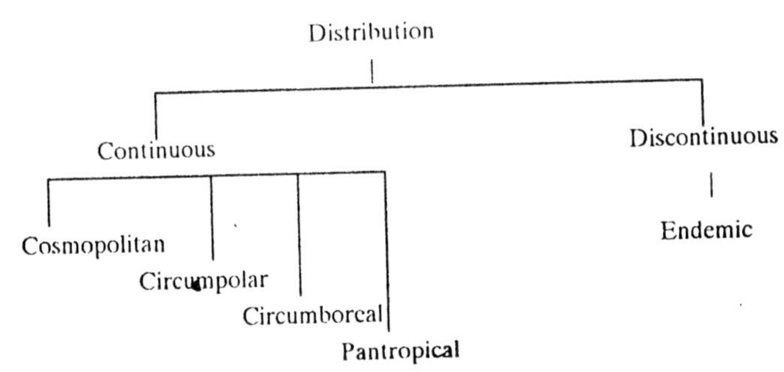

Continuous Distribution

A taxon distributed throughout the world in all climatic zones or in all the continents in atleast one climatic zone is said to have continous distribution. It is of the following four types :

(1) Cosmopolitan Distribution

Such a taxon is distributed throughout the world in all climatic zones. The common plant species of nearly cosmopolitan distribution are *Chenopodium album, Taraxcum officinale, Phragmites communis, Urtica dioica Poa annua* etc. The common cosmopolitan animal species are *Tyto alba, Ardea cineria, Falco pereginus, Vanessa cardui, Nonnophila noctuella* and *Macrobiotus hyfelandi.* Among the plants Gramineae, Compositeae, Cyperaceae, Caryophyllaceae, Orchidaceae, Papilionaceae, Labiateae, Liliaceae are cosmopolitan in distribution (Fig. 90).

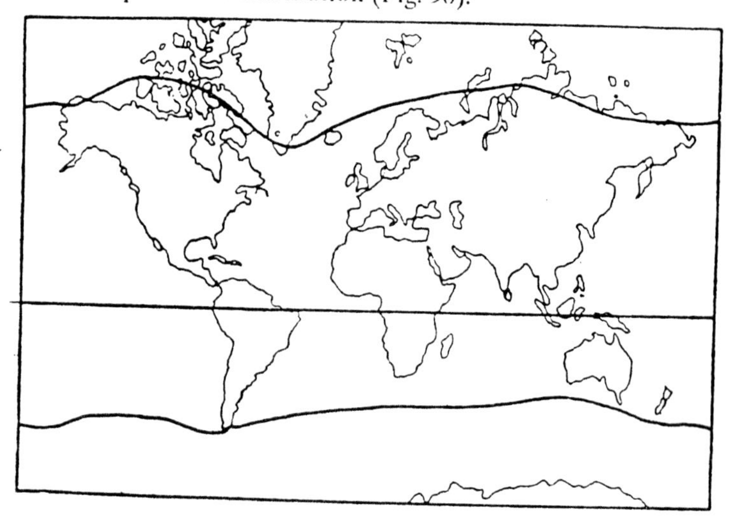

Fig. 90 : Cosmopolitan distribution of family Papilionaceae.

(2) Circumpolar Distribution

Certain species such as *Saxiphraga oppositaefolia, Carex lapponica, Ranunculus nivalis* etc. are distributed in a belt around the north pole.

(3) Circumboreal Distribution

Or circumaustral distribution includes taxon which are distributed in a near continuous belt in the temperate region of northern and/or sourthern hemisphere. The genera like *Alnus, Draba, Acer* etc. are circumboreal in distribution while *Danthonia* is circumaustral in distribution.

(4) Pantropical Distribution

Certain plant species like *Bauhinia, Dalbergia, Dioscorea, Corchorus, Ocimum, Cassia, Eugenia, Phyllanthus* etc. are distributed throughout the tropical belt.

Discontinuous Distribution

A taxon distributed in two or more widely separated geographical areas, is said to have discontinuous distribution. Such

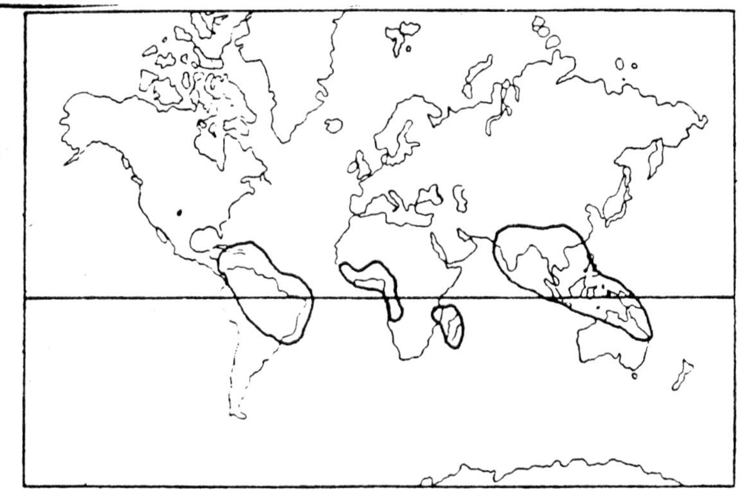

Fig. 91 : Discontinuous distribution : Family Lecythidaceae.

a taxon may occur in several small areas in the same continent or in two different continents of the same or different hemispheres. Fig. 91 shows the discontinuous distribution of family Lecythidaceae. Among plants *Saxifraga* and *Silene* are distributed in Arctic region and high altitudes. Likewise among animals *Epiceratodus, Protopterus* and *Lepidosiren* are distributed in Australia, Africa and South America. Other examples of discontinuous distribution have been indicated in Table 77.

In theory, of course all species are discontinuous to some extent insofar as they rarely, if ever, cover their general range so completely that the individual plants are actually in contact, and the greater the detail in which distribution is considered the more this point of view will become. In general however, and especially in considering the whole ranges of species and general, it is impossible to take into account, or indeed to mark, this degree of discontinuity, and the term is restricted in practice to ranges which on a large and obvious scale consist of two or more parts (Good, 1953).

Table 77. Examples of different types of discontinuous distribution

Discontinuity	*Distribution*	*Examples*
Arctic-alpine	Arctic region and high altitudes	*Saxifraga, Silene*
North Atlantic	North America and Western Europe	*Spiranthes, Eirocaulon, Polygonum*
North Pacific	Western North America and Eastern Asia	*Liriodendron, Torreya*
North-South American	North and South America	*Larrea divaricata, Sarraceniaceae, Koeberlenia*
Eurasiatic	Different parts of Europe and Asia	*Wulffenia, Cimifuga foetida.*
Mediterranean	Around Mediterranean sea in Europe and Africa	*Olea, Cistus, Ceratonia*
Tropical	In two or more parts of tropics	*Pandanus, Coffea arabica, Viola obyssinica, Nepenthes, Anona, Agathis, Dacrydium, Adansonia digitata*
South Pacific	Australia, New Zealand and South America	*Hebe, Drimys, Pernettia, Laurelia*
South Atlantic	South America and Southern Africa	*Saccoglotis, Asclepias*
Antarctic	South America, New Zealand and Some Islands	*Nothofagus*
Intracontinental	Several places within the same continent	*Drosera, Rubia, Erica, Daboecia*

To explain the cases of disjunction, two reasons readily come to mind: (*a*) The range of distribution has become discontinuous because individuals simply disappeared from the intervening areas, and (*b*) The range of species was never contiguous but diaspores (seed, spores, vegetatively reproducing organs) from the original area reached distant site suitable for its growth through the agency of wind, animals, especially migratory birds, ocean currents, etc., overcoming the barriers imposed by the unfavourable intermediate localities (long distance dispersal). Successful establishment of species through long distance dispersal need not involve frequent introductions. Several seeds carried by a bird at a time or a single seed introduction followed by a few more over a relatively long period may provide populations with sufficient heterzugosity to permit long-term survival (Carlquist, 1967). Mostly, the two disjunct areas are climatically and ecologically analogous. Mention may also be made of the action of man in recent years who has considerably speeded up the spread of weeds (Meher-Homji, 1974).

When a given case of disjunction is due to long-distance dispersal having occurred within the last 500 years could be known by the isoenzyme studies. A single plant that has been accidently transported to a far away place does not appear to build up within this short time a variable pool for isoenzymes as is likely to be possessed by the original populations (Meher-Homji, 1974).

The theory of continental drift has opened possibilities of explaining many intriguing cases of distribution patterns. The movement of continents could well segregate the populations of a species in space and time. *Polytopic* or simultaneous origin of species in two separated areas (double creation or parallel evolution) resulting in morphologic convergence of species has also been suggested since the time of Linnaeus.

Fossil record is of immense value in that it points out the past distribution but its absence does not necessarily mean that a species did not exist in a given region. The paleontological evidence that abounds in certain areas is either totally lacking or fragmentary in other parts. Fossilized leaves have atleast in some cases led to erroneous identifications (Meher-Homji, 1974).

Discontinuity is closely related to two matters : the first is the problem of *monophylatic origin of species*. If this view is maintained, then obviously the phenomenon of discontinuity take on a very great interest and importance because it may be assumed that whatever is the present separation between the constituent areas, they must have been continuous, or atleast the individuals contained in them must once have come from one ancestral plant. This being so, then the discontinuity has to be explained, and there must be taken into account all the factors which might possibly have caused it. On the other hand if plants had *polyphyletic origin of species*, then discontinuity loses much of its potential importance because it can always be explained on the supposition that the same species has arisen independently in each of the separated portions of its total range. It is unlikely that all species had a polyphyletic origin. On the contrary, facts generally indicate very strongly that discontinuity is the result of real disjunction (Good, 1953).

Hence, discontinuity is only to be regarded as a normal phenomenon of distribution. It is of course affected by all kinds of external factors and extreme discontinuity can arise only in definite circumstances. Discontinuity can be water discontinuity, land discontinuity or a mixture of both (Good, 1953).

Endemic Distribution

The endemics are taxa of very restricted distribution in small areas. Numerous plant families, genera and species are found to be highly endemic. Most of the angiosperm species belong to this group. They are restricted to a small region or country (Table 78).

Table 78. Some Endemic Taxa

Endemic Families	North America	Kameriaceae, Garryaceae, Fouquieriaceae, Tropeolaceae, Lactoridaceae, Heterostylaceae.
	Africa	Sphenocleaceae, Pandaceae, Scytopetalaceae.
	Asia	Sarcospermataceae, Siphonodontaceae, Pentaphragmataceae
	Malagache	Barbeuiaceae, Humbertiaceae
	Seychelles Island	Medusagynaceae.
	Australia	Austrobaileyaceae, Brunoniaceae, Cephalotaceae.
Endemic Genera	Europe	*Stratiotes, Pulmonaria, Physospermum.*
	China	*Litchi, Xanthoceras, Hovenia, Kolkwitzia*
	Western and Central Asia	*Cannabis, Spinacia.*
	Mediterranean region	*Atropa, Ulex, Verbascum, Drosophyllum.*
	Tropical Africa	*Cola, Khaya, Monotes, Ricinodendron*
	North African-Indian Desert	*Anastatica, Fortuynia, Reptonia, Omania, Xerotia, Leptadenia,*

(*Contd.*)

	region	*Daemia*
	India (incl. Ceylon)	*Amphicome, Dittoceras, Dodecania Ulteria, Cruddasia, Heylandia, Lagenandra, Zeylandium, Hitchenia, Blepharistemma.*
	Malaysian region	*Elettariopsis, Dryobalanops, Rafflesia, Panguium, Lunasia, Hallieracantha.*
Endemic Species	China	*Ficus pumila, Livistona, chinensis, Morus alba, Primula sinensis, Rosa banksiae, Lonicera nitida.*
	Tibet	*Poa altaica, Myricaria prostrata,*
	West Africa	*Coffea liberica, Elaeis guineensis, Erythrina excelsa, Uncaria africana.*
	Madagascar	*Apongeton fenestralis, Cryptostegia grandiflora, Calanchoe blosfeldiana, Ravenala madagascarensis, Delonix regia, Euphorbia fulgens.*
	Hawaii Island	*Acacia koa, Gunnera petoloides, Edwardsia grandiflora, Santalum pyrularium.*
	Amazon (South America)	*Theobroma cacao, Victoria regia, Arundo sacchariodes, Hevea brasiliensis, Leopoldiana pulchra.*
	Australia	Numerous species of *Acacia, Eucalyptus, Callistemon, Casuarina, Castanospermum.*
	India	*Ficus religiosa, Ficus bengalensis, Aegle marmelos, Artocarpus nobilis, Crotalaria juncea, Datura metel, Indigofera tinctoria, Elettaria repens, Eleusine coracana, Piper nigrum, Piper longum, Sesamum indicum, Hibiscus abelmoschus, Butea monosperma, Beaumontia grandiflora, Memecylon umbellatum, Holmskioldia sanguinea, Feornia elephantum, Saraca indica, Shorea robusta, Caryota urens.*

The word endemic should be properly used with due regard for the size of the taxonomic unit. Table 78 gives a list of important eneemic plant taxa of different regions. Although it is generally indescribable in words, there is an average range of families, an average range of genera, and an average range of species, these being progressively smaller, and the best practical limitation of the use of the word endemic is to restrict it to units whose ranges are obviously less than the average of their kind. For example, it is appropriate and valuable to consider families which are found in only one continent as endemic because the average distribution of families in greater than this. On this contrary, it is almost meaningless to apeak of species in terms of continental endemism, because comparatively few species are as wide or more widely distributed.

Endemism may be particularly useful in the recognition of different floristic regions, and also in determining the degree in which floras are peculiar.

In the first case, it is often to be noticed that while one part of a large region possesses a high proportion of endemics another and adjacent region may have comparatively fewer, and this is often a useful guide to the delimitation of the two. This is seen for instance between south-west Australia and other parts of that continent.

The second case may well be illustrated by three island groups—The Galapagos, Juan Fernandez, and Hawaii. The first has many endemic species, but very few endemic genera. Juan Fernandez has numerous endemic genera and even one endemic family. The Hawaii islands are very isolated and have a flora much larger than those of the other groups. Moreover, it has a very high degree of endemism, about 90 per cent of its species being confined to it. There are also many endemic genera.

It is clear that species will be of such narrow range as to permit the name endemic at two distinct period of their existence, namely at the beginning and at the end, when they are very young and when they are very old. A third possibility that species may retain a very small area without expansion or contraction for a very long time, is a hypothetical case, difficult to prove or disprove.

It is one of the principle features of plant geography that endemic species are much more numerous than others and this may be explained in two ways—according to one's view of evolution : (1) the *selectionists* propose that these endemics are mostly extinct species, and regard their large number as evidencing that widespread extinction of forms which their theory requires. (2) the *mutationists* believe that endemics are mostly new species and that their superior numbers point to the kind of exponential increase that should be associated with such a process. That is why the discussion of endemism enters so largely into the presentation of evolutionary theories, each author trying to find, among the facts of endemism, the support which he seeks for his opinions (Good, 1953).

In young species, the distribution is narrow in the beginning and it is likely to grow in its area in course of time. Such species are *progressive, expanding* or *neoendemics.* On the other hand, certain species on account of their gradual dwindling gets restricted to a small region and are called as *retrogressive endemics.* Some species which had extensive distribution in the past are narrowly distributed today. This change may be due to geographical and climatic changes. Such species is known as *relic endemics.* Even among endemics some are restricted to a very localised spot and these are called *local endemics.* Sometimes, a few mutants appear which do not compete successfully, and therefore disappear quickly. These are often called as *pseudo endemics.*

Phytogeography

It is the study of present and past geographical distribution of plants on the earth. According to Wulff (1943) phytogeography is a study of present and past areas and the elucidation of origin and history of development of floras. Leon Croizat (1952) considered phytogeography as the study of migration and evolution of plants in time and space.

The distribution of any taxonomic unit (Species, Genus or Family) of the plant world, is due to the biological peculiarities of plants and to their adaptations to local habitat conditions, which even within the limits of a small territory may vary to a considerable extent. The classical approach to phytogeography has been towards enumeration of the taxa of a region and on the basis of broad floristic differences, botanical regions have been recognised. With the aid of such information the causes and mechanism of evolution of different types of floras in different regions are also being studied. Thus phytogeography has two major approaches of study :

1. *Static* or *descriptive* phytogeography which involves assembling of floristic and vegetational data.

2. *Interpretive* phytogeography which seeks to understand the dynamics of migration and evolution of plants and floras.

Static or Descriptive Phytogeography

The term static is rather unfortunate in the sense that vegetation and the environment are changing at every place with increasing biotic activities. The following four broad vegetational belts are recognised : (1) Arctic, (2) North temperate, (3) Tropical, and (4) South temperate.

Arctic

This zone is divided into the following two types :

(*a*) *Arctic proper*

This zone occurs around the north pole and remains covered with ice throughout the year. It is infact Tundra biome and its chief components of vegetation include some algae, annual flowering plants, mosses and lichens.

(*b*) *Subarctic*

This is less defined zone which extends from southern arctic to the northern limits of temperate zone. This region is very cold, contains abundant bogs and have vegetation including small trees, shrubs and herbs in the months of June and July. Ground vegetation often includes some Pteridophytes, orchids, insectivorous plants, mosses and lichens.

North Temperate

This zone extends between 30° N Latitude and 55° Latitude. It has two major zones :

(*a*) *North temperate of the eastern zone*

This zone is further subdivided into the following four zones; (*i*) *Western and Central Europe.* This zone is demarcated in north by the subarctic and in south by Alps and British islands. Its forests are dominated by several gymnospermous tall trees like *Picea, Pinus* and *Abies* and angiospermous trees like Oak, mapple, and chestnuts. Ground vegetation comprises orchids, wild roses, buttercups, *Viola, Salvia, Dianthus* etc. At high altitudes of this zone trees are replaced by grasses with some herbaceous flowering plants. (*ii*) *Mediterranean.* This zone extends between 30°N and

40°N latitudes, south of mountain ranges in Europe and in Asia around Mediteranean sea and is characterised by warm temperate type climate. Vegetation is chiefly composed of fruit trees, olives, nut trees, oranges, and also some foreign palms, cacti, acacias etc. In the Asian region of Mediterranean as in Arab countries, rainfall is low, deserts are common and sparce vegetation includes species like *Atriplex, Alhagi, Polygonum* and *Phoenix dactylifera.* (*iii*) *North* 1
Africa. This zone includes the northern part of Morocco, Algeria, Libya and Egypt. It is characterised by scanty rain and sparce vegetation. In cooler areas some conifers and broad leaves oaks are common. Its deserts (including some portions of Sahara desert) have herbs, shrubs, woody acacias, and succulent xerophytes. (*iv*) *Himalayas, eastern Asia* and *Japan*. Tibet China and Japan have different type of vegetation. In China and Japan conifers like *Cryptomeria, Sciadopitys, Cemphalotaxus, Gonkgo biloba* and *Cycas* and angiosperms like *Rhododendrons, Cinnamomum camphora* and *Begonia* are common. The vegetation of Himalayas will be described later in this chapter.

(*b*) *North temperate of the wester hemisphere*

It includes the parts of United States and Canada lying mostly between north latitudes 30° to 55°. The eastern coastal region of these countries in the temperate belt have some very characteristic plant species like tropical fern (*Schizaea pusilla*). The forest communities are composed of conifers and deciduous trees. On lower altitudes some wild cherries, plums, roses and orchids are abundant. In the New England region trees of *Ulmus americana* and *Castania dentata* are abundant. Forests of conifers are common in southern parts and on western parts of Rocky mountains of USA. In north California grows *Sequoia sempervirens*, the tallest tree of world. Ground vegetation is composed of *Salicornia herbacea, Rumex maritima, Monotropa uniflora, Saxifragea, Primula* etc. In the Colorado desert of Arizona and southeastern California, there are several types of xerophytic plants.

Tropical Zone

This zone is divided into two sub-zones :

(*a*) *Palaetropic*

It comprises old world or eastern tropics and has the following two botanical regions : (*i*) *Tropical Africa*. This is a large landmass.

of varied topography. It includes high altitudes and Sahara desert which have little or no rainfall. In Africa, most remarkable plant is *Welwitschia mirabilis*. Eastern part of Central Africa has India like vegetation of *Borassus flabelliformis, Cassia fistula, Arythrina, Acacia, Albizzia, Zizyphus, Bauhinia* etc. (*ii*) *Tropical Asia*. It includes Arabia, Pakistan, India, Burma (Myanmar), Ceylon, Thailand, Indonesia, Philippines and island of Indian sea. In Arabia the rainfall is low and most of the desert species are found. *Coffea arabica* is a native plant of Arabia. Ceylon is rich in species diversity and ferns are the chief components of its sparce natural vegetation. Malaya, Java and Sumatra are characterised by heavy rainfall and their vegetation comprises varied palms, ferns, tall trees, lianas and insectivorous plants. The common plants of Miama and Thailand are jack fruit, orange, mango, banana, betalnut etc.

(*b*) *Neotropics*

It comprises Mexico and major part of Sourth America. The low rainfall areas of Mexico are rich in xerophytes. At higher cooler altitudes the coniferous forests have *Pinus, Spruce, Quercus* and *Populus*. On mountain peaks grasses are most common. West areas of Mexico have mosses, palms, bamboos, orchids etc. In south America there are extensive forests of *Bertholletia excetsa, Maximiliana regia* and Mangrove plants. There are also found many epiphytes.

South Temperate Zone

This zone includes extreme southern region of Africa, Australia and New Zealand. In African region the vegetation is chiefly made up of ferns and gymnosperms. On the hills conifers are common. Its lower wet regions have *Salix* and *Phragmites* while the dry regions have grasses like *Andropogon* and trees like *Acacia*. The vegetation of northern part of Australia is similar to that of south east Asia and includes trees like palms, nuts, *Eucalyptus, Acacia,* and *Casuarina*. Some Pteridophytes are also found in the ground vegetation. In the southern Australia Araucarias are common. The forests of New Zealand are mostly composed of conifers together with ferms, palm like *Rhopalostylis, Metrosideros* etc. The flora of New Zealand is rich in Bryophytes.

Phytogeography of India

Indian subcontinent lying between 8° and 37°N, and 68° and 97°E has its own peculiar physiographic, climatic and biotic features. It is surrounded on its south, east and west by oceans and in the north by mountain chains highest in the world. The subcontinent stretches out between tropical and subtropical belts. its climate is chiefly modified by oceans and mountains. In the south and in the far east the climate is typically tropical, while in the north it is temperate, and in the north west it is highly arid. The general climate of India is of monsoon type. In India the following three distinct seasons occur : 1. Cold weather season during December to February when the mean temperature falls very low. In the north snow fall occurs on several days and the temperature remains below 0°C for many days, while in the south the temperature may remain above 20°C. Some rainfall occurs during season. 2. From March to Mid June, the cold weather season is followed by the hot summer as the temperature gradually rises and relative humidity decreases. The change of season is marked by leaf fall and blooming of many trees particularly in north India. The summer in the north often becomes quite severe with average temperature above 40°C or even up to 50°C in the arid western regions. Hot dusty winds commonly blow at high velocity during this season. 3. Towards the end of the summer by late June, the monsoon winds bring rain and the temperature gradually falls down. The rainfall starts little earlier in the eastern and southern parts of India.

Although the Average precipitation over the country as a whole is about 1000 mm, this is very unevenly distributed in space and time. The west coast and the Assam region are areas of heavy rainfall, receiving 2500 mm and above annually. The eastern part of the peninsula and the northern plains receive moderate rainfall of 1000-2500 mm annually. The Punjab plains and upper western parts of the Deccan plateau receive low rainfall of 250-1000 mm. While the Rajasthan desert and Ladakh plateau of Jammu and Kashmir are regions of very low precipitation of less than 250 mm. The bulk of the precipitation occurs in the south west monsoon period covering 4 to 5 months of June to October. A large part of the country experience acute water shortage in the other months. It is only the south eastern part of peninsular India that receive the major share of the precipitation in November and December from the north east monsoon (Murthy, 1975).

The variability in temperature and rainfall patterns produces a great diversity in the vegetation of the subcontinent.

Vegetation of India

Champion (1936) had described the forest types of India in detail. His classification has been further modified by Champion and Seth (1967) who recognised 16 major types of forests in India. They have considered climate as the main factor in distinguishing broad types. Edaphic and biotic factors have been taken into account for recognising subgroups with each main type (Figure 92).

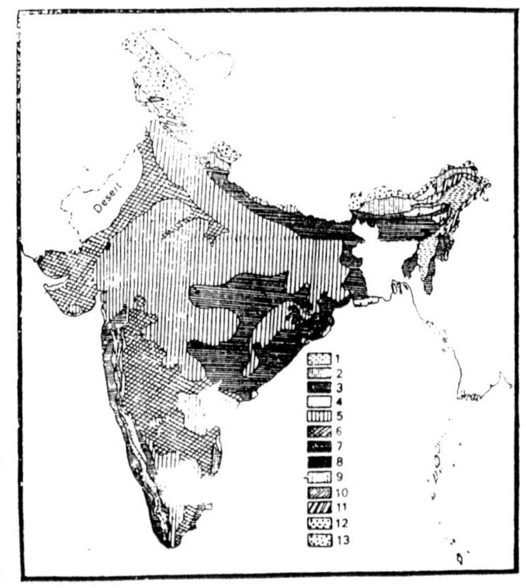

Fig. 92. Forest types of India (after Champion and Seth).

(A) Moist Tropical Forests

1. *Tropical wet evergreen forests*

These forests occur on the Western Ghats, Assam, Cachar, parts of Bengal, parts of Karnataka and Andamans. These are climatic climax forests with very dense growth of tall trees which are more than 45 meters high. The shrubs, lianas, climbers and epiphytes are abundant. The grasses and herbs are rare because the dense leaf canopy does not allow enough light penetration to the ground level. The rainfall exceeds 250 cm in these regions. The flora on the west coast includes *D. indica, Palaquium* and *Cellenia* and in assam it includes *D. macrocarpus, D. turbinatus, Shorea assamaica, Mesua ferrea* and *Kayea*. Besides these, the common flora includes species of *Mangifera, Eugenia, Mesua, Myristica, Pterospermum, Memecylon, Cinnamomum, Bobax, Dendracalamus, Calamus, Veteria, Calophyllum, Vitex, Pandanus, Cedrela, Tetrameles, Strobilanthes* and several ferns and orchids etc.

2. *Tropical moist semi-evergreen forests*

These forests occur along western Ghats, in parts of upper Assam and Orissa. The rainfall of these regions is usually high above 2000 mm per year. The deciduous species grow intermixed with the evergreen species, and therefore, the forest are called semi-evergreen. The leaves fall only for a short period. The trees and dense and tall (25-35 meters) but shrubs are common. In Andamans, major species are *Dipterocarpus alatus, Terminalia, Salmalia* etc. In east Himalayas *Schima wallichii* and *Bauhinia* are prominent. In Orissa *Artocarpus, Michelia* and *Mangifera* are common, while in Assam *Phoebe* and *Ammora* are abundant. Other taxa in these forests include species of *Xylia, Schleichera, Bambusa, Ixora, Webera, Storbilanthes, Cedrela, Actinodaphne, Garcinia, Legerstroemia, Sterculia, Lophopetalus, Mallotus, Vernonia, Dendrocalamus, Calamus, Pavetta, Elettaria, Pothos, Vitis, Shorea, Garuga, Bauhinia, Albizzia, Dillenia* etc. Orchids, ferns as well as grasses such as *Andropogon, Crotolaria, Imperata, Inula, Leea, Desmodium, Fambosa, Woodfordia* etc. are common.

3. *Tropical moist deciduous forests*

These forests are distributed in a narrow belt along the foot of the Himalayas, on the eastern side of the western Ghats, Chhota

Nagpur and Khasi hills. These forests have high rainfall (1500 - 2000 mm) throughout the year except very short dry period. The trees are deciduous, remaining leafless for one or two months only. The lowest story of short trees and shrubs sometimes is evergreen. High intensity of biotic factor have converted them to open savannas in several areas. The most common species in teak (*Tectona grandis*) in the south and sal (*Shorea robusta*) in the north. Other common species of these forests are *Dalbergia, Cedrela, Salmalia, Albizzia, Terminalia, Cordia, Melia, Dillenia, Eugenia, Dendrocalamus* etc.

4. *Littoral and swamp forests*

These forests occur along sea coast in wet and marshy areas and also in deltas of lerger rivers on the eastern coast. They comprise mostly of evergreen species. In saline swamps mangroves are chiefly composed of *Rhizophora, Bruguiera, Ceriops, Nipa* etc. In less saline areas *Phoenix, Ipomoea, Phragmites, Casuarina, Manilkaka, Calophyllum* etc. are found. Other swamp forests include species of *Barringtonia, Syzygium, Myristica, Bischofia, Trewia, Lagertoremia, Sophora, Pandanus, Enteda, Premna* etc.

(B) Dry Tropical Forests

5. *Tropical dry deciduous forest*

These are most widely distributed forests in India covering up to 40 per cent of the total forest area. These occur in areas with annual rainfall ranging from 750 to 1250 mm, and with a dry season of about 6 months. They are distributed almost throughout the country except Kashmir, beyond Bengal, Rajasthan and Western Ghats. The trees are small (10-15 meters) and the canopy is open. Shrubs are abundant. These forests are usually divided into northern deciduous forests dominated by *Shorea robusta* and southern deciduous forests dominated by *Tectona grandis*. The common species in the northern forests include *Anogeissus, Terminalia, Semecarpus, Buchanania, Carissa, Emblica, Madhuca, Acacia, Zizyphus, Lannea, Sterculia, Dendrocalamus, Salmella, Adina, Diospyros, Bauhinia, Eugenia, Aegle, Grewia, Adathoda, Helicteres* etc. The common species in the southern forests are *Dalbergia, Kydia, Dillenia, Terminalia, Pterospermum, Acacia, Diospyros, Anogeissus, Boswellia, Bauhinia, Chloroxylon, Hardwickia, Soymida, Gymnosporia, Zizyphus, Dendrocalamus and Holarrhena* etc.

6. *Tropical dry deciduous forests of south east deccan region*

In some parts of Tamil Nadu and the Karnatic coasts evergreen forests of short (10-15 meters) but dense trees are found though the annual rainfall is relatively small. There are no bamboos but grasses are abundant. The common flora of this region includes species of *Memecylon, Maba, Pavetta, Feronia, Terminalia, Ixora, Sterculia, Mesua, Schleichera* etc.

7. *Tropical thorn forests*

These forests are distributed in areas with very low rainfall as in western Rajasthan, parts of southern Punjab, Bundelkhand, Gujarat, Maharashtra, Madhya Pradesh and Tamil Nadu. The trees are short (8-10 meters), sparsely distributed, mostly thorny; shrubs are more common than trees. Plants remain leafless throughout the year. During rains grasses and herbs become abundant. The common plants of these forests are *Acacia nilotica, Prosopis spicigera, Capparis, Albizzia, Anogeissus pendula, Erythroxylon, Euphorbia, Cordia, Randia, Balanites, salvadora, Gymnosporia, Leptadenia, Suaeda, Grewia, Asparagus, Butea, Calotropis, Adathoda, Madhuca, Salmalia, Crotalaria, Tephrosia, Indigofera* etc.

(C) Montane Subtropical Forest

8. *Subtropical broad leaved hill forests*

These forests are dense, have predominantly evergreen species, and occur in relatively moist areas as lower slopes of eastern Himalayas, Bengal, Assam and on the hill ranges as Khasi, Nilgiri, Mahabaleshwar etc. In the north occur *Quercus, Castanopsis, Schima* and some temperate species. In eastern Himalayas due to higher humidity, bamboos, many epiphytes including orchids and ferns become abundant. In the south common trees are *Eugenia, Actinodaphne, Randia, Glochidion, Terminalia, Olea, Eleagnus, Murraya, Atylosia, Ficus, Pittosporum, Saccopetalum, Carreya, Alnus, Betula, Phoebe, Cedrela, Garcinia, Populus* etc., are common.

9. *Subtropical pine forests*

In western Himalayas and Khasi Jayantia hills of Assam, open forests composed mainly of *Pinus roxburghi* and *P. khasya* are found up to an altitude of 1875 meters.

10. *Subtropical dry evergreen forests*

Lower elevations of Himalayas have xerophytic forests containing mainly thorny and small leaved evergreen species. These areas are characterised by low rainfall and low temperature. The common species are *Acacia modesta, Dodonea viscosa, Olea cuspidata* etc.

(D) Montane Temperate Forests

These forests are found at 1800 to 3800 meters altitude in the Himalayas, where the climate is comparable to that of the temperate regions of high latitude in respect of low temperature. However, the humidity is comparatively lower. These forests are sub-divided into wet, moist and dry forests depending upon the moisture regime.

11. *Montane wet temperate forests*

These forests are found in eastern Himalayas with high rainfall and also in some parts of South India (Nilgiris). The forests in the south are evergreen and are known as *Sholas*. The trees are 15 to 20 meter high with a dense growth forming closed canopy. Epiphytes are abundant. The common plant species of these forests are *Rhododendron nilagircum, Hopea, Balanocarpus, Artocarpus, Pterocarpus, Myristica, Hardwickia, Gordonia, Salmella, Mucuna, Dioscorea* etc.

The wet temperate forests in the north extend from eastern Nepal to Assam at an altitude of 1800 to 2900 meters and sometimes above. These forests are also evergreen or semi-evergreen, dense and high (up to 25 meters). The common plant species include *Quercus, Acer, Prunus, Ulmus, Machilus, Eurya, Symplocos, Mahonia, Begonia, Michelia, Thunbergia, Rhododendron, Tsuga, Abies, Arundinaria, Bucklandia, Pittosporum, Loranthus* etc.

12. *Himalayan moist temperate forest*

The central and western Himalayas have the forests of conifers and/or oaks at an altitude of 1500 to 3000 meters. The trees are tall (up to 45 meters) but undergrowth is thin and deciduous. These forests have plants like *Cotoneaster, Berbaris, Spiraea* etc. In the lower zone the common species are *Quercus incana, Q. dilatata,*

Cedrus deodara, Pinus wallichiana, Picea smithiana, and *Abies pindrow*. At higher levels *Q. semecarpifolia* and *Abies pindow* are common. In the eastern Himalayas species like *Tsuga dumosa, Q. lineata, Picea spinulosa, Abies densa* and *Q. pachyphylla* are common. In some areas *Cupressus* is dominant due to presence of limestone.

13. *Himalayan dry temperature forests*

These forests occur in a nerrow belt in the western Himalayas extending from parts of Uttar Pradesh through Himachal Pradesh and Punjab to Kashmir. They are open, evergreen forests with scrub undergrowth. Oaks and conifers are dominant. In western drier areas *Pinus gerardiana* and *Quercus ilex* are common while in the eastern relatively moist zone *Abies, Picea, Larix griffithia* and *Juniperus wallichiana* are common. Other associated plant species belong to *Daphne, Artemisia, Fraxinus, Alnus, Cannabis, Plectranthus* etc.

(E) Sub Alpine Forests

14. *Sub alpine forests*

These evergreen forests are open and occur throughout the Himalayas at high altitudes (above 3000 meters up to tree lines). They are composed of mostly conifers like *Abies spectabilis* and broad leaved species like *Betula* and *Rhododendron*, shrub layer is composed of *Rosa, Lonicera, Strobilanthes, Smilex* etc.

15. *Moist alpine scrub*

Throughout the Himalayas above the timberline up to the height of 5500 meters, dwarf evergreen shrubby growth of conifers (*Juniperus*) and the broad leaved species (*Rhododendron*) are found. Other plants include *Arenaria, Saxifraga, Rheum, Thalicrum, Lonicera, Bergia, Sedum, Primula* etc.

16. *Dry alpine scrub*

In areas with low rainfall (below 35 cm) open xerophytic scrubs like *Juniperus, Caragana, Eurotia, Salix* etc. are found up to 5500 meters.

Grassland Vegetation

In India, natural grasslands (as climax formation) are not present but occur only during succession. The following three types of grasslands occur in India : (1) *Xerophilous* that occur in dry regions of north west India under semi-arid conditions; (2) *Mesophilous* which are extensive grass flats or savannas and occur in moist deciduous forests of Uttar Pradesh; and (3) *Hygrophilous* which are called wet savannas. These three basic types of Indian grasslands have been further subdivided into the following eight subtypes, each of which is named on its dominant species : *Sehima-Dicanthium, Dicanthium-Cenchrus, Phragmites-Saccharum, Bothriochloa, Cymbopogon, Arundinella, Deyeuxia-Arundinella* and *Deschampsia-Deyeuxia*.

Botanical Provinces of India

Hooker (1855) recognised 9 Botanical provinces of the then British India which included the Malayan Peninsula, Burma, Sri Lanka, Pakistan and Bangladesh as well as Nepal and Bhutan. Later Clarke (1898), Prain (1903) Candler (1938) and Chatterji (1939) made minor modification in the boundaries of the botanical provinces. In the post independence period, Razi (1955), classified the floristic provinces largely on the basis of climate, physiography and migration routes. Legris (1963) gave a detailed description of the Indian vegetation and floristic regions. From the phytogeographic viewpoint the classification of Chatterjee (1939) is still acceptable and largely followed. The country has been divided into following floristic regions (Figure 93).

1. Western Himalayas

It extends from Kumaon to Kashmir and has annual rainfall up to 2000 mm. Altitudinally there are following three zones of vegetation corresponding to three climatic belts :

(*a*) *Submontane zone*

This extends up to 1500 meters altitude and comprises mostly of Siwalik ranges. The forests are tropical and subtropical having trees like *Shorea robusta, Dalbergia sissoo, Cedrela toona, Ficus glomerata, Eugenia jambolana, Acacia catechu, Butea monosperma, Zizyphus* and thorny succulent *Euphorbias* on the slopes.

(b) *Temperate zone*

Abnove submontane zone extend montane temperate forest up to 3500 meters altitude. They are dominated by several species of *Quercus, Acer, Ulmus, Rhododendron, Betula, Salix, Populus, Cornus, Prunus, Fraxinus, Pinus, Cedrus, Picea* and *Taxus*.

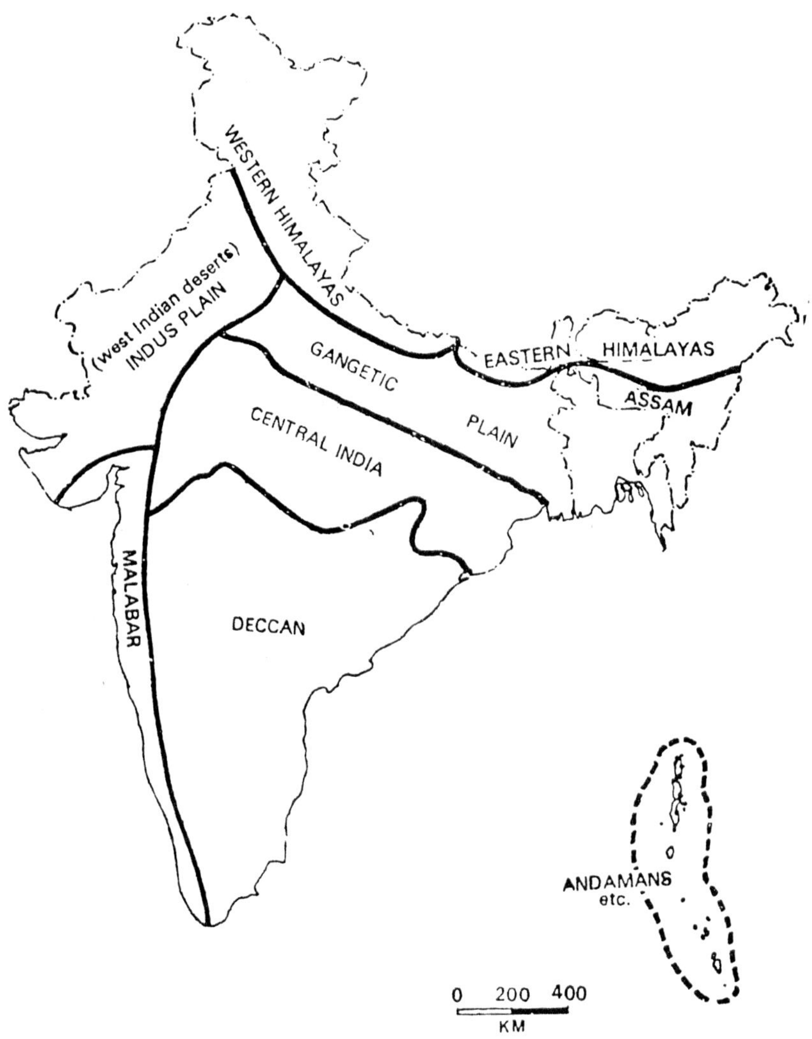

Fig. 93. Floristic regions of India.

(c) *Alpine zone*

Above 3500 meters up to about 4500 meters (snow line) the vegetation is alpine forest and scrub merging into meadows. Most common tree species are *Abies, Betula, Juniperus* and bushy *Rhodo-*

dendron. The herbs which occur near the snowline include species like *Primula, Potentilla, Polygonum, Geranium, Saxifraga, Aster* etc.

2. Eastern Himalayas

It includes regions of Sikkim and NEFA and is characterised by more rainfall, less snow and higher temperature. It is also sub-divided into three zones altitudinally :

(*a*) *Tropical zone*

Up to about 1800 meters altitude, this zone has tropical semi-evergreen or moist deciduous forests. The important species are *Shrea robusta, Acacia catechu, Dalbergia sissoo, Terminalia, Albizzia, Cedrela, Dendrocalamus* etc.

(*b*) *Temperate zone*

This zone extends between 1800 to 3800 meters altitude and has typical montane temperate forests which are dominated by Oaks like *Michelia, Quercus, Pyrus, Symplocus, Eugenia* etc., at lower levels; and by conifers such as *Abies, Pinus, Larix, Tsuga* and *Juniperus* and also *Salix, Rhododendron, Arundinaria* etc. at higher levels.

(c) *Alpine zone*

Beyond the temperate zone, extends the alpine zone up to 5000 meters altitude. It has alpine vegetation including *Juniperus* and *Rhododendron* with its other typical flora. Only herbaceous vegetation remains near snowline.

3. Indus Plains

This zone includes the arid and semi arid regions of Punjab, Rajasthan, Kutch, part of Gujarat and Delhi. The rainfall is less than 700 mm. The vegetation is tropical thorn forest in semi arid region and is typical desert in the arid region. The plants of this zone are xerophytes, for example, *Acacia nilotica, A. senegal, A. leucophloea, Anogeissus pendula, Gymnosporia montana, Prosopis spicigera, Salvadora, Capparis, Dalbergia, Albizzia, Euphorbia, Grewia, Tephrosia* and *Calotropis*.

4. Gangetic Plains

This region extends over Uttar Pradesh, Bihar, Bengal and part of Orissa. It is characterised by moderate amount of rainfall and most fertile soils. Vegetation of this zone is chiefly of tropical moist and deciduous and dry deciduous forest type. The common plants of this zone are *Dalbergia sissoo, Accacia nilotica, Saccharum munja, Butea monosperma, Madhuca indica, Terminalia arjuna, Buchanania lanzan, Diospyros melanoxylon, Cordia myxa, Acacia catechu, Azadirachta indica*, weeds and grasses like *Xanthium, Cassia, Argemone, Amaranthus* etc. In Gangetic delta (South Bengal) mangrove vegetation is common.

5. Central India

It comprises Madhya Pradesh, part of Orissa and Gujarat. The rainfall is 1500-2000 mm and its vegetation is thorny, mixed deciduous and teak type. The chief plants of this region are *Tectona grandis, Madhuca indica, Diospyros, Butea, Dalbergia, Terminalia, Carissa, Zizyphus, Acacia, Mangifera* etc. These forests in many areas have been degraded to a grassland vegetation due to heavy grazing, burning and other biotic interferences.

6. Malabar (West Coast)

This region include western coast of India from Gujarat to Cape Comorin and has heavy rainfall. The forests are tropical evergreen towards extreme west, semi-evergreen towards interior subtropical or montane temperate evergreen forests in Nilgiris and mangroves near Bombay and Kerala coast. The species composition of these forests has already been discussed.

7. Deccan Plateau

This region extends all over peninsular India, that is, Andhra Pradesh, Tamil Nadu and Karnataka, and has rainfall up to 1000 mm. Its central hilly plateau has tropical dry deciduous forests of *Boswellia serrata, Tectona grandis* and *Hardwickia pinnata*, while the low eastern dry Coromandal coast has tropical dry evergreen forests of *Santalum album, Cedrela toona* and plants like *Acacia, Prosopis, Euphorbia, Capparis, Phyllanthus* etc.

8. Assam

This region is characterised by heavy rainfall (2000-10000 mm). The vegetation is either dense evergreen forest or subtropical. The evergreen forest include trees like *Dipterocarpus macrocarpa, Mesua ferrea, Shorea robusta, Ficus elastica, Bambusa patlida, Dendrocalamus hamiltonil*, grasses like *amperata cylindraca, Sachharum, Themeda*, insectivorous plants like *Nepenthus*, epiphytes and orchids. In the northern cooler region, wet hill forest include plants like *Almus, Betula, Rhododendron, Mugnolia* etc. The hilly tracts also have pine forests of *Pinus khasiya* and *P. insularis*.

9. Andmans

This region posses a varied type of vegetation : mangroves and beech forest at its coast and evergreen forests of tall trees in the interior. Important plant species of the island are *Rhizophora, Mimusops, Calophyllum, Lagerstroamia* etc.

Interpretive or Dynamic Phytogeography

The explanation of why a particular taxon has come to occupy a particular part of the world requires a knowledge of where it originated, how it dispersed, and how it evolved its present adjustments and characteristics. When a population invade new regions and environments and geographically isolated they commonly differentiate into new species. Each species thus becomes the product or visible expression of a particular combination of environmental factors, interactions and locality.

Dispersal Dynamics

Dispersal is the spread of individuals away from their home sites. Dispersal movements are usually slow, and cover relatively short distances in the life time of an individual. The cumulative result of short dispersals by successive generations, however, may become conspicuous in the course of years, decades or centuries, when it constitutes *range expansion* of the species into a new habitat or area.

Manner and Means of Dispersal

According to Ridley (1930) dispersal means an active process of transportation of disseminules from place to place. Dispersal involves the dissamination from the parent and the distribution to a new place. Migration is also closely connected with the dispersal. It implies not only dispersal but also a successful growth and establishment. Thus, dispersal is a successful forerunner of migration which is actually accompanished only on establishment in a new place. In nature only a small number of plants which become dispersed actually become established and effect migration.

Not only many of them die prematurely or fall on barren land or come to rest where they cannot even start a new life or fail to survive the struggle with stranger compititors. Dispersal of plants takes place through dissemules or units of dispersal, also known as propagules or germules, which are mostly reproductive structures, for example, spores, nude embroy (*Seleginella*), Nude seed (Gymnosperms), seed (Angiosperms), Fruit (simple or aggregate) vegetative parts e.g., bulbils, rhizomes, runner, buds, stems etc.

Often the same plant species produces more than one type of dissemules - thereby increasing its chances of colonising vast areas. For example, common reed (*Phragmites communis*), which is often claimed to be the most widely distributed vascular plant in the world, has the advantage of a wind dispersed parachute like fruit and a water dispersed rhizome. Ridley classifies the main methods of dispersal as follows : (1) dispersal by wind, (2) dispersal by water and ice, (3) dispersal by animals, (4) dispersal by mechanical means, and (5) dispersal by human agency.

Dispersal by wind is also known as *anemochery*. The dissemules have various adaptations such as lightness, winged shape, hairyness etc. which helps them to be carried away by the wind to a shorter or longer distance from the parent plants. Spores are minute, light in weight and float to vast distances and to great heights, because they are easily blown away by gentle breeze. Some light seeds occur in a number of families but they are characteristic of the orchids. They may easily be carried away by the gentle breeze. Seeds and fruits of many plants develop one or more appandages in the form of thin flat membranous wings. This device help them to float in the air and facilitate their dispersal by wind. Thus we find that fruits of *Heleptelia, Hiptage, Acer, Shorea* etc. are provided with wings. In many plants of Compositeae the calyx is modified into hair like structure known as pappus. This pappus is persistent in fruit and open out in an umbrella like 'parachute'. Thus the seed float in the air and are carried by air current to a great distance. Seeds of *Calotropis, Holarrhena* and *Gossipium* are provided with either one or two tufts or all over their body. These hairs aid the dispersal of seeds by wind. In *Clematis* and *Naravelia* the styles are persistent and feathery. The fruits are thus easily carried away by wind.

Dispersal of seeds by water is also known as *hydrochory*. This is believed to be the earliest mode of dispersal as the primitive plants were aquatic. Plants which are hydrophytes and those which grow on the banks of water courses are dispersed by water. In *Cocos nucifera, Lodoicea schellarum, Potamogiton, Segittaria* etc. the pericarp is modified as floating organs. Seeds and fruits are light in weight in *Hibiscus, Ipomea* etc. Testa is buoyant in *Iris, Lamna, Anona, Euphorbia* etc. *Neptunia, Trapa, Eichhornia, Salvinia, Azolla* etc. are floating plants which are carried as a whole by flowing streams.

The most effective means of dispersal after wind is *zoochory*. It is effective where dispersal by wind any water fails. Animals tend to disperse disseminules in three ways : (1) by swallowing them, passing them through and out of the digestive tract, (2) by carrying them attached to their outer surface and (3) by carrying them in mud adhering to their feet. Locally restricted animals or relatively immobile animals do not play a large role. Birds are of vastly greater potential importance because of their greater range of action. Grazing animals are responsible for a more intensive type of dispersal, because their food will always contain a certain number of disseminules and these will be almost continuously passed out of the body. Attractiveness to animals and birds is greatly due to either colour or palatability. This type of dispersal is also known as *endozoic*. On the other hand in many seeds structural modifications are associated with wind dispersal. Many fruits are provided by hooks, spines, bristles, stiff hairs etc. on their body by means of which they adhere to the body of the wooly animals and are carried by them to distances for example, *Tribulus, Xanthium, Achyranthus, Martynia, Aristida* etc. Such a dispersal is known as *eccozoiz*.

In all the cases so far mentioned the fruits has been a relatively passive agent in dispersal. However, in some instances the dehiscence of the fruit is so sudden that the contained seeds are shaken out more or less violently, sometimes to a considerable distance. This is due to unequal strain set up in the ripening fruit and relieved by explosive rupture of the fruit wall for example, *Balsam, Castor* etc.

Mechanical dispersal involving rhizomes or runners which do infact play a kind of subsidiary role in dispersal because they give the plants some kind of mobility. In fact they enable new fruiting branches to arise at some distance from the parent axis.

The accidental or deliberate dispersal of plants is widespread. This type of dispersal extends over a long period of time and many of its early stages are now beyond elucidation. Although dispersal into the more distant parts of the world may be regarded as a fairly recent process, more localised dispersal of plants has been going on ever since mankind first begain to move freely about the world.

Barriers of Dispersal

Any feature of the physical or biological agency that restricts or prevents dispersal of plants is a barrier. Barriers are always relative and a barrier for one species may well be a main dispersal route for another. For example, water might be a barrier for a terrestrial species but normally is not for a aquatic species. Three classes of barriers are generally recognised : (1) Physiographic or physical barriers such as land, water, elevation, soil; (2) Climatic barriers such as temperature, humidity, rainfall, sunlight etc., and (3) Biotic or biological barriers such as lack of food and the presence of enimies or effective compititors.

Dispersal Pathways

An understanding of how plant groups have dispersed over the face of the earth involves the controversy as to whether all the continental land masses were at one time intimately connected and if so, how long ago they drifted apart to their present locations. Separated continents need atleast temporary land bridges if there is be dispersal and interchange of plants. Various theories have been put forward in this regard :

(1) *Continental drift theory*

This theory postulates that throughout the Palaeozoic and much of the Mesozoic the presently distributed continents were grouped into two great land masses: a northern ***Laurasia*** was separated from a southern *Gondwana* by the vast sea of ***Tethys***. In the Jurassic, the land masses were frangmented, and the fragments subsequently drifted apart. Laurasia is supposed to have split into

North America, Greenland, Europe and most of Asia; Gondwana into South America, Africa, Arabia, Madagascar, India, Australia and Antarctica (Wegener, 1924). This theory has been supported by various evidences.

(2) *Land bridges*

An alternative theory is that the continents have been situated in their present positions throughout geological time. They have changed their boundaries with rise and fall of oceans and with upheaval and subsidence of land masses but there has been no significant lateral drifting.

The separated continents however, have been connected by atleast temporary land bridges at various times during the period. A land bridges now connects North and South America, Eurasia and Africa. One of the best known and generally accepted land bridge of the past, but not now in existance, connected Asia and North America across the bering strait.

Land bridges serve as important dispersal routes for those land plants which are able to cross them. When a land bridge allows free passage of most plants in either or both directions, it is called *corridor*. It the land bridge is narrow, has on unfavourable climate, a lack of suitable niche, or too many compititors, it is called *filter*, since only a few species are able to pass over it. Because of its general unfavourableness, it is generally only one way.

Migration of Plants

During the evolution of our earth, plants have migrated from one place to another, they have been occupying new areas. Their movement to new areas was possible through dispersal of spores, seeds buds and bulbils or any other part which served as plant propagule. On reaching a new place the propagules established themselves. The plants moving to new areas has to acclimatise through morphological and physiological adaptations. Having established in the new locality plant reproduce and regenerate and form somewhat stable populations. Hence migration involves transport and establishment or plants from one place to the other (Misra, 1980).

Migration begins when the germule leaves the parent area and ends when it reaches its final resting place. The factors entering into migration are mobility, agent, distance and topography.

Mobility is the ability of a species to move out of the parent area. In landplants it is indicated mainly by the size, weight and surface of the diseminules, especially those carried by wind and water. Man and animals also help in the mobility of fruits in many ways. Mobility is most marked in those plants which are themselves mobile such as bacteria, diatoms, volvox etc. or possess motile spores as found in some of the green algae. On the other hand mobility is little or not at all developed in flowering plants with large heavy seeds or fruits.

Medlicott and Blanford (1879) suggested the Himalayan glaciation theory to explain the occurrence of the Himalayan plants and animals on the higher ranges of southern India. During the Pleisticene, the glaciation in the Himalayas lowered the temperature throughout India as a result of which plants and animals migrated southwards towards the equator. Subsequently, when the temperature increased these organisms sought refuge in the higher parts of hills of southern India. Burkill (1924) too discussed the importance of glaciation for migration of species : the spread of the Himalayan elements to the south India hills was affected by the glaciation and that the Malayan elements in these hills was much older and had migrated through Bay of Bengal or Ceylon-Deccan route. The glaciation also led to the disappearance of the rain forest Malayan element from the Himalayan ranges.

Blasco (1970, 1971a, b) disregards the hitherto commonly held view that Pleistocene glaciation is responsible for pushing the Himalayan flora southwards. He is in favour of intermountain migration through long distance dispersal either by the agency of birds or wind.

Meher-Homji (1972) has raised the questions that would arise if the view of migration in present times is accepted. For example, if the seeds of the Himalayan *Rhododendron, Mahonia* etc. are carried to the south in recent times, why sould the southern species differ from the Himalayan ones ? Foe example, ***Mahonia nepalensis*** occurs in the Himalayas, while *Mehonia* ***leschenaultii*** in south India.

Puri (1947) is of the opinion that the reason for the destruction of the oak-laurel community from the western Himalayas and later invasion of conifers from the Mediterranean region, from two different directions as shown by Biswas (1937) is that the western Himalayas were strongly glaciated and also uplifted during the Pleistocene. In the eastern Himalayas, the glaciation was less intense and the rainfall hingher; consequently, the oak-laurel type of vegetation and the Indo-Malayan species have been conserved. The intense glaciation caused destruction of the Malayan elements from the western Himalayas and conifers gained dominance (Gupta, 1962).

The second theory postulates land bridges across the Indian ocean. A continental bridge is proposed between Malaysia, Andaman-Nicobar, Ceylon and southern India. Clarke (1898) found in this route a possible explanation of the Malayan species into India, notable Dipterocarpaceae, Ericarceae and *Podocarpus nerifolia*. Land connections have also been suggested between India, Australia and New Zealand.

Towards Africa another route of migration has been offered by the theory of the Lemuria continent joining Madagascar with India and Ceylon. However, Millot (1952, 1953) shows that the present Malagasy fauna does not resemble the Indian fauna and that the concept of lemuria is not at all useful in explaining the distribution. Furon (1959) is of the view that if this connection existed at all it ceased in the Upper Cretaceous. According to Legris (1963) the notion is hypothetic and in no way useful in explaining the migration of the African elements into India.

The continuous range hypothesis postulates that the localities at which species are found at present, must have one time formed part of continuous range of distribution of the species. The satpura hypothesis (Hora, 1949) envisaged that during Pleistocene, the Satpura and the Vindhyan ranges had an altitude of 1500-1800 meter forming a continuous mountain chain between the eastern Himalayas and the western Ghats, to facilitate migration from the former to the latter areas. Auden (1949) and Dey (1949) have questioned the continuity of the Satpura ranges on geological grounds and have supported the theory of climatic changes in wake of the Pleistocene glaciation.

The importance of the Coromandel route has been emphasised by Legris (1963). This route joins Assam to the western Ghats passing through Bengal, the hills of Orissa, the eastern Ghats, the Karnataka plateau and the Nilgiris. Legris favours this hypothesis on climatic, geological, paleobotanical grounds and from present day distribution.

There seems to be one more route of migration across the Arabian sea. There was an important trade in horses between Arabia, Africa and Indian ports like Tuticorin, Pamban (Madurai) and Porbandar (Kathiawar). Along with this trade *Acacia spirocarpa* accidently migrated to India where over centuries it has emerged as a species only slightly distinct from the African member *A. planifrons*.

Similarly, the distribution of *Capparis decidua* is discontinuous. It is spread over northern tropical Africa, Egypt, Arabia, Persia, Pakistan and India. In India it covers Rajasthan, northern Gujarat and the deccan from Nasik up to Sholapur, Bijapur. After a considerable discontinuity it reappears in the extreme southeast part of the peninsula. It introduction in this southeast corner via the ship route from Africa may be suggested as a possibility.

Age and Area

If a species migrates at some time and some place on the surface of the earth, either by diverging from another part of the population, its starting point may be designated as some particular area. If such a species continues to maintain itself in the face of competition with other species, it has the opportunity of expanding its range or moving to some other areas if conditions permit. A successful species, then would have opportunity to occupy more area from time to time as opportunities offered. An unsuccessful species would be more likely gradually to restrict its range and finally to perish.

This general problem was considered by J.C. Willis (1951) from which he came to the conclusion : the area occupied at any given time, in any given country, by any group of allied species at least ten in number, depends chiefly, so long as conditions remain reasonably constant, upon the age of the species of that group in that country, but may be enormously modified by the presence of barriers such as seas, rivers, mountains, changes of climate from one region to the next or other ecological boundaries, and the like, also by the action of man.

This general idea of Willis has much merit in it, but in his statement it was hedged with so many details that the main point was lost in the confusion of the statement. His general idea might be stated to mean : on the whole, *spreading species tend to occupy increasing area with increasing age* (Woodbury, 1954).

This does not imply that all species will spread at the same rate. It is almost certain that some will spread faster than others and Willis tried to account for these differences by taking the average of atleast ten in number.

The statement of Willis needs further thinking with reference to species that are not increasing. Those that can hold their own may remain stationary, but this is not likely for long periods of time. Competition between species must be visualized as dynamic, some increasing and some decreasing. A corollary to his statement might be appended as follows : *On the whole, decreasing species tend to occupy decreasing area with increasing age and that the area occupied may be merely widely scattered remnants of the former area.*

This leads to the consideration of vigorous and decadent species in relation to present day range. It is sometimes difficult to interpret the present status of a species without looking backward in time to find its history. The general idea that vigorous species tend to occupy more area as age increases is useful in understanding widespread species. But it sometimes gives clues to young and vigorous species that have just begun their career. It is also useful, to use the general idea that decadent species tend to have remnants scattered here and there over large areas, but it is more difficult to use when the remnants have been reduced to one.

Apart from the short period during which a new species is expanding into a previously unexplored area, the pattern of distribution of plants are the results of the interactions of so many unrelated factors, such as geographical accidents of the accessibility of new exploitable environments, or the evolutionary accidents of contact with other, ecologically superior competitors that the concept of age and area is valueless and misleading (Cox, Healey and Moore, 1976)

20. *Famous Indian Ecologists*

(1) Professor R. Misra

Professor Ramdeo Misra, widely, variously and astonishingly read scholar of Ecology is not an individual but an institution. He is regarded a *Father of Indian Ecology*.

Professor Misra was borne on August 26, 1908. He had his schooling and university education at Varanasi. In the service of his *Alma Mater*, he joined as a Demonstrator in Botany in 1931. In 1935 he proceeded to the University of Leeds, London, to work for two years on "Ecology of British Lakes", under the guidance of Late Professor W.H. Pearsall. On returning to India he studied many ecosystems. In 1939, he went to Bhagalpur, after two years he again returned to Banaras Hindu University in 1941 as Assistant Professor. He moved to Sagar University as Reader and Head in 1946. After a period of nine years at Sagar, he again returned to Banaras Hindu University to occupy the covered chair of the Botany Department.

Prof. Misra's main research emphasis has been understanding of form, function and factor interaction at all levels of ecological inquiry with a strong experimental bias. Under his able guidance more than 30 of his students have secured the Ph. D. Degrees during 1953-67.

Prof. Misra has travelled far and wide. After his return from England in 1937, he made another big tour of Europe in 1955. He gave a series of lectures in U.A.R., Italy and England. At Rome in 1955 he represented India at the XII General Assembly of International Union of Biological Sciences. Wherever he has gone, his scholarship and affability have been appreciated by all those who came in contact with him.

He has organised many National and International Seminars/Conferences. He is the founder of the International Society for Tropical Ecology and the Journal Tropical Ecology.

Besides publishing several research papers, he is the author of the Indian Manual of Plant Ecology.

(2) Professor F.R. Bharucha

Professor Fardoom Pustomji Bharucha, Professor and first Director of the Institute of Science, Bombay was borne on May 4, 1904, educated in Bombay till graduation. He received M.Sc. Degree from Cambridge University (U.K.) in 1930. For his Ph. D programme, he joined professor J. Braun-Blanquet, a pioneer phytosociologist at Montpellier University in France. His first book on Vocabulary of Plant Sociology was a translation from the original French one by Braun-Blanquet and Pavillard. His D.Sc. Thesis has been credited as a first class model in the relevant area of research and presentation.

He introduced the subject of Plant Ecology at Bombay University. In spite of the administrative responsibilities at the Institute he supervised various research projects and thesis a large number of students, and published seventyfive odd research papers. His contribution in the field of phytosociology are the pioneering ones in the country. Prof. Bharucha studied mangrooves, grasslands, weeds of cropland, calcicolous, nitrophillous and ruderal vegetation types. His well-known contributions pertain to grasslands productivity with reference to trace mineral deficiencies is soil, life-forms of tropical plants and vegetation of arid and semi-arid zones of India.

He assessed the nutritive value of fodders and emphasized the role of trace elements in the soil. Professor Bharucha's application of the plant indicator concept in locating the 'gypsophytes' in Rajasthan was the particular significance.

One of his most notable contribution was on Crassulacean metabolism. He brought out for the first time that the same plant species may behave differently in trems of storage and reconversion of their metabolic products under tropical and temperate conditions. That the plants growing on rocks and calcium rich substrate survive in the precarious habitats, because of their ability to absorb and retain elements like calcium and iron in large quantities and secrete organic acids in a complementary way.

Professor Bharucha made his mark at many International Conferences, and was closely associated with *Vegetatio, Excerpta Botanica,* and *Socitas Phytogeographica*. His contribution to science, earnes him a number of honours for example, Fellowship of the National Academy of Science of India, place for his potrait in the Gallery of Eminent Botanists of the World at the Hunt Botanical Library of the Carnegie Mellon University at Pittsberg, USA.

After the end of his long teaching carrer in Botany, he was invited to work as U.N. Expert in Bombay at the University of Syria in Damascus (1956-59), and as Professor of Botany at the University of Baghdad, Iraq (1959-61).

Prof. F.R. Bharucha passed away on **March 30, 1981**. He spoke half a dozen languages fluently. He even taught French and German at Pune University. People who had personal contact with Prof. Bharucha, will always remember him.

Bibliography

Abbot, D., Biever, L., Flebbe, P., Kothari, A., Pfeiffer, W., and Seastedt, T. Model of potassium in Okefenokee swamp decomposition (unpublished).

Agarwal, S.K. 1971. Floristic and ecological studies on the deciduous forests of Gungunda and Prasad (Udaipur) Southeast Rajasthan. Ph. D. Thesis, University of Udaipur, Udaipur.

Agarwal, S.K. 1971. Germination regulation in *Indigofera tinctoria*. (unpublished).

Agarwal, S.K. 1971. Observation of dry matter production by deciduous trees in Rajasthan. IV. *Tectona grandis* Liff. f. In "*Trop. Ecol. Emphas. Org. Prod.* Ed. P.M. Golley and F.B. Golley. Georgia University, Athens, pp. 185-194.

Agarwal, S.K. 1974. The biological spectrum of the flora of Gogunda and Prasad (Udaipur Rajasthan). *J. Biol. Sci.* 17 : 67-71.

Agarwal, S.K. 1975. Biological spectra of the flora of western Rajasthan. *J. Biol. Sci.* 18 : 43-44.

Agarwal, S.K. 1977. Importance of woody species in the deciduous forests of Prasad, Rajasthan. *Ibid.* 20 : 24-27.

Agarwal, S.K. 1978. Observation on net above ground biomass production in *Butea monosperma*. *Comp. Physiol. Ecol.* 3 : 164-166.

Agarwal, S.K. 1980. Community structure of the deciduous forests at Prasad. *Acta Ecol.* 2 : 36-41.

Agarwal, S.K. 1980. Biomass production relation in four species of the deciduous forests of Udaipur. In "*Progress in Ecology*" Eds, V.P. Agarwal and V.K. Sharma. Today and Tomorrow, New Delhi. pp. 75-86.

Agarwal, S.K., and D'souja, R. 1980. Net primary production of *Calotropis* community at Kota. *Acta Ecol.* 2 : 62-63.

Agarwal, S.K., and D'souja, R. 1983. Net above ground biomass production of two shrub species at Kota. *Ibid.* 5 : 36-38.

Agarwal, S.K., and Kasat, M.L. 1979. Net primary production in *Bothrioclova pertusa* (L.A. Camus) grazingland at Kota. *Ibid.* 1 : 25-27.

Agarwal, S.K., and Vyas, L.N. 1970. Studies in the extent and role of dormancy in the seeds of *Indigofera astragalina* Dc Prodr. *J. Indian Bot. Soc.* 49 : 158-163.

Allee, W.C., Emerson, A.E., Park, O., Park, T., and Schmidt, K.P. 1949. *Principles of animal ecology.* W.B. Saunders Co., Philadelphia.

Allee, W.C., and Park, T. 1939. Concerning ecological principles. *Science*, 89 : 166-169.

Ambasht, R.S. 1963. Ecological studies on *Alhagi Camelorum* Fisch. *Trop. Ecol.* 4 : 72-82.

Ambasht, R.S. 1964. Ecology of the underground parts of *Cyperus rotundus* L. *Trop. Ecol.* 5 : 67-74.

Ambasht, R.S. 1980. *Plant ecology*. Students Friends & Co. Varanasi.

Ambasht, R.S., Singh, A.K., and Misra, K.N. 1982. Energy conserving efficiency and productivity of a gradient of community in Chakia forest ecosystem. *Proc. Internat. Forestry Seminar Univ.* Pertanian, Malaysia. Ed. Srivasta et al. pp. 209-218.

Amen, R.D. 1968. A model of seed dormancy. *Bot. Rev.* 34 : 1-31.

Andrewartha, H.G. 1961. *Introduction to the study of animal population*. University of Chicago Press, Chicago.

Ansari, M.Y. 1956. Ecological and phytosociological studies of the plant associations of mountain screens. M.Sc. Thesis, Bombay University, Bombay.

Arditti, J. 1967. Factors affecting the germination of orchid seeds. *Bot. Rev.* 33 : 1-77.

Arora, R.K. 1966. On the biological spectrum of North Kanara flora. *Indian For.* 91 : 85-88.

Arwidsson, T.H. 1928. Bizentrische Arten in Skandinavien-eine terminologische Erorterung. *Bot. Nortiser.* No. 1.

Ashby, W.R. 1956. *An introduction to cybernetics*. Chapman & Hall, Ltd.

Attenberg, A. 1909. Die Nachreife des Getreides. *Landio. Vers. Stat.* 67 : 127-143.

Aubreville, A. 1963. Classification des plants vascularires en milieu tropical. *Adansonia*, 3 : 221-226.

Auden, J.B. 1949. A geographical discussion on the Satpura hypothesis and Garo-Rajmahal gap. *Proc. Natn. Inst. Sci. India.* 15 : 315-340.

Babu, R.V., and Joshi, M.C. 1970. Studies on physiological ecology of *Borreria articulata* (lian) P.N. Will., a common weed of bajra (*Pennisetum typhoides* (Burm. f) Stapf et C.E. Hubb) fields. I. seed production, germination and seeling survival. *Trop. Ecol.* 11 : 126-139.

Bachelard, E.P. 1967. Effects of gibberellic acid, Kinetin and light on the germination of dormant seeds of some Eucalypt species. *Aust. Jour. Bot.* 15 : 393-402.

Bacquerel, P. 1932. *Compts. Revd. Acad. Sci. Paris.* 194 : 2158.

Bacquerel, P. 1934. Ibid. 199 : 1662.

Baes, C.E. Jr. Goeller, H.H., Olson, J.S., and Rotly, R.M. 1977. Carbon dioxide and climate : The uncontrolled experiment. *Amer. Sci.* 65 : 310-320.

Bakshi, T.S. 1952 a. The effect of a preliminary period of darkness on the percentage germination of the seeds of *Anisochilus eriocephalus* Benth. *Curr. Sci.* 21 : 108.

Bakshi, T.S. 1952 b. The autecology of *Anisochilus eriocephalus Benth. Jour. Indian Bot. Soc.* 31 : 269-280.

Bakshi, T.S., and Kapil, R.N. 1952. The autecology of *Mollugo nudicaulis* Lam. *Bull. Bot. Soc. Bengal*, 6 : 45-48.

Bakshi, T.S., and Kapil, R.N. 1954. The morphology and ecology of *Mollugo cerviana. Jour. Indian Bot. Soc.* 33 : 309-328.

Barton, L.V. 1961. *Seed preservation and longivity.* Leonard Hill Ltd., London.

Barton, L.V. 1965. Seed dormancy: General survey of dormancy types in seeds and dormancy imposed by external agents. In "*Encyclopedia of plant physiology.*" Ed. W. Ruhland. Springer-Verlag, Berlin. 15/2 : 619-720.

Barton, L.V., and Crocker, W. 1948. *Twenty years of seed research*. Faber and Faber, London.

Basile, R.M. 1971. Conservation of the atmosphere. In "*Conservation of natural resources*". Ed. G.M. Smith. John Wiley & Sons, New York. pp. 133-158.

Bennett, E. 1965. *Genecological aspects of plant introduction and genetic conservation.* Scottish Plant Breeding Station. Record.

Bhandari, M.C., and Sen, D.N. 1971. Productivity studies in some arid zone Cucurbitaceous species. *Trop. Ecol.* 14 : 155-159.

Bhandari, M.C., and Sen, D.N. 1973. Phytochrome and seed germination in *Citrullus colocythis* (L) Schrad. *Sci. & Cult.* 39 : 458-459.

Bharucha, F.R., and Ferreira, D.B. 1941 a. The Biological spectrum of the flora. *J. Univ. Bomb. N. S. Biol Sci.* 9 : 93-100.

Bharucha, F.R., and Ferreira, D.B. 1941 b. The biological spectrum of the Matheran and Mahabaleshwar flora. *Jour. Indian Bot. Soc.* 20 : 195-211.

Bharucha, F.R., and Dave, R.N. 1944. The biological spectrum of grassland association. *J. Univ. Bomb. N.S. Biol. Sci.* 13 : 15-16.

Bharucha, F.R., and Dubash, P.J. 1951 a. The problem of nitrophily. *Vegetatio*, 3 : 183-194

Bharucha, F.R., and Dubash, P.J. 1951 b. Studies in nitrophily. II. Nitrophilous plants of Bombay *Jour Indian Bot. Soc.* 30 : 83-87.

Bharucha, F.R., and Satyanarayan, Y. 1954. Calcicolous associations of the Bombay state. *Vegetatio*, 5-6 : 129-134.

Bharucha, F.R., and Shankarnarayana, K.A. 1958. Studies on the grasslands of the western Ghats, India. *J. Ecol.* 46 : 681-705.

Bhat, J.L. 1968. Study of ecological life history of two weeds of crop fields and methods of their control. Ph. D. Thesis., Vikram University, Ujjain.

Bhatia, K.K. 1954. Factors in distribution of teak (*Tectona grandis*) and a study of teak forests in Madhya Pradesh. Ph. D. Thesis, Sagar University, Sagar.

Billings, W.D. 1938. The structure and development of old field short leaf pine stands and certain associated physical properties of the soil. *Ecol. Monogr.* 8 : 437-499.

Billore, S.K. 1973. Net primary production and energetics of a grassland ecosystem at Ratlam. Ph. D. Thesis, Vikram University, Ujjain.

Biswas, K. 1937. The distribution of wild conifers in the Indian Empire. *J. Indian Bot. Soc.* 12 : 24-47.

Black, M., and Naylor, J.M. 1959. Prevention of the onset of seed dormancy by Gibberellic acid. *Nature*, 184 : 468-469.

Black, M., and Wareing, P.F. 1955. *Physiol. Plant.* 8 : 300-316.

Blackman, F.F. 1905. Optima and limiting factors. *Ann. Bot.* 19 : 2 : 1-198.

Blasco, F. 1970. Aspects of the flora and ecology of Savannas of the south Indian hills. *J. Bomb. Nat. Hist. Soc.* 67 : 522-534.

Blasco, F. 1971 a. Montagnes du Sud de l'Inde Forests, Savanes, Ecologie. *Inst. Fr. Pondichery Tr. Sec. Sci. Tech.* 11 : 1-436.

Blasco, F. 1971 b. Orophytes of south India and Himalayas *J. Indian Bot. Soc.* 50 : 377-381.

Blumenthal-Goldschmidt, S., and Land, A. 1960. The presence of Gibberellin like substances in nature lettuce seeds. *Nature*, 186 : 815-816.

Bohra, P.N. 1976. Soil-plant-atmosphere system : a study of plant water relations in Indian aird zone. Ph. D. Thesis, Jodhpur University, Jodhpur.

Bohra, P.N., and Sen, D.N. 1977. Behaviour of some arid zone plants under water stress. All Indian Symposium of Advancement of Ecology. Meerut University (Abstract).

Borthwick, H.A., Hendricks, S.B., and Parker, M.W. 1952 The reaction controlling floral initiation. *Proc. Natl. Acad. Sci. USA*. 38 : 929-934.

Borthwick, H.A., Hendricks, S.B., and Parker, M.W., 1956. Photoperiodism. In "*Radiation biology*" Ed. A. Hollander. McGraw Hill Book Co., New York. 3 : 479-517.

Borthwick, H.A., Hendricks, S.B., Parkar, M.W., Toole, E.H., and Toole, V.K. 1952. A reversible photoreaction controlling seed germination. *Proc. Natl. Acad. Sci. USA*. 38 : 662-668.

Borthwick, H.A., Hendricks, S.B., Parkar, M.W., Toole, E.H., and Toole, V.K. 1954. Action of light on lettuce seed germination. *Bot. Gaz.* 115 ; 205-225.

Borthwick, H.A., Toole, E.H., and Toole, V.K. 1964. Phytochrome control of Paulonia seed germination. *Isreal Jour. Bot.* 13 : 122-133.

Braun-Blanquet, J. 1932. *Plant sociology*. Engl. Transl by G.D. Fuller and H.S. Conard. Macgraw Hill, New York.

Bray, J.R. 1960. The chlorophyll content of some native and managed plant communities in central Minnesota. *Canad. J. Bot*. 38 : 313-333.

Bray, J.R. 1963. Root production and the estimation of net productivity. *Canad. J. Bot*. 41 : 65-72.

Bray. J.B., and Gorham, E. 1964. Litter production in forests of the world. *Adv. Ecol. Res.* 2 : 101-157.

Brummitt, L.W. 1962. Cypripediums. *Orch. Rev.* 70 : 181-183; 249-251; 384-387.

Bunning, E. 1947. Die endogene Ruheriode der Samen. *Planta* (Berlin), 35 : 352-359.

Burkholder, P.R. 1952. Cooperation and conflict among primitive organisms. *Amer. Sci.* 40 : 601-631.

Burkill, I.H. 1927. The botany of the Abor expedition. *Rec. Bot. Surv. India*. 10.

Butler, W.L. 1961. In "*Progress in photobiology*". Eds. B.C. Christenson and B. Buchmann. Proc. Third Internat. Congr. Photobiology, Amsterdam. pp. 569-571.

Butler, W.L., Morris, K.H., Siegelman, H.W., and Hendricks, S.B. 1959. Detection assey and priliminary purification of the pigment controlling photoresponsive development of plants. *Proc. Natal. Acad. Sci. USA*. 45 : 1703-1708.

Butler, W.I., Lane, H.C., and Siegelman, H.W. 1963. Non-photochemical transformation of phytochrome in vivo. *Plant Physiol*. 38 : 514-519.

Butler, W.L., Siegelman, H.W., and Miller, C.O. 1964 a. Denaturation of phytochrome. *Biochem*. 3 : 851-857.

Butler, W.L., Hendricks, S.B., and Siegelman, H.W. 1964 b. Action spectra of phytochrome in vivo. *Phytochem. Photobiol*. 3 : 521-528.

Cain, S.A. 1944. *Fundamentals of plant geography*. Harper and Bros. New York.

Cain, S.A. 1945. A biological spectrum of the flora of great smoky mountains national park. *Butler Univ. Bot. Stud*. 7 : 1-14.

Cain, S.A. 1950. Life forms and phytoclimate. *Bot. Rev*. 16 : 1-32.

Capon, B., and Van Asdall, W. 1967. Heat pretreatment as a means of increasing germination of desert annual seeds. *Ecology*, 48 : 305-306.

Carles, J. 1948. Le spectra biologique reel. *Extr. Bull. Soc. Bot. France,* 95 : 340-342.

Carlozzi, C.A. 1965. *Conservation of natural resources*. Cornell University, Ithaca, New York.

Carlquist, S. 1967. *Bull. Torrey, Bot. Club*. 94 : 129-162.

Cantifanto, Y.M., and Silver, W.S. 1964. Leaf-nodules symbiosis. I. Edophyte of *Psychotria bacteriophila*. *J. Bacteriol*. 88 : 776-781.

Champion, H.G. 1936. A preliminary survey of forest types of India and Burma, *Indian. For. Rec. N.S.* 1 : 1-286.

Champion, H.G. 1939. The relative stability of Indian vegetational types. *Jour. Indian Bot. Soc*. 18 : 1-12.

Champion, H.G., and Seth, S.K. 1965. Forest types of India. Mimeo. Forest Research Institute, Dehra Dun.

Champion, H.G., and Seth, S.K. 1967. *A revised survey of the forest types of India*. Manager of Publications, New Delhi.

Candler, C.C. 1939. An outline of the vegetation of India. In "*An outline of the field science of India.*" Ed. S.L. Hora. Calcutta.

Chaphekar, S.B. 1967. Ecological studies of Bombay weeds. Ph. D. Thesis, Bombay University, Bombay.

Chatterjee, D. 1939. Studies on the endemic flora of India and Burma, *Jour. As. Soc. Bengal.* 5 : 19-67.

Chaudhary, V.B. 1967. Seasonal changes, annual net production and energetics of *Dicanthium annulatum* stands. Ph. D. Thesis, Banaras Hindu University, Varanasi.

Chorley, R.J., and Kennedy, B.A. 1971. *Physical geography : a systems approach*. Prentice Hall, Englewood, Cliffs.

Choudhuri, G.N. 1966. Seeds germination and flowering in *Vallisnaria spiralis, Northwest Sci.* 40 : 31-35.

Chourey, S. 1953. A study of the reproductive capacity of six dominant forest trees. M. Sc. Thesis, Sagar University, Sagar.

Clapham, Jr. W.B. 1973. *Natural ecosystems*. The Macmillan Co., New York.

Clarke, C.B. 1898. On the subareas of British India. Illustrated by the detailed distribution of the Cyperaceae in that Empire. *J. Linn. Soc. Bot.* (London). 34 : 1-146.

Clark, G.L. 1973. *Elements of ecology*. John Wiley & Sons, New York.

Clausen, J., Keck, D.D., and Hiesey, W.M. 1940. Experimental studies on the nature of species. I. Effect of varied environments on western north American plants. *Carnegie Inst. Wash. Publ.* 520.

Clausen, J., Keck, D.D., and Hiesey, W.M. 1948. Experimental studies on the nature of species. III. Environmental responses of climatic races of *Achillea. Carnegie Inst. Wash. Publ.* 581.

Clements, F.E. 1905. *Research methods in Ecology*. Nebraska University Publ. Co, Lincoln.

Clements, F.E. 1916. *Plant succession : an analyusis of the development of vegetation*. Carnegie Institution, Washington.

Clements, F.E. 1920. *Plant indicators*. Carnegie Institution, Washington.

Clements, F.E. 1936. Nature and structure of the climax. *J. Ecol.* 24 : 252-284.

Clements, F.E., and Shelford, V.E. 1939. *Bio-Ecology.* John Willey & Sons, New York.

Cloudsley-Thompson, J.L. 1975. *Terrestrial environment.* Crom Helm, London.

Cockrum, E.L., and McCauley, M.J. 1965. *Zoology.* W.B. Saunders Co., Philadelphia.

Coffey, G.N. 1912. A study of the soils on the United States. *U.S.D.A. Bull. Soil Sci.* 85.

Cooper, W.S. 1926. The fundamentals of vegetational changes. *Ecology.* 7 : 391-413.

Cowles, H.C. 1899. The ecological relations of the vegetation on the sand dunes of lake Michigan. *Bot. Gas.* 27 : 95-117.

Cowles, H.C. 1911. The cause of vegetation cycles. *Bot. Gaz.* 51 : 161-183.

Cox, C.B., Healey, I.N., and Moore, P.D. 1976. *Biogeography an ecological and evolutionary approach.* Blackwell Sci. Publ. Oxford.

Crocker, W. 1936. Effect of the visible spectrum upon the germination of seeds and fruits. In "*Biological effects of radiation*". Ed. B.M. Duggar. McGraw Hill, New York. pp. 791-828.

Crocker, W.H. 1907. Germination of seeds of water plants. *Bot. Gaz.* 44 : 375-380.

Crocker, W. 1948. Life-span of seeds. *Bot. Rev.* 4 : 235-274.

Crocker, W., and Barton, L.V. 1931. After ripening, germination and storage of certain Rosaceous seeds. *Contr. Boyce Thompson Inst.* 3 : 385-404.

Crocker, W., and Barton, L.V. 1953. *Physiology of seeds.* Chronica Botanica Co, Waltham, Mass.

Croizat, L. 1952. *Mannual of phytogeography.* W. Junk, Hague.

Curtis, J.H. 1959. *The vegetation of Wisconsin.* University of Wisconsin Press, Madison.

Curtis, J.T. 1955. A praire continuum in wisconsin. *Ecology*. 36 : 558-566.

Curtis, J.T., and McIntosh, R.P. 1950. The relationship of certain analytic and synthetic phytosociological characters. *Ecology*, 31 : 434-455.

Dansereau, P. 1945. Essai de correlation sociologique entre les plantes superieurs et les poissons de la beine du lac Saint-Louis. *Rev. Canad. Biol.* 4 : 369-417.

Das, R.B., and Sarup, S. 1951. The biological spectrum of the Indian desert flora. *Univ. Raj Stud. Biol.* Sci. 1 : 36-42.

Datta, S.C. 1961. Studies of oleander seed germination. *Bull. College Sci.*. 6 : 71-79.

Davis, D.E., and Golley, F.B. 1963. *Principles in mammology*. Reinholt, New York.

Davis, P.A. 1928. The effect of high pressure on the percentage of soft and hard seeds of Medicago sativa and Melilotus alba. Amer. J. Bot. 15 : 433-436.

Davis, P.H., and Haywood, V.H. 1963. Principles of *Angiosperm Taxonomy*. D. Van Nostrand Co., Inc., New York.

Davis, W.E. 1939. An explanation of the advantage of alternating temperatures over constant temperatures in the germination of certain seeds. *Amer. J. Bot.* 26 : 175-185.

D'Amato, F., and Hoffman-Ostenhof, O. 1956. Metabolism and spontaneous mutations in plants. *Advanc. Genet.* 8 : 1-28.

Deininger, R.A. 1980. Systems analysis for environmental pollution control. *Indian J. Environ. Hlth.* 22 : 4 : 263-277.

DeLint, P.J.A.L., and Spuit, G.J.P. 1963. Phytochrome destruction following illumi-nation of mesocotyls of *Zea mays L. Medgll. Landbouwhogeschool*. 63 : 17.

Delwiche, C.C. 1965. The cycling of carbon and nitrogen in the biosphere. In "*Microbiology and soil fertility Proc. 1964*". Biol. Coll. Oregon University.

Delwiche, C.C. 1970. The nitrogen cycle. *Sci. Amer.* 223 : 137-146.

Dey, A.K. 1945. The age of the Bengal gap. Symposium - Satpura hypothesis of the distribution of Malayan fauna and flora to Peninsular India. *Proc. Natn. Inst. Sci. India.* 15 : 409-410.

Dixit, A.P. 1966. Autecological studies of three weeds. Ph. D. Thesis, Banaras Hindu University, Varanasi.

Dokuchayev, V.V. 1900. *Pektise o Pohavedenie*, Moscow. 7 : 257-296.

Downs, R.J., and Piringer, A.A. 1958. Seed germination in the Bromeliaceae. *Bromel. Soc. Bull.* 8 : 36-38.

Downs, R.J., *et al.* 1961. *Light and plants.* USDA Agric Res Ser.

Drude, O. 1897. *Manual de geographie botanique.* Paris.

Drude, O. 1913. Die oekologie der pflanzen. *Die Wissenschaft.* 50 : 308.

Dubey, P.S. 1990. The ecological modelling (memio).

DuReitz, G.E. 1931. Life-forms of terrestrial flowering plants. *Acta Phytogeogr. Suecica,* 3 : 1-95.

Edwards, C.A. 1974. Dynamics of pesticide residues in the environment. In *"Environmental pollution by pesticides"*. Eds. C.A. Edwards. Plenum Press, New York. pp. 457-493.

Edwards, T.I. 1932. Temperature relations of seed germination. *Quart. Rev. Biol.* 7 : 428-443.

Ehrenfield, D.W. 1970. *Biological conservation.* Holt Rinehart and Winston, New York.

Elton, C. 1927. *Animal ecology.* Sidwick & Jackson, London.

Elton, C. 1947. *Animal ecology.* McMillon, New York.

Engelmann, M.D. 1966. Energetics, terrestrial field studies and animal productivity. *Adv. Ecol. Res.* 3 : 73-115.

Ethrington, J.R. 1975. *Environment and plant ecology.* John Wiley and Sons, London.

Evenari, M. 1949. Germination inhibitors. *Bot. Rev.* 15 : 153-194.

Evenari, M. 1956. Seed germination. In "*Radiation biology*". Ed. A. Hollaender. McGraw Hill, New York. 3 : 519-549.

Evenari, M. 1957. The physiological action and biological importance of germination inhibitors. *Symp. Soc. Exptl. Biol.* 11 : 21-43.

Evenari, M. 1965. Light and seed dormancy. In "*Encyclopedia of plant physiology*". Ed. W. Ruhland. Springer-Verlag, Berilin. 15/2 : 804-847.

Ewart, A.J. 1908. On the longivity of seeds. *Proc. Roy. Soc. Victoria N.S.* 21 : 1-203.

Fassett, N.C. 1930. The plants of some northeastern Wisconsin lakes. *Trans. Wisconsin Acad. Sci.* 25 : 157-168.

Ferreira, D.B. 1940. The vegetable life-forms of central and southern Deccan in peninsular India. M.Sc. Thesis, Bombay University, Bombay.

Forbes, E. 1944. Report on the mollusca and radiata of the Aegean Sea. *Rep. Brit. Assoc. Adv. Sci.* 130-193.

Forbes, S.A. 1887. The lake as a microcosm. *Bull. Peorio Sci. Ass.* 1887 : 77-87.

Fosberg, F.R. 1967. a classification of vegetation for general purpose. In "*Guide to the check list for IBP atlas*". Ed. G.F. Peterken. Blackwells, Oxford.

Flint, L.H., and McAlister, E.D. 1937. Wavelength of radiation in the visible spectrum promoting the germination of light sensitive seeds. *Smithsonian Inst. Publs. Misc. Coll.* 96 : 1-8.

Forrester, J.W. 1968. *Principles of systems*. Wright-Allen Press, Cambridge, Mass.

Furon, R. 1959. *La paleogeographie : Essai sur 1' evolution des continents et des Oceans*. Paris.

Galston, A.W. 1968. Microspectrophotometric evidence for phytochrome in plant nuclei. *Proc. Natl. Acad. Sci, USA*. 61 : 2 : 454-460.

Gandhi, S.M., and Bhatnagar, M.P. 1961. Effect of certain hormones on germination, flowering, fruiting, branching and yield of cumin (*Cuminum cyminum*). *J. Indian bot. Soc.* 42 : 628-634.

Garg, H.P. 1977. *Urja, energy affairs monthly*. 24-29.

Garg, R.K. 1973. Floristic and ecological studies on deciduous forests of Kewaranal (Udaipur) South Rajasthan. Ph. D. Thesis, Udaipur University, Udaipur.

Garg, R.K., Ranawat, M.P.S., and Vyas, L.N. 1972. Plant biomass and net production of *Butea monosperma* (Lamk) Teub. *Forstw. Chl.* 91 : 357-364.

Garg, R.K., and Vyas, L.N. 1975. Litter production in deciduous forests near Udaipur (South Rajasthan) India. In "*Trends in terrestrial and aquatic research*". Eds. F.B. Golley and E. Medima. Springer-Verlag, New York. pp. 131-136.

Gates, D.M. 1968. Energy exchange and ecology. *Bioscience*, 18 : 2 : 90-95.

Gause, G.F. 1934. *The struggle for existence*. Waverly, Baltimore.

Geiger, R. 1961. *The climate near the ground*. 4th ed. Harward University Press, Mass.

Gelhar, L.W. 1972. The aqueus underground. *Tech. Rev.* 75 : 5 : 3-11.

Gimingham, C.H. 1975. *An introduction to heathland ecology*. Oliver and Boyd, Edinburg.

Gleason, H.A. 1926. The individualistic concept of the plant association. *Amer. Midl. Nat.* 21 : 92-110.

Golley, F.B. 1960. Energy dynamics of a food chain of an old-field community. *Ecol. Monogr.* 30 : 187-206.

Golley, F.B. 1961. Energy values in in ecological materials. *Ecologyl*, 42 : 581-584.

Golley, F.B. 1965. Structure and function of an old-field Broomsedge community. *Ecol. Monogr.* 25 : 113-131.

Golley, F.B., and Leith, H. 1972. Basis of organic production in the tropics. *Trop. Ecol. Emphas. Org. Prod.* eds. P.M. Golley and F.B. Golley. Georgia University, Athens. pp. 1-26.

Good, R. 1953. *The geography of flowering plants.* Longmans Green & Co., London.

Gopal, B. 1977. Principles of environmental science. In "*Current trends on Indian environment*". Eds. D. Bandhu and E. Chauhan. Today and Tomorrow, New Delhi. pp. 19-24.

Gopal, B., and Bandhu, D. 1972. Distribution of foliar chlorophyll at different strata in a dry deciduous forest stand at Varanasi, India. *Photosynthetica*, 6 : 219-224.

Gregor, J.W., Davey, V. McM., and Lang, J.M.S. 1936. Experimental taxonomy. I. Experimental garden technique in relation to the recognition of small taxonomic units. *New Phytol.* 35 : 323-350.

Grisebach. A. 1884. *Die vegetation der Erde*. Bd. I and II. Leipzing.

Grisebach, A. 1938. *Linnea*, 12 : 159-200.

Grinnel, J. 1917. Field test of theories concerning distributional control. *Amer. Nat.* 51 : 115-128.

Gundeson, A., and Hasting, G.T. 1944. Interdependence in plant and animal evolution. *Sci. Monthly*, 59 : 63-72.

Gupta, K.C. 1966. Some aspects of the autecology of *Dactyloctenium aegypticum* (Linn) P. Beauv. M.Sc. Thesis, Vikram University, Ujjain.

Gupta, R.K. 1962. Some observations on the plants of the south Indian hill tops (Nilgiri) and Palani plateau and their distribution in the Himalayas. *J. Indian Bot. Soc.* 41 : 1-15.

Gupta, P.K., and Srivastava, A.K. 1969. Ecotypic differentiation in *Eleusine indica* (L) Gaertn. *Curr. Sci.* 38 : 15 : 373-374.

Gupta, R.K., and Prakash, I. 1975. *Environmental analysis of Thar desert*. English Book Depot, Dehra Dun.

Gupta, R.K., Saxena, S.K., and Sharma, S.K. 1971. Effect of weed competition on dry matter production in *Cenchrus ciliaris L. J. Indian Bot. Soc.* 50 : 382-388.

Gupta, V.C., and Kachroo, P. 1983. Life-form classification and biological spectrum of the flora of Yusmarg, Kashmir. *Trop. Ecol.* 24 : 22-28.

Gutterman, Y., Witzium, A., and Evenari, M. 1969. Physiological and morphological differences between populations of Ble *Blepharis persica* (Burm) Kuntz. *Israel Jour.* Bot.. 18 : 89-95.

Haberlandt, G. 1884. *Physiologische pfianzenantomie.* Engelmann, Leipzig.

Hedley, E.B., and Bliss, L.C. 1964. *Ecol. Monogr.* 34 : 331.

Haeckel, E. 1869. *Generelle morphologie der organismen.* Georg. Reimer, Berlin (2 vols).

Hale, M.E. 1967. *The biology of lichens.* Edward Arnold, London.

Hanson, H.C., and Churchill, E.D. 1965. *The plant community.* Reinhold Publ. Co., New York.

Harberd, D.J. 1961. The case for extensive rather than intensive sampling in genecology. *New Phytol.* 60 : 325-338.

Harper, J.L. 1969. The role of predation in vegetational diversity. In "*Diversity and stability in ecological systems*". Brookhaven Symposium in Biology. No. 22 : 48-62.

Harrington, G.T. 1916. Further studies on the germination of Johnson grass seeds. *Proc. Ass. Off. Seed. Anal. N. Amer.* 1916-1917 : 71-76.

Haskell, E.F. 1949. A classification of social science. *Main Currents in Modern Thought.* 7 : 45-51.

Hendricks, S.B. 1964. Photochemical aspects of plant photoperiodicity. In "*Photophysiology*". Ed. A.C. Giese. Academic Press, New York. *pp.* 305-331.

Hendricks, S.B., and Borthwick, H.A. 1959. Photocontrol of plant development by the simultaneous excitations of two interconversible pigments. *Proc. Natn. Acad. Sci. USA.* 45 : 344-349.

Hendricks, S.B., Butler, W.L., and Diegelman, H.W. 1962. A reversible photoreaction regulating plannt growth. *J. Physical Chem.* 66 : 2550-2555.

Hendricks, S.B., Toole, V.K., and Borthwick, H.A. 1968. Opposing actions of light in seed germination of *Poa pratensis* and *Amaranthus arenicola*. *Plant Physiol.* 43 : 2023-2028.

Hey, G.L. 1962. The mixed orchid house. Orch. Rev. 70 : 315-316.

Hiremath, S.C. 1965. Ecological study of genus Cleome at Ujjain. M.Sc. Thesis, Vikram University, Ujjain.

Hooker, J.D. 1855. Introductory essay. In *Flora Indica*. Ed. J.D. Hooker. Th. Thomas, London.

Hora, S.K. 1949. Torrential fishes and the significance of their distribution in zoogeographical studies. *Bull. Nat. Geo. Soc.* India. 7 : 1-10.

Hult, R. 1885. *Medd. Soc. Fenn.* 122 : 161.

Hult, R. 1887. *Meddel. Soc. Pro. Fenna Flora Fennia Fennica*. 14 : 153-228.

Humboldt. A. 1805. *Essai sur la geographie des plantes.* Paris.

Humm, H.S. 1944. Bacterial leaf nodules. *J.N. Y. Bot Garden*. 45 : 193-199.

Humphreys, J.L. 1960. Help wanted. *Orch. Rev*. 68 : 141-142.

Hutchinson, G.E. 1944. Nitrogen and biochemistry of the atmosphere. *Amer. Scient*. 30 : 178-195.

Hutchinson, G.E. 1951. *A treatise of limnology*. Vol. III. *Limnological Botany*. John Wiley, New York.

Hutchinson, G.E. 1958. Concluding remarks. *Cold Spring Harbor Symp. Quart Biol*. 22 : 415-427.

Isikawa, S. 1954. Light sensitivity against germination. I. Photoperiodism of seed. *Botan. Mag*. (Tokyo). 67 : 51-56.

Jain. S.K. 1971. Production studies in some grasslands of Sagar. Ph. D. Thesis. Sagar University. Sagar.

Jindal, K.B. 1956. Phytosociological and ecological studies of rederal vegetation of Bombay. M.Sc. Thesis, Bombay University, Bombay.

Joshi, H.K., and Acry, N.C. 1989. Studies on the biomass dynamics, productivity and energy contents of *Crotalaria medicaginea* dominated communities in the environs of Udaipur. *Acta Ecol.* 11 : 124-132.

Joshi, M.C., and Kambhoj, O.P. 1959. Studies on the autecology of *Gisekia pharmaceoides* Linn. *Jour. Indian Bot Soc*. 38 : 8-34.

Joshi, M.C., and Nigam, B.C. 1970. Autecological studies on Rajasthan desert plants. III. Seed output and seed germination in *Trianthema portulaastrum* L. *Trop. Ecol.* 11 : 140-147.

Joshi, M.C., and Verghese, T.M. 1967. Autecological studies on Rajsthan desert plants. I. *Anticharis linearis* Hochst. *Jour. Indian Bot. Soc.* 46 : 63-75.

Joshi, M.C., Kahate, S., and Bishnoi, S. 1967. Autecological studies on Rajasthan desert plants. *Jour. Indian Bot. Soc.* 46 : 169-184.

Kasperbauer. M.J., Borthwick, H.A., and Hendricks, S.B. 1963. Inhibition of flowering of *Chenopodium rubrum* by prolonged far-red radiation. *Bot. Gaz.* 124 : 444-451.

Kaul, N., and Sarin, Y.K. 1976. Life-form classification and the biological spectrum of the flora of Bhaderwah. *Trop. Ecol.* 17 : 132-138.

Kaul, R.N. and Manohar, M.S. 1966. Germination studies on arid zone tree seeds. I *Acacia senegal* Willd. *Indian For.* 92 : 8 : 499-503.

Kaul, V. 1959. Physiologico-ecological studies of *Xanthium strumarium* and *Croton sparciflorus* Morong. Ph. D. Thesis, Banaras Hindu University, Varanasi.

Kaul, V.N., and Misra, R. 1957. Seed germination and photoperiodic behaviou of *Xanthium strumerium*. *Indian Sci. Congr.* (Abstracts).

Kaul, A. 1968. Ecology of two species of *Alternathera*. Ph. D. Thesis, Vikram University. Ujjain.

Kendeigh, S.C. 1961. *Animal ecology*. Prentice Hall, Englewood Cliffs.

Kendeigh, S.C. 1974, *Ecology with special reference to animals and man*. Prentice Hall of India, New Delhi.

Kerner, V.M.A. 1875. Vorlaufige Mittercilung uber die Bedeutung der Asyngramie fur die Entstehung der Arten. *Ber. Naturw. Med. Ver. Innsbruck*. 5.

Ketkar, M.V. 1965. Some aspects of autecology of *Alysicarpus* Neck. M.Sc. Thesis, Vikram University, Ujjain.

Klipple, G.E., and Costello, D.F. 1960. Vegetationa and cattle responses to different intensities of grazing on short grass range on the Central great plaing *U.S.D.A. Tech. Bull.* 1216 : 82 p.

Koller, D., and Roth, N. 1964. Studies on the physiological and ecological significance of amphicarpy of *Gymnorrhena micrantha* (Compositae). *Amer. J. Bot.* 51 : 26-35.

Killer, D., Sachs, M., and Negbi, M. 1964. Spectral sensitivity of seed germination in *Artemisia monosperma. Plant and Cell Physiol.* 5 : 79-84.

Komareck, E.V. 1966. The meteorological basis for fire ecology. In "*Tall timber fire ecology conference*". 5 : 85-126.

Komareck, E.V. 1967. The nature of lightening fires. *Tall Timber fire ecology conference.* 7 : 5-41.

Kormondy, E.J. 1965. *Readings in ecology*. Prentice Hall, New Jersey.

Kormondy, E.J. 1969. *Concepts of ecology*. Englewood Cliffs, New Jersey.

Koul, M. 1965. Ecology of three medicinal plants. Ph. D. Thesis, Banaras Hindu University, Varanasi.

Kourtz, P. 1967. *Lightening behaviour and lightening fires in Canadian forests.* Canada Deptt of Forestry and Rural Development Publ. No. 1179, Ottawa.

Kraus, E.H.L. 1891. Die eintheilung der pflanzen nach ihrer Daur. *Ber. d. dt. bot. Ges.* 9 : 233-237.

Krebs, C.J. 1972. *Ecology*, Harper and Row, New York.

Kruckeberg, A.R. 1951. Interspecific variability in the response of certain native plant species to serpentine soil. *Amer. J. Bot.* 38 : 208-419.

Kuera, C.L., Dahlman, R.C., and Koelling, M.R. 1967. Total net productivity and turnover on an energy basis for tallgrass prairie. *Ecology*, 48 : 536-541.

Kumar, A., and Joshi, M.C. 1972. The effect of grazing on the structure and productivity of vegetation near Pilani, Rajasthan. *Indian J. Ecol.* 60 : 665-675.

Kumar, H.D. 1981. *Modern concepts of ecology*. Vikas, New Delhi.

Lack, D. 1944. Ecological aspects of species formation in passerine birds. *Bot. Not.* 86 : 260-286.

Lahiri, A.N., and Kharbanda, B.C. 1963. Germination studies on arid zone plant II. Germination inhibitors in the spikelet glumes of *Lasiurus sindicus, Cenchrus ciliaris, Cenchrus setigerus*. *Ann. Arid Zone*, 1 : 114-125.

Lakshmanan, N.K. 1962. The application of Raunkiarer's life-forms. *J. Indian Bot. Soc.* 41 : 585-589.

Lamb, B. 1984. Gaseous pollutants characteristics. In "*Handbook of air pollution technology*". Ed. S. Calvert and H.M. Englund. John Wiley and Sons, New York. p. 65-96.

Lane, H.C., Siegelman, H.W., Butler, W.L., and Firer, E.M. 1963. Detection oif phytochrome in green plants. *Plant Physiol.* 38 : 414-416.

Lang, O.L. 1965. Leaf temperature and methods of measurement. *Aird Zone Res.* 25 : 203-209.

Langlet, O. 1971. Two hundred years of genecology. Taxon. 20 : 653-722.

Larsten, N.R., and Horner Jr. H.T. 1967. Development and structure of bacterial leaf nodules in *Psychotria bacteriophila* Val (Rubiaceae). *J. Bacteriol.* 94 : 6 : 2027-2036.

Legris, P. 1963. La vegetattion de l'Inde : Ecologie et Flora. *Inst. Fr. Pondicherry Trav. Sec. Sci. Tech.* 6 : 1-589.

Lewis, T., and Taylor, L.R. 1967. *Introduction to experimental ecology*. Academic Press, London.

Leith, H. 1965. Indirect methods of measurement of dry matter production. In "*Methodology of plant ecophysiology*". Ed. F.E. Eckardt. UNESCO, Paris, pp. 513-518.

Leith, H. 1970. Phenology in productivity studies. In "*Analysis of temperate forest ecosystems*". Chapman and Hall, London. pp. 29-46.

Leith, H. 1972. Modelling oif the primary productivity of the world. *Trop. Ecol.* 13 : 1 : 125-130.

Leith, H., Oswald., D., and Martens, H. 1965. Staffproduktion spor B/Wurzel-Verhaltnis, chlorophyll-gebaht und Blattflache von Jangappein-Mitt. Vereinnsforstt. *Standostosjunde Fors Pflanzenzucht.* 15 : 70-74.

Liebig, J. 1840. *Chemistry and its application to agriculture and physiology.* Taylor and Walton, London.

Lindman, R.L. 1942. The trophic dynamic aspects of ecology. *Ecology*, 23 : 399-418.

Linnaeus, C. 1939. Rom om Waxters plantering, grundat, Po. naturen. *Swenska Weltensk. Acad. Handl.* 1.

Lloyd, P.S. 1968. *J. Ecol.* 56 : 811-826.

Lute, A.M. 1928. Impermeable seeds of alfalfa. *Colorado Exp. Station Agric. Sec. Bull.* No. 326.

MacFadyen, A. 1957. *Animal ecology : Aims and methods.* Pitman, London.

MacFadyen, A. 1963. *Animal ecology.* Pitman, London.

Madge, D. 1966. How leaf litter disappears. *New Scientist,* 113-115.

Madina, E., and Leith, H. 1964. Die Beziehungen Zurischen chlorophyll-gehalt, assimilierender Flache und Trocken substanzprduktion in eingen Pflanzengemeinchaften. Beitr. *Biol. pflenzen.* 40 : 451-494.

Major, J. 1951. A functional, factoral approach to plant plant ecology. *Ecology*, 32 : 392-412.

Malcolm, W.M. 1966. Biological interactions. *Bot. Rev.* 32 : 243-255.

Mall, L.P. 1955. Ecology of drying pools and some common weeds. Ph. D. Thesis, Sagar University, Sagar.

Mall, L.P. 1956. Some aspects of the autecology of *Chrozophora rottleri* A. Juss. *Bull. Bot. Soc. Univ. Sagar.* 8 : 13-24.

Mall, L.P. 1957. Contribution to the autecology of *Cassia tora* Linn and *C. obtusifolia* Linn. *J. Univ. Sagar.* 8 : 13-24.

Mall, L.P. 1958. Germination studies of some seeds. In *Modern Developments in plant physiology*". Ed. P. Maheshwari. Proc. Delhi Univ. Seminar. pp. 140-141.

Mall, L.P. 1961. Ecology of temporary pools during dry phase. *J. Indian Bot. Soc.* 40 : 139-163.

Mall, L.P., and Arzere, K.C. 1956. Autecological study of *Achyranthus aspera Linn. Bull Bot. Soc. Univ. Sagar.* 8 : 69-76.

Mall, L.P., Billore, S.K., and Misra, C.M. 1973. Relations of ecological efficiency with photosynnthetic structure in some terrestrial herbaceous stands of Ujjain, India. *Trop. Ecol. Symp*. Venezuella. 29.

Mall, L.P., and Manilal, K.S. 1962. Some ecological and physiological studies on *Crypsis aculeata* (L) Cut. *Bull. Bot. Soc. Univ. Sagar.* 14 : 22-32.

Mall, L. P., and Raina, D.K. 1957. Autecological studies on *Achyranthus aspera* Linn. *Bull. Bot. Soc. Univ. Sagar.* 9 : 21-33.

Mall, S.L. 1972. Ecological study of *Sehima nervosum* (Stapt). M.Sc. Thesis, Vikram University, Ujjain.

Manohar, M.S. 1966 a. Measurement of the water potential of intact plant tissues. III. The water potential of germinating peas (*Pisum satium* L). *Jour. Exptl. Bot.* 17 : 231-235.

Manohar, M.B. 1966 b. Effect of osomtic systems on germination of peas (*Pisum sativum* L). *Planta* (Berlin). 71 : 81-86.

Manohar, M.S., and Heydecker,W. 1964 a Effect of water potential on germination of *Poa* seeds. *Nature*, 202 : 22-24.

Manohar, M.S., and Heydeker, W. 1964 b. Water requirements for seed germination. *Univ. Nottingham Sch. Agric Report.* 55-62.

Margalef, R. 1957. The food web in the pelagic environment, *Helgolander. Wiss. Meesunters.* 15 : 548-559.

Margalef, R. 1968. *Perspective in ecological theory.* Univ. of Chicago Press, Chicago.

Margalef, R. 1978. Stress in ecosystems : a possible approach. 2nd International Congress Ecology, Jerusalem. Abstract. Vol. 1 p. 227.

Martell and Calvin 1952. *Op. Cit.* Narayana and Shah 1966.

Maurya, APN., and Ambasht, R.S. 1973. Significance of seed dia-morphism in *Aysicarpus monilifer* DC. *Jour. Ecol.* 61 : 213-217.

Maxwell, L.R., and Kempton, J.R. 1939. Delayed killing of maize seeds X-rayed at liquid air temperature. *Jour. Washington Acad. Sci.* 29 : 368-374.

Mayer, A.M., and Poljakoff-Mayber, A. 1963. *The germination of seeds.* Pergamon Press, London.

McCormick, J.F. 1978. Pattern, processes and strategies of recovery following perturbations to terrestrial ecosystems. 2nd International Congress Ecology, Jerusalem. Abstract Vol. 1 p. 232.

McGinnies, W.J. 1960. Effect of moisture stress and temperature on germination of six range grasses. *Agron. J.* 52 : 159-162.

McIntosh, D.H. 1963. *Meteorological glossary* H.M.S.O.

McIntosh, R.P. 1958. Plant communities. *Science.* 128 : 115-120.

McLarsenn, A.D., and Peterson, G.H. 1967. *Soil biochemistry.* Vol 1 : Dekker, New York.

McNair, J.B. 1945. Plant fats in relation to environment and evolution. *Bot. Rev.* 11 : 1-59.

Medlicott, H.B., and Blanford, W.T. 1978. *A manual of geology of India.* 2 Vols, pp. 374-375. Calcutta.

Meher-Homji, V.M. 1960. Des bioclimate due sub-continent Indian et leures types analogues dans le monde. *Documents pour les productions vegetales* (Toulouse). Series General 4, 386 p. Inst. Fr. Pondicherry Tran. Sect. Sci. Tech. 7 : 1-386.

Meher-Homji, V.M. 1964. Life-forms and biological spectra as epharmonic criteria of ardity and humidity in the tropics. *J. Indian Bot. Soc.* 43 : 424-430.

Meher-Homji, V.M. 1972. Himalayan plants on the south Indian hills : Role of Pleistocene glaciation VS long distance dispersal. *Sci & Cult.* 38 : 8-12.

Meher-Homji, V.M. 1974. *Sci. & Cult.* 40 : 217-227.

Meher-Homji, V.M., and Misra, K.C. 1973. Phytogeography of the Indian subcontinent. In "*Progress of plant ecology in India*". Eds. R. Misra and B. Gopal. Today and Tomorrow, New Delhi. pp 9-11.

Meijar, G. 1959. The spectral dependence of flowering and elongation. *Acta Bot. Neerl.* 8 : 189-246.

Menhinick, E.F. 1967. Structure, stability and energy flow in plants and arthropods in a *Sericea lespedeza* stand. *Ecool. Monogr.* 37 : 255-272.

Meyer, B.S., Andersen, D.B., and Bohning, R.H. 1963. *Introduction to plant physiology.* Von Nostrand & East-West Press.

Midgley, A.I. 1926. Effect of alternate freezing and thawing on the impermeability of alfalfa and dodder seeds. *J. Amer. Soc. Agron.* 18 : 1087-1098.

Miller, J.H., and Miller, P.M. 1964. Blue light in the development of fern gametophytes and its interaction with far-red light. *Amer. J. Bot.* 51 : 329-334.

Millot, J. 1952. La faune malgache et la mythe gondwanien. *Mem. Inst. Sci. Madagascar*, Series A, 7 : 1-36.

Millot, J. 1953. Le continent de gondwana et les methodes de relsonnement de la biogeographie classique. *Ann. Sci. Nat.* Zool. lle Series 15 : 185-219.

Milne, L.J., and Maline, M. 1959. *Plant life.* Prentice Hall, Inc. New York.

Misra, C.N. 1973. Primary productivity of grassland ecosystems at Ujjain, India. Ph.D. Thesis, Vikram University, Ujjain.

Misra, K.C. 1974. *Manual of ecology.* Oxford & IBH, Calcutta.

Misra, K.C. 1980. *Manual of plant ecology.* Second edition. Oxford & IBH, Calcutta.

Misra, R. 1944. Variation in leaf form oif *Potamogetom perfoliatus L. J. Indian Bot. Soc.* 23 : 44-52.

Misra, R. 1957. Plant ecology : Progress of science in India. *Natl. Inst. Sci. India.* 6 : 7141-148.

Misra, R. 1959. The status of the plant communities in the upper Gangetic plain. J. *Indian Bot. Soc.* 38 : 1-7.

Misra, R. 1959 a. Environment, adaptations and plant distribution. 46th Indian Science Congress Presidential Address. Section of Botany. pp. 1-12.

Misra, R. 1962. The science of vegetation with reference to India. *Bull. Bot. Surv. India.* 4 : 115-118.

Misra, R. 1967. Form, function annd factors in ecology. *J. Indian Bot. Soc.* 46 : 144-153.

Misra, R. 1967. Ecology and development. Inaugural address. Symposium on advancement of Ecology. D.A.V. College, Muzaffarnagar.

Misra, R. 1968. *Ecology work book.* Oxford & IBH, Calcutta.

Misra, R. 1970. Save the tropical ecosystems. INTECOL. Bulletin No. 2.

Misra, R. 1976. Ecology and development. (Memiographed).

Misra, R. 1968 a. Energy transfer along terrestrial food chains. *Trop. Ecol.* 9 : 2 : 105-118.

Misra, R., and Kothari, A. 1971. Effects of population density on the growth behaviour of *Crotalaria medicaginea. Proc. Indian Sci. Cong. Ass.* (Abstract).

Misra, R., and Puri, G.S. 1954. *Indian Manual of plant ecology.* English Book Depot, Denhra Dun.

Misra, R., and Ramkrishnan, P.S. 1960. Ecological distribution of *Peristrophe bicalyculata* Neres. *Proc. Natl. Inst. Sci. India.* B 26 : 51-23.

Misra, R., and Rao, B.S. 1948. A study in the autecology of *Lindenbergia polyantha* Royle. *J. Indian Bot. Soc.* 27 : 186-199.

Misra, R., Singh, J.S., and Singh, K.P. 1967. Preliminary observations on the production of dry matter by Sal (*Shorea robusta* Gaertn. f). *Trop. Ecol.* 8 : 94-104.

Misra, R., Singh, J.S., and Singh, K.P. 1968. A new hypothesis to account for the opposite trophic biomass structure on land and in water. *Curr. Sci.* 37 : 13 : 382-383.

Mobius, K. 1877. *Die Auster und die Austern Wirtschaft,* Berlin.

Mohan, N.P., and Puri, G.S. 1955. The Himalayan conifers III. The succession of forest communities in Oak-conifer forests of the Bashahar Himalayas. *Indian For.* 81 : 465-487; 549-562; 646-652; 706-711.

Mohan, N.P., and Puri, G.S. 1956. The Himalayan conifers V. The succession of forest communities in chir-pine (*Pinus roxburghi*) forests of the Punjab and Himachal Pradesh. *Indian For.* 82.

Mohr, H. 1957. Der Einfluss monochromatischer strahiung auf des lungenwachestum des hypocotyls and aud die Anthocy-anbildung bei Keimlingen von *Sinapsis alba* L. (=*Brassica* alba Boiss). *Planta.* 49 : 389-405.

Mohr, H. 1959. The photochemical control of anthocyanin synthesis and correlated photomorphogenesis in dark grown seedlings of *Sanapsis alba* L. *Proc. IX Internat. Bot. Congr. Montreal.* 1 : 267-268.

Mohr, H. 1960. Photomorphogenetische Reaktionssy steme in pflamzen Teil.

Mohr, H. 1961. 1 : Des Reversible Hallrot-Dunkelrot-Reaktionssystem und des Blau-dun-kelrot-Reaktionssystem. *Ergb. Biol.* 22 : 67-102.

Mohr, H. 1962. Primary effects of light on growth. *Ann. Rev. Plant Physiol.* 13 : 465-488.

Mohr, H., and Haung, A. 1962. Die histologischen vorgange wahrend der lichtabhagigen schliessung und offnung des plumulahakens bei der Keimbingen vox *Lactuca sativa L. Planta*, 59 : 151-164.

Mohr, H., and Nes, E.V. 1963. Der einfluss sichtharer strahlung auf die flavonoid synthesis und morphogenese der Buchweizen-Keimlinge (*Fagopyrum esculentum* Moench). 1. Synthesis von Anthocyan. Zeit. Bot. 51 : 1-16.

Moir, W.H. 1969. Stepper communities in the foot hills of the Colorado Front range and their relative productivities. *Amer. Midl. Natur.* 81 : 331-340.

Mukerji, S.K. 1932. On the genus *Artemisia* - its species, varieties and ecads as found in Kashmir. Proc. 19th Indian Sci. Cong. (Abstracts).

Mukerjee, S.K. 1936. Contributions to the ecology of *Mercurialis perennis L. J. Ecol.* 24 : 38-81.

Muller, K. 1912. Versuche nit hartschaligen Kleesamen. Ber. Gnobherzogl-Badische. Landw. Vers. Anst. Augustenburg. 81-88.

Mullick, P., and Chatterji, U.N. 1967. Ecophysiological studies on seed germination : Germination experiments with seeds of *Clitoria ternata* Linn. *Trop. Ecol.* 8 : 117-129.

Muntzing, A. 1954. The cytological basis of polymorphism in *Poa alpina. Heriditas*, 40 : 451-516.

Murakami, Y. 1959. The occurrence of Gibberellins in mature dry seeds. *Bot. Mag. Tokyo*. 72 : 438-442.

Murthy, Y.K. 1975. Water for tomorrow. *Commerce*, 131 : 3372 : 9-12.

Narayanna, N., and Shah, C.C. 1966. *Physical properties of soils.* P.C. Manaktala & Sons, Bombay.

N.R.C. 1979. *Ammonia*. National Research Council, University Park Press, Baltimore, USA.

Newbold, P.J. 1967. Methods for primary production for forests. Handbook No. 2. Blackwell Scientific Publication, Oxord.

Nobs, M.A. 1961. *Gradual speciation and genecology.* Recent Advances in Botany, Toronto.

Nutman, P.S. 1976. *Symbiotic nitrogen fixation in plants.* Cambridge University Press, London.

Odum, E.P. 1963. *Ecology*. Holt Richart and Winston Inc. New York.

Odum, E.P. 1968. Energy flow in ecosystems; A historical review, *Amer Zool.* 8 :11-18.

Odum, E.P. 1969. The strategy of ecosystem development. *Science*, 164 : 262-270.

Odum, E.P. 1971. *Fundamentals of ecology* W.B. Saunders, New York.

Odum, H.T. 1957. Trophic structure and productivity of silver springs, Florida, *Ecol. Monogr.* 27 : 55-112.

Odum, H.T. 1971. *Environment, power and society*. Wiley, New York.

Odum, H.T., and Odum, E.P. 1957. Principles and concepts pertaining to energy in ecosystems. In "*Fundamentals of ecology*". Ed. E.P. Odum. W.B. Saunders Co, Philadelphia. pp. 43-87.

Olson, E.C. 1972. The habitat : Climatic change and its influence on life and habitat. In "*Biology of nutrition*". Eds. T.W. Frennes and R.N. Pergamon. Oxford. pp. 267-305.

Oosting, H.J. 1956. *Study of plant communities*. San Francisco.

Obsorn, H.F. 1897. The limits of organic selection. *Amer. Natural.* 31.

Ovington, J.D., and Lawrence, D.B. 1967. Comparative chlorophyll and energy studies of prairie, savanna, oakwood and maize field ecosystems. *Ecology*, 48 : 515-524.

Ovington, J.D., and Madgwick, H.A.I. 1959. Distribution of organic matter and plant nutrients in a plantation of Scots pine. *Forest Sci.* 5 : 4 : 344-355.

Owen, E.B. 1956. *The storage of seeds for maintenance of viability*. Commonwealth Burew of Pasture and field Crops. Bull No. 43. Hurley, Berks, England.

Oxley, T.A. 1949. *The scientific principles of grain storage*. Northern publishing Co., Liverpool.

Pandey, S.B. 1968. Ecological observations on *Anagallis arvensis* L with special emphasis on seed germination. Ph. D. Thesis, Banaras Hindu University, Varanasi.

Pandey, S.B. 1968 a. Germination and seedling mortality in *Echinops echinatus* Roxb. *Proc. Symp. Recent Adv. Trop. Ecol.* 237-244.

Pandey, S.B. 1968 b. Adaptive significance of seed dormancy in *Anagallis arvensis* Linn. *Trop. Ecol.* 9 : 171-193.

Pandey, S.B. 1969. Photocontrol of seed germination in *Anagallis arvensis* Linn. *Trop. Ecol.* 10 : 96-138.

Pandeya, S.C. 1952. Grasslands of Sagar, Madhya Pradesh. *Indian For.* 78 : 638-654.

Pandeya, S.C. 1953. Studies in the morphology and ecology of three species of *Dicanthium* Will. *J. Indian Bot. Soc.* 32 : 86-100.

Pandeya, S.C. 1954. Grassland ecology. Ph. D. Thesis, Sagar University, Sagar.

Pandeya, S.C. 1959. The concept of maturation in Indian soils. *Proc. Natl. Acad, Sci. India.* 29 : 262-271.

Pandeya, S.C. 1962. Ecology as an aid to floristics. I. Adaptation and ecotypes. *Bull. Bot. Surv. India.* 4 : 141-146.

Pandeya, S.C. 1964 a. Ecology of grasslands of sagar, M.P.II. Composition of the fenced grassland associations. *J. Indian Bot. Soc.* 43 : 577-605.

Pandeya, S.C. 1964 b. Ecology of grasslands of M.P. II b. Composition of the associations open to grazing or occupying special habitat. *J. Indian Bot. Soc.* 43 : 606-639.

Pandeya, S.C. 1977. An approach to systems analysis. In "*Desert ecosystems and its improvement*". Ed. H.S. Mann, CAZRI, Jodhpur. pp 172-177.

Pandeya, S.C., and Goswami, R.M. 1962. Further contribution to the autecology of *Acanthospermum hispidum* De. *Op. Cit.* Pandeya *et al.* 1968.

Pandeya, S.C., Pandeya, S.M., Murthy, M.S., and Kuruvilla, K. 1967. Forest ecosystems - classification of forest vegetation with reference to forests in the river Narmada Catchment area. J. *Indian Bot. Soc.* 46 : 412-427.

Pandey, V.N. 1979. Comparative study of nutrient dynamics of Savanna and teak (99999999Tectona grandis L) plantation of Chandraprabha region, Varanasi. Ph. D. thesis, Banaras Hindu University, Varanasi.

Patil, R.P., and Singh, J.S. 1963. On the ecology and anatomy of *Phyla nodiflora* Greene at Allahabad, U.P. *Trop. Ecol.* 4 : 83-88.

Patten, B.C. 1959. An introduction to the cybernetics of the ecosystem : the trophic dynamic aspect. *Ecology*, 40 : 221-231.

Patten, B.C. 1971. A primer for ecological modelling and simulation with analog and digital computers. In "*Systems analysis and simulation in ecology*". Vol. 1. Ed. B.C. Patten. Academic Press, New York. pp. 3-121.

Pavillard, J. 1928. Escapes et associations. *Arch. Bot. Bull. Mens.* 2 : 68-72.

Perkins, C.J. 1967. Prescribed burning on international paper company lands. *Tall Timber Fire Ecology Conference,* 6 : 1-6.

Peterken, G.P., and Newbold, P.J. 1966. Dry matter production by *Ilex aquifolium. J. Ecol.* 54 : 143-150.

Petrides, G.A. 1956. Densities and range carrying capacity in east Africa. *N.A. Wildl. Conf. Trans.* 21 ; 525-537.

Petrides, G.A. 1968. *Proc. Symp. Recent Adv. Trop. Ecol.* 705-709.

Phillips, J. 1965. Fire as master and servant : its influence in the bioclimatic regions of trans-saharan Africa. *Tall Timber Fire Ecology Conference,* 4 : 7-110.

Phillipson, J. 1966. *Ecological energetics.* Edward Arnold. Great Britain.

Pinaka, E. 1974. *Evolutionary ecology.* Harpur and Row, New York.

Poore, M.E.D. 1955. The use of phytosociological methods in ecological investigations I. Braun-Blanquet system. *Jour. Ecol.* 43 : 226-244.

Poore, M.E.D. 1956. The use of phytosociological methods in ecological investigations IV. General discussion of phytosociological problems. *J. Ecol.* 44 : 1 : 28-50.

Poore, M.E.D. 1962. The methods of successive approximation in descriptive ecology. In "*Advances in ecological research*". Vol. 1. Ed. J.B. Cragg. Academic Press, New York. pp. 35-68.

Porter, R.H. 1949. Recent developments in seed technology. Bot. Rev. 15 : 221-344.

Pound, R., and Clements, F.E. 1898. *The phytogeography of*

Nebraska. Lincoln.

Pratt, L.H., and briggs, W.R. 1966. Photochemical and non-photochemical reactions of phytochrome in vivo. *Plant Physiol.* 41 : 467-474.

Puri,G.S. 1950. The distribution of conifers in the Kulu Himalayas with special reference to geology. *Indian For.* 76 : 144-153.

Puri, G.S. 1960. *Indian forest ecology*. Oxford Book Co., New Delhi.

Puri, G.S., and Gupta, A.C. 1951. The Himalayan conifers. II. The ecology of humus in conifer forests of the Kulu Himalayas. *Indian For.* 77 : 55-63; 124-129.

Prain, D. 1903. *Bengal plants,* Calcutta.

Puri, G.S. 1947. Fossil plants and the Himalayan uplift. *J. Indian Bot. Soc.* 25 : 167-184.

Purohit, S.S. 1977. Ecophysiological observations on *Helinanthus annus* L with special reference to germination, productivity and energetics. Ph. D. thesis, University of Udaipur, Udaipur.

Radley, M. 1958. The distribution oif substances similar to gibberellic acid in higher plants. *Ann. Bot.* 22 : 297-357.

Ramadas, L.A. 1974. Wheather and climatic pattern. In "*Ecology and biography in India*". Ed. M.S. Mani. W. Junk Publ. Hague.

Ramakrishnan, P.S. 1960. Distribution of *Euphorbia thymifolia* Linn in relation to soil calcium. *Mem. Indian Bot. Soc.* 3 : 52-57.

Ramakrishnan, P.S. 1960 a. Studies in the autecology of *Euphorbia hirta* Linn. *Jour. Indian Bot. Soc.* 39 : 455-473.

Ramakrishnan, P.S. 1960 b. Ecology of *Echinnochloa colunum* Link. Proc. Indian *Acad. Sci.* B 52 : 73-90.

Ramakrishnan, P.S. 1960 c. Ecology of *Setaria glauca* Beauv. *Proc. Indian Sci. Cogr.* Abstract III. 410.

Ramakrishnan, P.S. 1961 a. Calcicole and calcifuge problems in *Euphorbia thymifolia* Linn. *Jour. Indian Bot. Soc.* 40 : 66-81.

Ramakrishnan, P.S. 1961 b. Studies in the ecological life history of *Euphorbia thymifolia* Linn. *Proc. Natl. Inst. Sci. India.* 27 : 347-358.

Ramakrishnan, P.S. 1963. Contribution to the ecological life-history of *Setaria glauca* Beauv. *J. Indian Bot Soc.* 42 : 118-129.

Ramakrishnnan, P.S. 1963. Ecology of *Mullugo hirta* Thunb. *Proc. Natl. Inst. Sci. India.* B 29 : 4 : 396-406.

Ramakrishnan, P.S. 1965 a. Studies on edaphic ecotypes in *Euphorbia thymifolia* L.I. Seed germination. J. Ecol. 53 : 157-162.

Ramakrishnan, P.S. 1965 b. Studies on edaphic ecotypes in *Euphorbia thymifolia* L. II. Growth performance, mineral uptake and inter ecotypic competition. *J. Eco.* 53 : 703-714.

Ramakrishnan, P.S., and Bisht, S.S. 1968. Adaptability of the edaphic ecotypes in *Adathoda vasica* Nees. *Proc. Recent Adv. Trop. Ecol.* 180-186.

Ramakrishnan, P.S., amd Jain, R.S. 1966. Germinability of the seeds of edaphic ecotypes in *Tridex procumbens* L. *Trop. Ecol.* 6 : 47-55.

Ramaley, F. 1940. The growth of science. *Colorado Univ. Studies.* 26 : 3-14.

Ramam, S.S. 1966. Organisation of grass communities on western and Vindhyan uplands of Varanasi district. *J. Indian Bot. Soc.* 45 : 266-276.

Ramirez. W.B. 1969. Fig. wasps mechanism of pllen *traffer. Science,* 163 : 580-581.

Ranawat, M.P.S. and Vyas, L.N. 1975. Litter production in deciduous forests of Koriat, Udaipur (South Rajasthan), India. *Biologia,* 30 : 41-47.

Ranganathan, C.R. 1958. Studies in the ecology of the flora of Sind. *J. Indian Bot. Soc.* 8 : 283-286.

Rao, C.C. 1968. Biological spectrum of Kamnasa flora (Varanasi district). *Proc. Recent Adv. Trop. Ecol.* 458-465.

Rathore, J.S. 1971. Whether and climate as functions of solar radiations aznd other external influences. In "*Proceedings of the school of plant ecology*". Eds. R. Misra and B. Gopal. Oxford & IBH, Calcutta. pp. 4-16.

Raunkiaer, C. 1908. Liveformernes statistik som Grundlag for Biologiskche Plantegeographi. *Betanisk Tideskr*, 29 : 1.

Raunkiaer, C. 1911. Statislik der Labenformeno als Grundlag fur die Biologiskche Geographie *Beih. bot. Zbl* 27 : 171-205.

Raunkiaer, C. 1934. *The life-form of plants and statistical plant geography.* Clarendon Press, Oxford.

Razi, B.A. 1955. The phytogeogrphy of Mysore hill tops. *J. Mysore Univ.* Section B. 14 : 10 : 87-107; 15 : 11 : 109-144.

Reiter, H. 1885. *Die consolidation der Physiognomik.* Graz.

Revelle, R., Brocker, W., Greig, H., Keeling C.D., and Smagorinsky, J. 1965. Atmospheric carbon dioxide. In "*Restoring the quality of our environment*". Ennviron. Poll. Panel Presidents Science Advisory Comm. Washington. pp. 113-133.

Rice, E.L. 1974. *Allelopathy.* Academic Press, New York.

Ridley, H.N. 1930. *Dispersal of plants throughout the world.* L. Reeve & Co., London.

Riehl, H. 1965. *Introduction to the atmosphere.* McGraw Hill Book Co., New York.

Robertson, J.H. 1947. Responses of range grasses to different intensities of competition with sagebrush (*Artimisea tridentata* Nutt). *Ecology*, 28 : 1-16.

Robinowitch, E.I. 1945. *Photosynthesis and related processes.* Vol 1. Interscience Publications, New York.

Rodin, L.E., and Bazilevich, N.I. 1965. *Production and mineral cycling in terrestrial vegetation.* Oliver and Boyd, London.

Sachs, J. 1860. Abhangigkeit der Keimung vor der temperature. *Jahrb. Wiss. Bot.* 2 : 338-377.

Salisbury, E.J. 1942. *The reproductive capacity of plants.* G. Bell & Sons, London.

Salisbury, F.B. 1963. *The flowering plants.* Pergamon, New York.

Sant, H.R. 1964. Seasonal variations in coverage of selected grasses and forbs in relation to grazing intensities in India. *J. Range Manag.* 17 : 74-76.

Sant, H.R. 1964. Ecological studies with special reference to reproductive capacity in relation to grazing of *Panicum psilopodium* Trin. *Jour Indiann Bot. Soc.* 45 : 476-486.

Sant, H.R. 1966. Grazing effects on grassland soils of Varanasi, India. *J. Range Manag.* 19 : 362-363.

Sant, H.R. 1972. Ecological studies with special reference to reproductive capacity in relation to grazing of *Indigofera linnifolia* Linn. *Botanique*, 3 : 107-110.

Sant, H.R. 1974. Ecological studies in *Bothrioclova pertusa* A. Camus as a component of pastures on the Upper Gangetiic plain. *Botanique*, 5 : 67-84.

Sarup, S. 1952. The biological spectrum of the flora of Mount Abu *Univ. Raj. Stud. Biol. Sci.* 23 : 1-7.

Satoo, T. 1966. *Production and distribution of dry matter in forest ecosystems.* The Tokyo University Forest Miscellaneous Information. No. 16.

Saxena, M.C. 1963. Germination studies in *Hyptis suaveolens* Poit. Part I. Breaking of dormancy and effect of certain physical treatment on the percentage germination. *Proc. Natl. Acad. Sci. India.* 33 : 70-76.

Saxena, M.D.1964. Studies on the genus Datura linn of Ujjain. (M.P.) with special reference to the autoecology of *Datura innoxia* Mill. M. Sc. Thesis. Vikram University, Ujjain.

Saxena, S.K. 1977. Vegetation and its succession in the Indian desert. In "*Desertification and its control.* Ed. P.L. Jaiwswal, ICAR, New Delhi. pp. 176-192.

Saxena, S.K., and Sharma, S.K. 1981. Productivity of a desert shrub (*Zizyphus nummularia*). In "Bordi (*Zizyphus nummularia*) : A shrub of the Indian arid zone - its role in silvipasture. Ed. H.S. Mann and S.K. Saxena. CAZRI. Jodhpur. Monograph. No. 13 : 25-30.

Scheibe, J. and Lang, A. 1965. Lettuce seed germination : evidence for a reversible light induced increase in growth potential and Fr phytochrome mediation of the low temperature effect. *Plant Physiol.* 40 : 485-492.

Schjelderup-Ebbe, T. 1936. Uber die labensfahigkeit alter Samen. *Skr. norske Vidensk. Acad.* 1 Mat-nat Kl. 1936. 1-178 (1936) Abstr. in Biol. Abstr. 11 : 1114, No. 10520.

Sellack, G.W. 1960. The climax concept. *Bot. Rev.* 26 : 625-631.

Sellers, W.D. 1965. *Physical climatology.* University of Chicago Press, Chicago.

Semenik, R.A. 1954. Micoflora. In Anderson and Alcock. pp. 77-151.

Sen, D.N. 1977. *Environment and seed germination of Indian plants.* Chronica Botanica Co., New Delhi.

Sen, D.N. 1988. *Concepts in Indian ecology.* Shoban Lal Nagin Chand & Co, Delhi.

Sen, D.N. and Chatterji, U.N. 1966. Temperature relations of *Ruellia tuberosa* Linn. *Osterr. Botan. Zeit.* 113 : 390-394.

Schimper, A.F. 1898. *Pflanzen geographie auf physiologische gundiage.* Fishcher Jene.

Sen, D.N. 1969. Action of light in the germination of seeds and seedling growth in some Asclepiadaceae. *Acta Bot. Acad. Hung.* 15 : 327-335.

Shankar, V. 1968. Ecology of seed germination in *Trichoderma amplexicaule* Roth. Trop. Ecol. 9 : 201-207.

Shah, C.B. 1956. Phytosociological and ecological study of weeds of rice field. M. Sc. Thesis, Bombay University, Bombay.

Shankar, V., and Dadhich, N.K. 1977. Effect of long term enclosure on changes in dune vegetation. Ann. Arid Zone, 16 : 381-386.

Shankarnarayan, K.A. 1977. Impect of overgrazing on the grasslands. *Ann Arid Zone,* 16 : 349-359.

Sharma, I.K. 1978. Mutualism between flora and fauna of the Indian desert *Quarter. Inform. Bull. Ecol. Indian Soc. Ecol.* Ludhiana. 11 : 4 : 18.

Sharma, S. 1976. Ecological studies on the herb layer of hills at Udaipur (South Rajasthan). Ph. D. Thesis, Udaipur Univ., Udaipur.

Sharma, S.S., and Sen, D.N. 1975. Effect of light on seed germination and seedling growth of *Merremid* species. *Folia Geobot. et Phytotax.* 10 :

Sharma, V.B. 1953. Autecological study of *Sida acuta* burm. M. Sc. Thesis, Satugar University, Sagar.

Sharma, V.B. 1955. An autecological study of *Boswellia serrata Roxb.* Ph. D. Thesis, Sagar University, Sagar.

Shelford, V.E. 1913. *Animal communities in temperate America.* University of Chicago Press, Chicago.

Shirjaev, G. 1932. Generis *Onionis* revisio critica. *Beih. Z. Bot. Centracol.* 49.

Shrimal, R.L., and Vyas, L.N. 1975. Net primary production in grassland at Udaipur, India. In "*Trends in terrestrial and aquatic researches.* Eds. F.B. Golley and E. Medina. Springer Verlag, New York. *pp.* 265-271.

Shrivastava, A.K. 1953. Ecotypic concept and the meothod of establishing ecotypes. In "*School of plant ecology*". Eds. R. Misra and R.R. Das. Oxford & IBH, Calcutta, pp. 168-176.

Shrivastava, A.K. 1966. Ecology of *Boerhaavia diffusa* Linn and *Gomphrena celosioides* Mart. Ph. D. Thesis, Banaras Hindu University, Varanasi.

Shrivastava, A.K. 1967. Ecological studies of *Chenopodium album* L. *Ann Arid Zone,* 6 : 211-214.

Shrivastava, A.K., and Misra, K.C. 1968. Ecotype differentiation in *Boerhaavia diffusa* Linn. *Trop. Ecol.* 9 : 52-63.

Shrivastava, G.D. 1944. The biological spectrum of the Allahabad flora. *J. Indian Bot. Soc.* 23 : 1-7.

Shukla, S.P. 1964. Some aspects of the autecology of *Cyperus rotundus* L (Motha). M. Sc. Thesis, Vikram University, Ujjain.

Siddiki, M.O. 1972. Biological spectrum of the forests of Gorakhpur. *Trop. Ecol.* 13 : 234-235.

Siegelman, H.W., and Hendricks, S.B. 1957. Photocontrol of anthocyanin in turnip and red cabbage seedlings. *Plant Physiol.* 32 : 393-398.

Siegelman, H.W., Hendricks, S.B., and Firer, E.M. 1964. Purification of phytochrome from oat seedlings. *Biochem.* 3 : 418-423.

Siegelman, H.W., and Hendricks, S.B. 1956. Two photochemically distinct controls of anthocyanian formation in *Brassica* seedlings. *Plant Physiol.* 31 (Suppl) XIII.

Siegelman, H.W., Turner, B.C., and Hendricks, S.B. 1966. The chromophore of phytochrome. *Plant Physiol.* 41 : 1289-1292.

Silvertown, J.W. 1987. *Introduction to plant population. Ecology,* IInd Ed. ELBS, London.

Simpson, E.H. 1949. Measurement of diversity. *Nature*, 163-688.

Simps, P.L., and Singh, J.S. 1971. Herbags dynamics and net primary production in certain ungrazed and grazed grassland in north America. In "*Primary analysis of structure and function in grasslands*". Range Science Department, Science Series No. 10. Ed. N.R. Freisch. Colorado State University, USA. pp. 59-124.

Singer, S.F. (Ed). 1975. *The changing global environment.* D. Reidel Publ. Co., Boston.

Singh, J.S. 1967. Growth performance of *Cassia tora* L plants of different seed origins at different temperatures. In *PL 480 Annual Report.* pp. 12-23.

Singh, J.S. 1967a, Seasonal variations in composition, plant biomass and net community productivity in the grasslands at Varanasi. Ph. D. thesis, Banaras Hindu University, Varanasi.

Singh, J.S. 1968. Net above ground community productivity in the grassland at Varannasi. *Proc. Symp. Recent Adv. Trop. Ecol.* 2 : 631-653.

Singh, J.S. 1968. Comparison of growth performance of seeds of *Cassia tora* L and *C. obtusifolia* L. *Trop. Ecol.* 9 : 64-71.

Singh, J.S. 1969. Underground plant biomass, annual production and turnover inthe grassland at Varanasi. *Symp. Plant* Prod. Stud. India. 57th Indian Science Congress, Kharagpur. p. 6 (Abstract).

Singh, J.S. and Misra, R. 1968. Efficiency of energy capture by grassland vegetation at Varanasi. *Curr. Sci.* 37 : 636-637.

Singh, K.P. 1966. Seed dormancy and its control by germination inhibitors in *Anagallis arvensis*. In PL 480 annual Report 1965 -1966 : pp 89-102.

Singh, K.P. 1967. Distribution and growth performance of *Anagallis arvensis* L plants raised from seed of different localities. In PL 480 Annual Report. pp. 96-105.

Singh, K.P. 1968. Thermoresponses of *Potulaca oleracea* seeds. *Curr Sci.* 37 : 506-507.

Singh, S.N. 1961. Germination if grape (*Vitis vinifera* L) hybrid seeds by chilling. *Curr. Sci.* 30 : 62.

Slobodkin, L.B. 1960. Ecological energy relationship at population level. *Amer. Nat.* 94 : 213-236.

Smith, R.L. 1974. *Ecology and field biology*. Harper and Row, New York.

Smith, R.L. 1977. *Elements of ecology and field biology* harper and Row, New York.

Sorenson, E. 1909. *Op. Cit.* Narayana and Shah, 1966.

Southwick, C.H. 1976. *Ecology and the quality of our environment.* D. Van. Nostrand Co. New York.

Spragg, S.M., and Yemm, E.W. 1959. Regulatory mechanisms annd the changes of glutathione and ascorbic acid in germinating peas. *J. Exptl. Bot.* 10 : 409-425.

Srivastava, R.P., and Saxena, R.C.1989. *A text book of insect toxicology*. Himanshu Publications, Udaipur.

Stapledon, R.G. 1928, Cocks-foot grass (*Dactylis glomerata L*) ecotypes in relation to the biotic factor. *J. Ecol.* 16 : 71-104.

Stern, W.L., and Buell, M.F. 1951. Life-form spectra of New Jersey pine barren forest and Minnessota Jack pine forest. *Bull. Torrey Bot. Club.* 78 : 61-65.

Stewart, W.D.P. 1966. *Nitrogen fixation in plants.* The Athione Press, London.

Stocks, P. 1965. Temperature and seed dormancy. In "*Encyclopedia of plant physiology*". W. Ruhland. Springer-Verlag, Berlin. 15/2 : 746-803.

Sukachev, V.N. 1944. On principles of genetic classification in *biocenology* (in Russion). *Zur. Obschecheri, Biol.* 5 : 213-227.

Sussman, A.S. 1965. Physiology of dormancy and germination in the propagules of Crypotogamic plants. In "*Encyclopedia of plant physiology*". Ed. W. Ruhland. Springer-Verlag, Berlin. 15/2 : 933-1025.

Sussman, A.S. and halvorson, H.O. 1966. *Spores their dormancy and germination.* Harper and Row, New York.

Sweeney, J.R. 1956. Responses of vegetation to fire. *University of California Publication Botany.* 28 : 193-249.

Szymkiewicz, D. 1937. contributions a la geographic des plants. IV. Une nouvelle methode pour la recherche des centres de distribution geographique des genres. *Kosmos.* 62 : 1-15.

Tadaki, Y. 1966. Some discussions on the leaf biomass of forest stands and trees. *Bull. Govt. Forest Expt. Station,* (Tokyo). No. 184 : 135-161.

Takimota, A., and Hammer, K.C. 1965 a. Effect of far red light and its interactions with red light in photoperiodic response of *Pharbitis nil. Plant Physiol.* 40 : 859-864.

Tansley, A.G. 1935. The use and abuse of vegetational concepts and terms. *Ecology.* 16 : 284-307.

Tansley, A.G. 1939. *The British Isles annd their vegetation.* The Cambridge University Press, London.

Tansley, A.G., and Rankin, M.M 1911. The plant formation of calcarious soils. B. The sub-formations of the chalk. In "*Types of British vegetation*". Cembridge University Press, Cambridge. pp. 161-166.

Taylor, W.P. 1936. *Ecology*, 17 : 333-336.

Thimann, K.V. 1956. Promotion and inhibition, twin themes in physiology. *Amer. nat.* 90 : 145-162.

Tiagi, Y.D., and Ketewa, S.S. 1987. Net primary production of three important grassland communities in the environs of Udaipur.

In "*Environmental issues and researches in India*". Eds. S.K. Agarwal and R.K. Garg, Himanshu Publications, Udaipur. pp. 147-162.

Thompson, P.A. 1968. Germinating primula seeds. *J. Roy. Hort. Soc.* XCIII : 3 : 133-138.

Thompson, P.A. 1968. Germination of Caryphyllaceae at low temperatures in relation to geographical distribution. *Nature,* 217 : 1156-1157.

Throp, J., and Smith, G.D. 1949. *Soil Sci.* 67 : 117-126.

Timson, J. 1966. The germination of *Polygonum convovulus* L. *New Phytol.* 65 : 423-428.

Toole, E.H., Borthwick, H.A., Hendricks, S.B., and Toole, V.K. 1953. Physiological studies of the effects of light and temperature on seed germination. *Competes rendus de 1' Association Innternationale D'assais de Semences*, 18 : 267-276.

Toole, E.H., and Borthwick, H.A. 1968. The photoreactionn controlling seed germination in *Eragrostis curvula. Plant & Cell Physiol.* 9 : 125-136.

Toole, V.K. Toole, E.H., Borthwick, H.A., and Snow, Jr. A.G. 1962. Responses of seeds of *Pinus taeda* and *P. strobus* to light. *Plant Physiol.* 37 : 228-233.

Toole, E.H., Hendricks, S.B., Borthwicks, H.A., and Toole, V.K. 1956. Physiology of seed germination. *Ann. Rev. Plant Physiol.* 7 : 299-324.

Toole, E.H., Toole, V.K., Brothwick, H.A., and Hendricks, S.B. 1956. Photcontrol of Lepidium seed germination. *Plant physiol.* 30 : 15-21.

Toole, E.H., Toole, V.K., Borthwick, H.A., and Hendricks, S.B. 1955 a. Photocontrol of Lepidium seed germination. *Plant Physiolo.* 20 : 5-21.

Toole, E.H., Toole, V.K., Borthwick, H.A., and Hendricks, S.B. 1955 b. Interaction of temprature and light in germination of seeds. *Plant Physiol.* 30 : 473-478.

Toole, V.K., and Toole, E.H., Hendricks, S.B., and Borthwick, H.H. 1961. Responses of seeds of *Pinus virginiana* to light. Plant Physiol. 36 : 285-290.

Toy, S.J., and Willingham, B.c. 1966. Effect of temperature on seed germination of ten species and varieties on *Limnanthes*. *Eco. Bot.* 20 : 71-75.

Trench, R.K. 1978. The cell biology of plant-animal symbiosis. *Ann. Rev. pt. Physiol.* 30 : 485-531.

Tripathi, R.S. 1965. The resource ratio hypothesis of plant succession. *Amer. Natur.* 125 : 283-302.

Tripathi, R.S. 1968 a. Certain autecological observations on *Asphodelus tenuifolius* Cav : A troublesome weed of Indian agrivulture. *Trop. Ecol.* 9 : 208-219.

Tripathi, R.S. 1968 b. Ecology of *Cyperus rotundus* L. I. Effect of exposure of sub-freezing temperature on survival and subsequent growth. *Trop. Ecol.* 9 : 239-242.

Tripathi, R.S., and Srivastava, P.P. 1970. Preliminary studies on the germination behaviour of *Desmondium pullchelium* (Linn) Benth. *Trop. Ecol.* 11 : 75-79.

Trivedi, B.S., and Sharma, P.C. 1965. Biological spectrum of Lucknow flora. Nat. *Acad. Sci. India*. 35 : 15-20.

Turesson, G. 1922 a. The species and variety as ecological units. *Heriditas,* 3 ; 100-113.

Turesson, G. 1922 b. The genotypical response of the plant species to the habitat. *Ibid.* 3 : 211-350.

Unger, G. 1936. *Op. Cit.* Narayana and Shah 1966.

Van Dyne, G.M., Smith, F.M., Czaplewski, R.L., and Woodmansee, R.G. 1978. Annalysis and synthesis of grassland ecosystems dynamics. In "*Glimpses of ecology*" Eds. J.S. Singh and B. Gopal. Internatioal Scientific Publications, Jaipur. pp. 1-79.

Varshney, C.K. 1964. Ecology of wall flora of Varanasi. Ph. D. Thesis, Banaras Hindu University, Varanasi.

Varshney, C.K. 1966. Ecological life-history of *Bidens biternata* (Lour) Merr and Sherff. *Trop. Ecol.* 7 : 14-24.

Varshnney C.K. 1968. germination of the light sensitive seeds of *Ocimum americanum* Linn. *New Phytol.* 67 : 125-130.

Varshney, C.K. 1972. Propductivity of Delhi grasslands. In *Trop. Ecol. Emphas. Org. Prod.* Eds. P.M. Golley and F.B. Golley. Georgia Univ., Athens. *pp*. 27-42.

Vyas, L.N. 1962. Ecological studies on the vegetation of Alwar Rajasthan. Ph. D. Thesis, University of Rajasthan, Jaipur.

Vyas, L.N. 1964. Studies on the grassland communities of Alwar. J. *Indian Bot. Soc.* 33 : 3 : 490-494.

Vyas, L.N. 1965. Vegetation of hills around Alwar, northeast Rajasthan - Phytosociological studies. *J. Indian Bot. Soc.* 35 : 3 : 305-313.

Vyas, L.N., and Agarwal, S.K. 1970. Germination behaviour of *Verbena bipinnatifida* Nutt. Seeds. *J. Biol. Sci* 13 : 57-63.

Vyas, L.N., and Agarwal, S.K. 1970. Studies in the germinability of the dormant seeds of *Indigofera linnaei* Ali. *J. Biol. Sci.* 13 : 15-25.

Vyas, L.N.., and Agarwal, S.K. 1970. Temperature and spectral sensitivity of germinating *Dalbergia sissoo* seeds. *Phyton*, (Argentina). 27 : 41-45.

Vyas, L.N., Agarwal, S.K., and Garg, R.K. 1971. Responses of *Verbena bipinnatifida* Nutt to soil types. *Phyton*. 28 : 161-164.

Vyas, L.N., and Agarwal, S.K. 1972. Studies in the germination behaviour of Indigofera cordifolia Heyne ex Roth. *Japanese Jour. Eco.* 22 : 161-169.

Vyas, L.N., and Agarwal, S.K. 1973. Studies in the seed dormancy of desert plants 1. *Abrus precatorius. Ibid* 29 : 237-242.

Vyas, L.N., Agarwal, S.K., and Garg, R.K. 1972. Biomass production by *Erythrina suberosa* Roxb. *Trop. Ecol. Emphas.* Org. Prod. Eds. P.M. Golley and F.B. Golley. Georgia University, Athens, pp. 195-200.

Vyas, L.N., Agarwal, S.K., and Ranawat, M.P.S. 1971. Relation between above ground biomass, girth and number of growth rings in three species of the deciduous forests of Udaipur, Rajasthan, *Japanese Jour. Ecol.* 21 : 52-54.

Vyas, L.N., Garg, R.K. 1970. Studies on the seed dormancy of desert plants. 3. Interaction of GA and light on the germination and growth of seedlings of *Cleome viscosa* Linn. *Acta Botanica Scientiaria Hungaricae.* 15 : 273-279.

Vyas, L.N., and Garg, R.K. 1971. Responses to GA of light requiring seeds of *Verbena bipinnatifida* Nutt. *Zeitschrift. fur Pflanzenphysiologie.* 65 : 189-194.

Vyas, L.N., Garg, R.K., and Ranawat, M.P.S. and Das, R.R. 1971 b. Method for estimation of biomass and productivity of a tropical tree. *Japanese Jour. Ecol.* 21 : 144-246.

Vyas, L.N., Garg, R.K., and Ranawat, M.P.S. 1971 C. Observations on the dry matter production in *Mitrogyna parvifolia* Korth. *Ibid.* 21 : 227-230.

Vyas, L.N., Garg, R.K., and Agarwal, S.K. 1972. Net above ground biomass production in the monsoon vegetation at Udaipur. *Troip. Ecol. Emphas. Org. Prod.* Eds. P.M. Golley and F.B. Golley, Gergia University, Athens. pp. 95-99.

Vyas, L.N., Garg. R.K., and Ranawat, M.P.S. 1972. Observations on the net biomass production in *Schrebera switetenoides* Roxb. *Japanese Jour. Ecol.* 22 : 245-249.

Vyas, L.N., and Shrimal, R.L. 1973. Studies on the effect of thiourea. IAA and gibberellic acid onthe germination of the dormant seeds of *Celosia argentia* Linn. *Acta Botanica Scientieria Hungericae,* 18 : 423-430.

Vyas, L.N., and Shrimal, R.L. 1974. Promotion of kinetin induced seed germination by inorganic nitrogegenous compunds in *Amaranthus spinosus* Linn and *Celosia argentia* Linn. Biochem. *Physiol. Pflanzen.* 165 : 209-211.

Vyas, L.N., Billore, D.K., Sharma, A., and Jindal, K. 1980. Morphactin GA_3 interaction in *Verbena bipinnatifida* Nutt seeds. *Comp. Physiol. Ecol.* 5 : 334-337.

Vyas, L.N., and Garg, R.K. 1974. Reversal of the effect of growth inhibitors by GA and IAA. *Biochem. Physiol. Pflanzen.* 166 : 166 : 549-554.

Vyas, L.N., Garg, R.K., and Ranawat, M.P.S. 1973. b. Plant biomass and net production of *Anogeissus latifolia* Wall in forests of semiarid zone of Rajasthan, (India) *Biologia Plantarum.* 15 : 280-285.

Vyas, L.N., Garg, R.K., Ranawat, M.P.S., and Shrimal, R.L. 1973 c. Plant biomass and net production of *Soymida febrifuga* Juss. *Biologia* 28 : 499-506.

Vyas, L.N., Garg, R.K., and Ranawat, M.P.S. 1974. Branch dimensions and estimation of branch production in *Anogeissus latifolia* Wall. *Geobios* 1 : 36.

Vyas, L.N., Garg, R.K., and Vyas, N.L. 1976 b. Litter production and nutrient release in deciduous forests of Bansi, Udaipur (India). *Flora*, 165 : 103-111.

Vyas, L.N., Garg, R.K., Vyas, N.L., and Jindal, K. 1978 a. Plant biomass and net production relations of *Boswellia* serrata Roxb. ex Colber. In "*Environmental physiology and ecology of plants.*" Ed. D.N. Senn. Bishan Singh M.P. Singh, Dehra Dun. pp 289-296.

Vyas, L.N., Garg, R.K., and Jindal, K. 1979. Relation between photosynthetic area and above ground biomass in five tree species of deciduous forests near Udaipur (Rajasthan), India, *Biologia* 34 : 7 : 547-556.

Vyas, L.N., Jain, S.L., and Sharma, S. 1975. Reaction of gibberllin induced dark germination of *Borreria stricta* (Linn.f) Schuum. to Perature and N compounds. *Biochem. Physiol. Pflanzen.* 163 ; 343-347.

Vyas, L.N., Ranawat, M.P.S., and Garg, R.K. 1973 a. Studies on the production rekations to deciduous forests of semi-arid zone of Rajasthan (India) : plant biomass and net production in *Adina cordifolia* Hook. F. Fortow. *Chl.* 92 : 343-349.

Vyas, L.N., Sharma, S., and Jain, S.L. 1975. Reversal of the effect of 2,4- Dichlorophenoxy acetic acid by GA, Kinetin and Potassium nitrate. *Biochem. Physiol. Pflanzen.* 167 : 351-365.

Vytas, L.N., Shrimal, R.L., and Jindal, K. 1978 b. Plant biomass and net production relations of *Anogeissus pendula* at deciduous forests near Udaipur (Rajasthan) India. Flora. 167 : 6 : 457-465.

Vyas, N.L. 1975. Community structure and plannt productivity studies on Kalighat forest block, Udaipur (South Rajasthan). Ph. D. Thesis, Udaipur University, Udaipur.

Vyas, N.L. 1987. Structure, successional trend and above ground production in woody species of dry deciduous forests of Aravalli hills near Udaipur. In "*Environmental issues and researches in India*". Eds. S.K. Agarwal and R.K. Garg. Hinanshu Publication, Udaipur. pp. 215-228.

Vyas, N.L., Garg, R.K. Vyas, L.N. : 1976 a. Plant biomass and net production relations of *Terminalia tomentosa* Weight et Arn at deciduous forests near Udaipur (Rajasthan) India. *Flora*. 165 : 381-387.

Vyas, N.L., Garg, R.K., and Vyas, L.N. 1977. Plant biomass and production relations of *Diospyros melanoxylon* Roxb Cor at deciduous forests near Udaipur (Rajasthan), India. *Biologia*, 32 : 7 : 461-467.

Vyas, N.L., and Vyas, L.N. 1975. Distribution of chlorophyll and carotenoids in different canopy strata of dry deciduous tropical forest at Udaipur. *Photosynthetica*, 9 : 241-245.

Wali, M.K. 1966. Life-forms and biological spectrum of Lolab valley, Kashmir in relation to climate. *J. Bomb. Nat. Hist. Soc.* 63 : 115-122.

Walter, C.J. 1971. Systems ecology : the systems approach and mathematical models in ecology. In "*Fundamentals of ecology*". Ed. E.P. Odum, W.B. Saunders Co. Philadepphia pp. 276-292.

Ward, R.C. 1973. *Principles of hydrology*. 2nd Edn. McGraw Hill, London.

Wareing, P.F. 1965. Endogenous inhibitors in seed germination and dormancy. In "*Encyclopedia of plant physiology*". ed. W. Ruhland. Springer-Verlag, berlin, 15/2 : 909-924.

Warning, E. 1896. *Labuch der oekologischen*. Vid. Medd. Foren.

Warning, R. 1909. *Oekology of plants*. Oxford.

Watson, D.J. 1958. The dependence of net assimilation rate of leaf area index. Ann. Bot. 22 : 37-54.

Weaver, J.E., and Darland, R.W. 1947. A method of measuring vigour in range grasses. Ecology. 28 : 116-162.

Wegener, A. 1924. *The origin of continents and oceans*. London.

Wells, H.G. Huxley, J., and Wells, G.P. 1931. *The science of life,* Cassetl, London.

Went, F.W. 1957. *The experimental control of plant growth.* Chronica Botanica Co., Mass.

Whittaker, R.H. 1953. A consideration of the climax theory : The climax as a population and pattern. *Ecol. Monogr.* 23 : 41-78.

Whittaker, R.H. 1965. Dominance and diversity in land plant communities. *Science,* 147 : 250-260.

Whittaker, R.H. 1967. Gradient analysis of vegetation. *Biol. Rev.* 42 : 2 : 207-264.

Whittaker, R.H. 1970. *Communities and ecosystems.* MacMillan Co, London.

Whittaker, R.H., and Garfine, V. 1962. Leaf characteristics and chloropyll in relation to exposure and production in *Rhododendron maximum. Ecology,* 13 : 120-125.

Whittaker, E., and Gimingham, C.H. 1962. The effect of fire on regeneration of *Calluna vulgaris* (L) Hull from seed. *J. Ecol.* 50 : 815-822.

Wiegert, R.G. 1964. Population energetics of meadow spittle bugs (*Philaenus spurmarius*) as affected by migration and habitat. *Ecol. Monogr.* 34 : 217-241.

Willism, J.C. 1951. *Dictionary of flowering plants and ferns.* Cambridge University Press.

Woodbury, A.M. 1954. *Principles of general ecology.* McGraw Hill Book Co., New York.

Woodwell, G.M., and Whittaker, R.H. 1968. Primary production in terrestrial ecosoystems. *American Zoologist,* 8 : 19-50.

Wulff, E.V. 1943. *An introduction to historical plant geography.* Chronica Botanica, Waltham Mass.

Yeatom, R.I. 1978. A cyclic relationship between *Larrea tridentata* and *Opuntia leptocaulis* in the north Chihuahuan desert of Texas, USA. *Jour. Ecol.* 66 : 651-656.

Subject Index

Abiotic components, 3
Adiabatic changes, 56
Age and area, 387
Aggregation, 239
Air composition, 65
Albedo values, 46
Alluvial, 97
Ammensalism, 128
Anabolism, 257
Arctic, 146, 366
Assam flora, 379, 380
Association of species, 215
Atmospheric factor, 63
Autecology, 6, 137
Barriers, 383
Biogeochemical cycles, 309
Biological clocks, 184
Biomagnification, 34
Biological spectrum, 212
Biotic factor, 117
Biotic components, 2
Botanical provinces, 376
Catabolism, 257
Celluvial, 97
Centre of origin, 144
Chamaephytes, 207
Climax, 240
Circumpolar 358
Clouds, 80
Coaction, 118, 240
Commensalism, 126
Competition, 119, 239
Community dynamics, 233
—Structure, 198
Compartment model, 333
Continental drift, 383
Continuum concept, 230
Cryptophytes, 208
Cycling of carbon, 312
—nitrogen, 314
—phosphorus, 318
—radioactive, 320
—sulphur, 317
Cyclones, 70
Day neutral plants, 52
Deccan plateau, 379
Decomposers, 2
Density, 217
Dispersal, 72, 148
—dynamics, 380
—pathways, 383
Dispersion, 214
Distribution, 143, 355, 357
Diversity, 224
—index, 224
Dominance, 223
—index, 223
Dormancy, 61
Dry matter dynamics, 335
Dynamism, 33

Ecology, 1

Ecology analytical, 15

—descriptive, 14

—experimental, 17

—production, 19

Ecads, 189

Ecais, 239

Ecological amplitude, 184

Ecological energetics, 7, 267

Ecological pyramids, 263

Ecosystem, 253

Ecosystem model, 327

Efficiency of energy capture, 267

Energy transfer, 276

Energy flow model, 277

—upright, 277

—schematic model, 279

—Y-shaped model, 279

Energy exchange in ecosystem, 46

Ecotypic differentiation, 194

Endemism, 147, 362

Eolin, 97

Exchange capacity, 108

—anion, 109

—cation, 108

Exosphere, 65

Famous Indian ecologists, 389

Father of Indian ecology, 389

Fidility, 221

Fog, 81

Food-chain, 259

Food-web, 262

Floristic composition, 200

Frequency, 216

Gangetic plains, 378

Genecology, 6, 187

Glacial, 97

Gleization, 90

Grassland, 375

Green house effect, 42

Gradient analysis, 231

Heliophytes, 50

Hemicryptophytes, 207

Hekistotherms, 62

History of ecology, 10, 12

Holocoenotic, 24

Homeostasis, 26

Homologous series, 228

Humus formation, 104

—function, 105

Hydrologic, cycle, 76, 77

Hydrosere, 242

Importance value index, 226

Individualistic concept, 230

Information theory, 242

Interdependence, 21

Inter-relatedness, 21

Invasion, 237

Land bridges, 384

Laterization, 90

Lapse change, 56

Levels of hierarchy, 322

Light factor, 39

Limiting factor, 28

Lithosere, 248

Litter production, 303

Maturation of soil, 90

Megatherms, 62

Melanisation, 90

Mesotherms, 62

Mesosphere, 63

Microtherms, 62

Migration, 187, 237, 384
Modelling, 323
Moisture holding capacity, 101
Monoclimax theory, 240
Mutulaism, 120
Neutralism, 118
Niche, 198
North temperate, 366
Nudation, 237
Organisation of communities, 197
Origin of ecotypes, 192
Pantropical 146, 359
Parasitism, 130
Plagiosere, 250
Podzolisation, 90
Polyclimax theory, 241
Porosity, 100
Primary producers, 2
Precipitation, 79, 84
Presence 222
Principles of ecology, 22
Production ecology, 7, 19, 283
Phytogeography, 365
—descriptive, 365
—interpretive, 366
Physiogromy, 204
Phanerophytes, 206
Protocooperation, 126
Pyramid of biomass, 265
—energy, 277
—number, 263
Pyric factor, 133
Radiation, 39
Reaction, 106, 239
Reproductive capacity, 181
Sciophytes, 50
Scope of ecology, 7
Seed dormancy, 176
—germination, 157, 162
Seed imbibition, 158
—output, 151
—viability, 154
Seedling growth, 181
Simulation, model, 338
—linear, 339
—non-linear, 340
Soil, 87
—Colour, 100
—classification, 96
—porosity, 100
—profile, 93
—skeleton, 95
—texture, 96
—reaction, 106
—colloids, 108
Stabilization, 240
Status of ecology, 3
Stratificationn, 201
Sub divisions of ecology, 6
Sub arctic, 366
Succession deflected, 250
—kinds, 234
—primary, 234
—secondary, 234
—autogenic, 235
—autotrophic, 234
—allogenic 235
—Hetrotrophic 234
—Cyclic 235
—induced, 235
—retrogressive, 235
Successive approximation, 229

Systems analysis, 341

—ecology, 321

—theory, 354

—transfer function, 275

Synecology, 6

Systematics, 142

Temperate, 146

Temperature, 55

—inversion, 59

Thermodynamic laws 36

Thermosphere, 63

Thermoperiodism, 60

Therophytes, 208

Thunderstomers, 133

Tolerance, 31

Topographic factors, 113

Troposphere, 63

Tropical 367

Turn over 225

Vegetation analysis 229

Vegetation of India, 370

Vitality, 215

Water potential 162

Water 75

Xerosere, 246

Zonal soils, 98